Microprocessor-Based Control Systems

International Series on
MICROPROCESSOR-BASED SYSTEMS ENGINEERING

Microprocessor-Based Control Systems

edited by

NARESH K. SINHA

Department of Electrical and Computer Engineering,
McMaster University, Ontario, Canada

D. REIDEL PUBLISHING COMPANY

A MEMBER OF THE KLUWER ACADEMIC PUBLISHERS GROUP

DORDRECHT / BOSTON / LANCASTER / TOKYO

Library of Congress Cataloging in Publication Data

CIP

Microprocessor based control systems.

(International series on microprocessor-based systems engineering)
Includes bibliographies and index.
1. Automatic control 2. Microprocessors. I. Sinha, N. K. (Naresh Kumar), 1927– . II. Series.
TJ223.M53M49 1986 629.8′95 86–11849
ISBN 90–277–2287–0

Published by D. Reidel Publishing Company
P.O. Box 17, 3300 AA Dordrecht, Holland

Sold and distributed in the U.S.A. and Canada
by Kluwer Academic Publishers,
101 Philip Drive, Assinippi Park, Norwell, MA 02061, U.S.A.

In all other countries, sold and distributed
by Kluwer Academic Publishers Group,
P.O. Box 322, 3300 AH Dordrecht, Holland

Printed in The Netherlands

CONTENTS

Chapter 4
ADAPTIVE CONTROL OF SYNCHRONOUS MACHINE EXCITATION
O. P. Malik

Chapter 5
MICROPROCESSOR-BASED SYSTEM IDENTIFICATION BY SIGNAL COMPRESSION METHOD
N. Aoshima

Chapter 6
DIGITAL SYSTEM CONTROLLERS FOR EXACT MODEL SIGNAL TRACKING
G. H. Hostetter

Chapter 7
MICROPROCESSOR-BASED CONTROL OF DC MOTORS
M. R. Stojic

Chapter 8
MICROPROCESSOR-BASED AC MOTOR CONTROL
M. H. Park

Chapter 9
MICROPROCESSOR-BASED NONLINEAR CONTROLLER FOR THE ROLL MOTION OF A MODEL AIRCRAFT
N. A. Krikelis

Chapter 10
A MICROPROCESSOR-BASED CONTROL SYSTEM FOR A GLASS FURNACE
H. N. Koivo and A. Setälä

Chapter 11
MULTI-PROCESSOR ROBOT CONTROL SYSTEM USING MICROPROGRAMMING
M. Jelšina

Chapter 12
COMPUTER CONTROL SYSTEM FOR THE TRAINING SIMULATOR
K. Furuta and Y. Ohyama

Chapter 13
MICROCONTROL-BASED DIRECT NUMERICAL CONTROL
R. Kitai and T. P. Taylor

Chapter 14
MICROPROCESSOR-BASED REPETITIVE CONTROL
M. Nakano and S. Hara

Chapter 15
REAL-TIME PROGRAMMING LANGUAGES FOR PROCESS CONTROL

Chapter 16
INDUSTRIAL LAN USING DISTRIBUTED MICROPROCESSORS
O. L. Storoshchuk and B. Szabados

Chapter 17
MICROCOMPUTER-EMBEDDED DISTRIBUTED CONTROL OF A SWITCHING AND COMMUNICATION SYSTEM
D. Tabak and J. Etkin

Chapter 18
MICROPROCESSOR-BASED DATA REDUCTION AND COMPRESSION SYSTEMS
S. G. Tzafestas and G. Papakonstantinou

PREFACE

Recent advances in LSI technology and the consequent availability of inexpensive but powerful microprocessors have already affected the process control industry in a significant manner. Microprocessors are being increasingly utilized for improving the performance of control systems and making them more sophisticated as well as reliable. Many concepts of adaptive and learning control theory which were considered impractical only 20 years ago are now being implemented. With these developments there has been a steady growth in hardware and software tools to support the microprocessor in its complex tasks. With the current trend of using several microprocessors for performing the complex tasks in a modern control system, a great deal of emphasis is being given to the topic of the transfer and sharing of information between them. Thus the subject of local area networking in the industrial environment has become assumed great importance.

The object of this book is to present both hardware and software concepts that are important in the development of microprocessor-based control systems. An attempt has been made to obtain a balance between theory and practice, with emphasis on practical applications. It should be useful for both practicing engineers and students who are interested in learning the practical details of the implementation of microprocessor-based control systems. As some of the related material has been published in the earlier volumes of this series, duplication has been avoided as far as possible.

I would like to express my sincere thanks to the authors of the different chapters who have agreed to share some of their valuable experience for the book. Without their generous contribution this book could not have been possible.

I am aware of the fact that there are some variations in the organization and flavour of the various chapters but I hope that these will not affect the usefulness of the book.

Hamilton 1986 — Naresh K. Sinha

CONTRIBUTORS

S.I. Ahson, Department of Electrical Engineering,
Indian Institute of Technology, New Delhi

N. Aoshima, Institute of Applied Physics,
University of Tsukuba, Ibaraki, Japan

D.W. Clarke, Department of Engineering Science,
Oxford University, Oxford, England

N.K. De, Department of Electrical Engineering,
Indian Institute of Technology, Kharagpur(W.B.), India

K. Furuta, Faculty of Control Engineering,
Tokyo Institute of Technology, Tokyo, Japan

S. Hara, Faculty of Control Engineering,
Tokyo Institute of Technology, Tokyo, Japan

G.H. Hostetter, Electrical Engineering Department,
University of California, Irvine, California, U.S.A.

M. Jelsina, Department of Technical Cybernetics,
Technical University, Kosice, Czechoslovakia

R. Kitai, Department of Electrical and Computer Engineering,
McMaster University, Hamilton, Canada

H.N. Koivo, Tampere University of Technology,
Tampere, Finland

N.A. Krikelis, Control Systems Division,
National Technical University, Athens, Greece

O.P. Malik, Department of Electrical Engineering,
The University of Calgary, Calgary, Canada

D.S. Naidu, Department of Electrical Engineering,
Indian Institute of Technology, Kharagpur(W.B.), India

M. Nakano, Department of Control Engineering,
Tokyo Inst. of Technology, Tokyo, Japan

Y. Ohyama, Faculty of Control Engineering,
Tokyo Institute of Technology, Tokyo, Japan

G. Papakonstantinou, Division of Computer Science,
Department of Electrical Engineering,
National Univ. of Athens, Athens, Greece

M.H. Park, Department of Electrical Engineering,
Seoul National University, Seoul, Korea

G.P.Rao, Department of Electrical Engineering,
Indian Institute of Technology, Kharagpur(W.B.), India

A. Setala, Tampere University of Technology,
Tampere, Finland

N.K. Sinha, Department of Electrical and Computer Engineering,
McMaster University, Hamilton, Canada

S. Sinha, Department of Electrical Engineering,
Indian Institute of Technology, Kharagpur (W.B.), India

M.R. Stojic, Faculty of Electrical Engineering,
University of Belgrade, Belgrade, Yugoslavia

O. Storoschchuk, General Motors of Canada Limited,
Oshawa, Ontario, Canada

B. Szabados, Department of Electrical and Computer Engineering,
McMaster University, Hamilton, Canada

D. Tabak, Department of Electrical and Computer Engineering,
George Mason University, Fairfax, Virginia, U.S.A.

T.P. Taylor, Atlantis Flight Research Inc.,
Woodbridge, Ontario, Canada

S. G. Tzafestas, Division of Computer Science,
Department of Electrical Engineering,
National Univ. of Athens, Athens, Greece

CHAPTER 1

MICROPROCESSORS IN CONTROL SYSTEMS

Naresh K. Sinha
Department of Electrical & Computer Engineering
McMaster University
Hamilton, Ontario
Canada L8S 4L7

1. INTRODUCTION

Recent developments in the large-scale integration of semiconductor devices and the resulting availability of inexpensive microprocessors, memory chips and analog-to-digital converters have made it possible to use computers as integral parts of control systems without any significant increase in cost. Consequently, it is now quite practical to implement complex control and decision strategies, which are required in optimal and adaptive control theory, in real time. This has led to a great deal of activity in both the theory and application of digital control. It can be said, without any exaggeration, that the impact of the microprocessor on the process control industry will be nothing short of revolutionary.

Several books have recently been published on digital control (Franklin and Powell 1980, Katz 1981, Isermann 1981, Phillips and Nagel Jr. 1983, VanLandingham 1985). All of these expect the use of a microprocessor as a component of a feedback control system. In the simplest schemes for digital control, the error signal in a closed-loop system is sampled and converted to the digital form by an analog-to-digital converter. A digital controller is then utilized to process the sequence of numbers representing the error signal and produce another sequence of numbers corresponding to the input that should be applied to the plant to be controlled. Finally, a digital-to-analog converter produces the actual analog input which is then applied to the plant. The digital controller can, therefore, be regarded as a digital filter which processes the error sequence to generate the sequence of numbers corresponding to the desired input. This can be easily implemented on a microprocessor using one of many possible realization methods (Ahson 1983, Nagel Jr. and Carroll 1974, Nagel Jr. and Nelson 1981). A PID controller may be implemented using a microprocessor in a similar way (Krikelis and Fassois 1984).

N. K. Sinha (ed.), Microprocessor-Based Control Systems, 1–5.

Although such schemes for digital control are affected by the finite word-length used in the microprocessors, the problem is not as severe with 16-bit processors of today as it was with the 4-bit processors used ten years ago. The flexibility that is obtained with microprocessors in the adjustment of the parameters of the controller has made them specially attractive for adaptive control (Clarke and Gawthrop 1981, Dexter 1983). These developments have given a great impetus to research in the field of adaptive control. Several schemes for the practical implementation of adaptive control using microprocessors have become quite practical. Some of these implementations require several microprocessors for carrying on simultaneously the tasks of on-line system identification, computation of the optimal parameters of the digital controller and implementing the controller itself.

In some large factories, computer control is utilized at a number of different locations for inter-related control systems. In general, one or more computers may be required for each of these systems. For proper coordination, it is necessary that all of these computers communicate with each other and with a central computer. Thus, with the development more sophisticated multi-computer control systems, we have to be concerned with local area networking at many levels.

An interesting feature of the microelectronics revolution has been the relationship between the relative cost of hardware. During the past ten years, we have witnessed a dramatic fall in the price of computer hardware. Within a decade, there have been four generations of microprocessors, with the price and size spiralling down while at the same time their computing power has been increasing rapidly. Today one can buy a microcomputer which fits inside a briefcase, has the same capability as a large main-frame computer available 20 years ago, at less than one hundredths of the cost of the latter. On the other hand, the cost of related software appears to have increased. At the present time, often the computer hardware costs much less than the software that is necessary to run it effectively. The situation is really critical in the case of software designed specially for applications in industry. The main reason has been due to the fact that process control software has been written almost exclusively in assembly language in the past. Although such software is usually very expensive, it has been preferred due to the fact that a control engineer could generate compact and efficient code. It is now being recognized that with the present day 16-bit microprocessors one can generate very efficient code using high-level structured languages at a fraction of the cost required for programming in assembly language. Some 32-bit microprocessors are now available which are even more suitable for this purpose. Hence, efficient use of microprocessors in control systems requires a good knowledge of high-level real-time programming languages and skills in software engineering.

2. OUTLINE OF THE BOOK

The present book is an attempt to put together various aspects of microprocessor-based control systems under one cover. The topics that have been discussed include some recent developments in control theory inspired by the availability of inexpensive microprocessors as well as important aspects of related software with emphasis on practical implementation for several specific applications. In *chapter 2* by Rao et al, there is a general discussion of microprocessor-based control, with special reference to the implementation of simple classical control techniques, motion control systems, on-line identification and control. The topic of *chapter 3* is the microprocessor implementation of adaptive control using self-tuning regulators. After describing the background to many current approaches to self-tuning design, Clarke presents an outline of the implementation of this self-tuner based on the Generalized Minimum Variance strategy. This is followed by a discussion of additional features required to make it work in practice. Finally, a new method, called Generalized Predictive Control is shown to overcome the limitation that the knowledge of the dead-time of the plant is required with the Generalized Minimum Variance approach. In *chapter 4*, Malik presents a discussion of microprocessor-based adaptive control of the excitation of synchronous machines. Interesting features of this chapter are the use of an adaptive controller with a variable forgetting factor and self-searching pole-shift. Experimental results are included. In *chapter 5*, Aoshima presents a new approach to microprocessor-based system identification. The signal compression method, proposed by the author, uses a special computer generated test signal. All signal processing is done by a microprocessor-based measuring system. The method is illustrated with several examples of practical application. In *chapter 6*, Hostetter describes new results on the design of digital controllers for discrete-time tracking systems. The transient performance of the resulting feedback system is exactly as desired by the designer. After the initial transient period, a specified transformation of the state of the plant tracks any reference input from a class chosen by the designer. The proposed method is specially suitable for control using a microprocessor.

Since servomotors are the "muscle arms" of most control systems, the next two chapters are devoted to the control of motors using microprocessors. In *chapter 7*, Stojic provides a detailed description of microprocessor-based control of d.c motors. The author has presented the design of a real digital position control system using a d.c. motor for the purpose of illustrating the proposed grapho-analytical method of pole-placement. In *chapter 8*, Park has described microprocessor-based control of a.c. motors, which have advantages in respect of cost, robustness, weight and maintenance. The special topics which are discussed include slip frequency control, pulse-width modulation inverter control, field oriented control and energy optimization control strategy.

Chapters 9 to 13 are devoted to the implementation of microprocessor-based control systems in practical industrial problems. In *chapter 9*, Krikelis describes a microprocessor-based controller for the roll motion of a model aircraft.. A nonlinear control scheme is proposed and results of simulation are included. In *chapter 10*, Koivo and Setala discuss a microprocessor-based control system for a glass furnace. The multi-microprocessor control system for robots is presented by Jelsina in *chapter 11*. It includes the architecture of the multiprocessor control system as well as algorithms for the microprogramme control of the robot control system executive level. In *chapter 12*, Furuta and Ohyama discuss a computer control system for the training simulator of a radio-controlled helicopter. This is followed by a description of the direct numerical control of machine tools using microprocessors in *chapter 13* by Kitai and Taylor. The authors have given a brief description of the basic principle of direct numerical control. This is followed by the description of their implementation of direct numerical control of a large milling machine for the metal machining industry, using an Intel 8031 microprocessor. In *chapter 14*, Nakano and Hara have discussed the application of microprocessor-based repetitive control systems. This is new control scheme, in which the control variables follow the periodic reference commands with high accuracy. The method is applied to the control of a single axis servomechanism in which the controller is implemented on a 16-bit microprocessor. It can be applied to the control of industrial robots.

In *chapter 15*, Ahson has given a survey of high-level languages used for real-time process control. This includes languages like concurrent Pascal, Ada, Forth and Modula-2. It is felt that there will be greater interest in this field as the use of microprocessors in process control becomes more common. Many industries are already using a large number of computers in their plants and a major concern has been the flow of information between these computers. This is very important for effective control action. In *chapter 16*, Szabados and Storoshchuck have described the organization of an industrial local-area network using distributed microprocessors. In *chapter 17* Tabak and Etkin have presented a related topic. The microcomputer-embedded distributed control of a switching and communication system described in this chapter presents a new methodology for controlling and managing a complex distributed communication system. This is followed by the description of microprocessor-based data reduction systems in *chapter 18* by Tzafestas and Papakonstantinou. This is a very important subject as it includes saving storage space as well as the time required for computation. Both of these considerations are of vital importance to engineers who want to improve the overall performance of control systems by using inexpensive microprocessors as components.

3. REFERENCES

1 Nagel Jr., H.T. and Nelson, V.P.: 1981, "Digital filter implementation on 16-bit microprocessors", IEEE Micro, vol. 1, pp.23-41.

2. Nagel Jr., H.T. and Carroll, C.C.: 1974, "Realization of digital controller ", Proc. IFAC Symp. Autom. Control in Space, Armenia, USSR.

3 . Phillips, C.L. and Nagel Jr., H.T.: 1984, "Digital control system analysis and design", Prentice-Hall inc., Englewood Cliffs, N.J.

4. Tabak, Daniel: 1983, "Hardware and software aspects of control applications of microprocessors" in "Microprocessors in signal processing, measurement and control", Tzafestas, S.G.(ed.), D.Reidel Publishing Company, Dordrecht.

5. Katz, P.: 1981, "Digital control using microprocessors", Prentice-Hall Inc., Englewood Cliffs, N.J.

6. Isermann, R.:1981, "Digital Control Systems", Springer-Verlag, Berlin.

7. Franklin G.F. and Powell, J.D.: 1980, "Digital Control of Dynamic Systems", Addison-Wesley, Reading, Mass.

8. Krikelis, N.J. and Fassois, S.D.: 1984, "Microprocessor implementation of PID controllers and lead-lag compensators", IEEE Trans. Industrial Electronics, vol. IE-31, pp.79-85.

9. Clarke, D.W. and Gawthrop,P.J.: 1981, "Implementation and application of microprocessor-based self-tuners", Automatica, vol. 17, pp.233-244.

10. Dexter, A.L.: 1983, "Self-tuning control algorithm for single-chip microcomputer implementation", Proc. IEE, vol. 130, Pt.D, pp.255-260.

11. Vanlandingham H.F.: 1985, "Introduction to digital control systems", Macmillan Publishing Company, New York.

CHAPTER 2

SOME ASPECTS OF MICROPROCESSOR-BASED CONTROL AND IDENTIFICATION

Ganti Prasada Rao, S. Sinha, D. S. Naidu and N. K. De
Department of Electrical Engineering
Indian Institute of Technology
Kharagpur (WB) 721302, India

ABSTRACT. This article describes microprocessor applications in control and identification. Beginning with a simple system of open loop control for a stepper motor, microprocessor based schemes are outlined for closed loop temperature control and for a more complex motion control system employing modern control techniques. A microprocessor system for parameter identification in continuous time dynamical systems is then presented. The motion control system and the parameter identification system which is based on the so called Poisson moment functional (PMF) method have been developed at I.I.T., Kharagpur.

I. INTRODUCTION

In the history of technology, few inventions like that of microprocessors, have merited the epithet "revolutionary" ever since their inception. In less than a decade we have seen four generations of microprocessors - a growth rate three times faster than that in the case of computers. As the cost and size spiral down with an exponential increase in their computing power, the number of applications is tremendously increasing. In control and automation the microprocessor has gained great importance [4,5]. The advent of microprocessors is a major step towards enabling us to implement complex control and decision strategies such as optimal control, identification and adaptive control thereby reducing the gap between theory and practice in the field of modern control [1,2,3,9,14].

The key to effective use of the great progress in microcomputer technology for this purpose is the understanding of software and methodology for developing suitable control strategies. After some experience with microprocessors we realize that the software turns out to be most critical, expensive and time consuming part of their application. Our discussion in this article begins with simple control schemes to aid in gradual development. Based on simple conventional control concepts we outline in Section II the cases of open loop control of a stepper motor and closed loop temperature control.

In Section III we discuss implementation of certain modern control

N. K. Sinha (ed.), Microprocessor-Based Control Systems, 7–34.

techniques in the so called area of motion control systems. Motion control engineering is a newly emerging area wherein a union of techniques of electric machine drives, power electronics and modern control play the main role. Recent advances in modern control theoretic techniques and the big strides in power electronics along with classical methods of control of electric machines give rise to new possibilities in the development of made-to-measure motion control systems. Beginning with a brief review of the necessary control theoretic background, we describe a programmable controller using an 8-bit-microprocessor culminating in the presentation of a casestudy of motion control.

Section IV begins with a brief outline of the role of microprocessors in identification and adaptive control. A specific casestudy implementing the so called poisson moment functional (PMF) technique [8,13] for parameter estimation in continuous-time models is then discussed. Review and assessment of the various cases presented here along with some possible future directions are given at appropriate places in the text.

II. IMPLEMENTATION OF SIMPLE CLASSICAL CONTROL TECHNIQUES

2.1. Openloop Control of a Stepper Motor

In several industrial applications, position control is required, particularly in automatic assembly lines, graphics and plotter operations. The importance of stepper motor in industrial applications spans a wide section of several engineering disciplines. The microprocessor is a natural choice for programmable control of a stepper motor wherein it is required to have a first starting, stopping, precise stepping to a destined position and remaining there for a programmable (hold) time before stepping to another destined position.

The basic scheme using a stepper motor in openloop position control is shown in Fig. 2.1. The motor receives a series of pulses for its accu-

Figure 2.1. Open loop control using a stepper motor

rate stepping to the desired position. The purpose of control electronics is to generate the pulses and direction signals. The microprocessor is used as a part of the control electronics as shown in Fig. 2.2. Port A is for input signals and Port B is for output to drive the stepper motor coils C_o-C_3. The output port activates the coils with proper delay and number of steps for a given direction. For a given stepper motor the manufacturer's data sheets should be consulted for specifications regarding the step angle, voltage, power etc. The diodes D_o-D_3 prevent the transients of the coils.

The software required for the stepper motor control consists of delay (wait) program, pulse generation program, and finally motor control program detailed as follows:

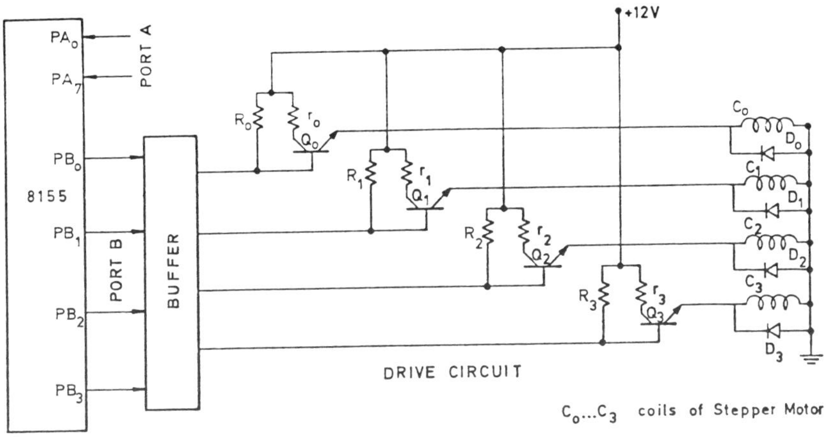

Figure 2.2. Simplified diagram of stepper motor drive and interface

2.1.(a): Delay (wait) operation

The two usual methods of generating the required delay are by decrementing a register count and by using no-operation (NOP) instructions. We describe the one using NOP instructions. Since each machine instruction takes a specific amount of time for execution (typically a few microseconds), it is possible to compute delay time in terms of a group of instructions a preset number of times. The flow chart for the delay operation is given in Fig. 2.3.

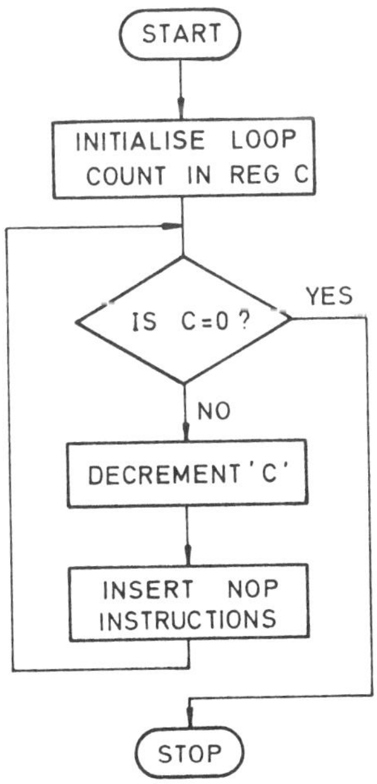

Figure 2.3. Flow chart for delay time

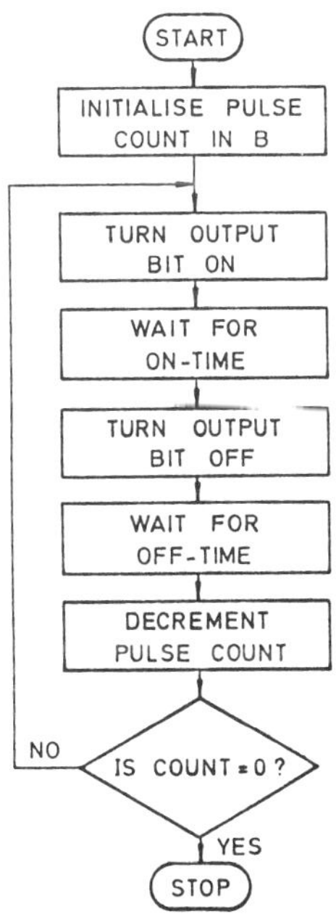

Figure 2.4. Flow chart for pulse generation

2.1.(b): Pulse generation

We need standard rectangular wave shaped pulses of proper height, on- and off-times. For stepper motor control the off-time is longer than the on-time. A simple flow chart for the generation of pulses is given in Fig. 2.4. The on-time and off-time delays are generated by repeated decrementation of the contents of a register.

2.1.(c): Motor control

The complete flow chart for motor control is given in Fig. 2.5. The direction of the motor is controlled by two independent output lines on the same output port. For further details of stepper motor control with microprocessors see [2,6].

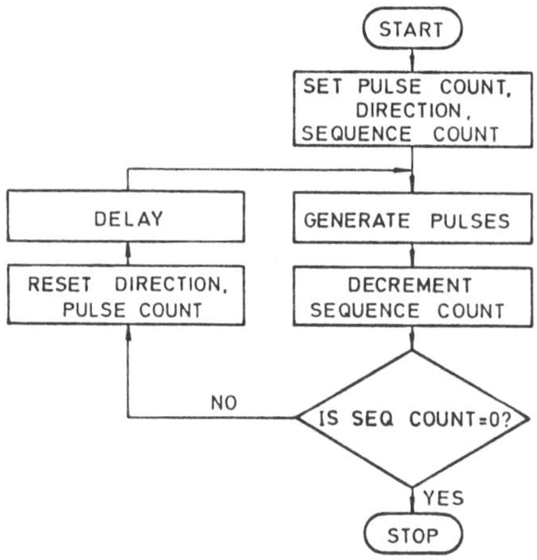

Figure 2.5. Flow chart for stepper motor control

2.2. Closed loop temperature control

One of the simplest and most common applications of feedback is in a typical temperature control system shown in Fig. 2.6. The reference value of the desired temperature is stored in a microcomputer and the actual temperature of the thermal process is measured and compared via analog-to-digital-converter (ADC). Any error in the temperature, expressed in

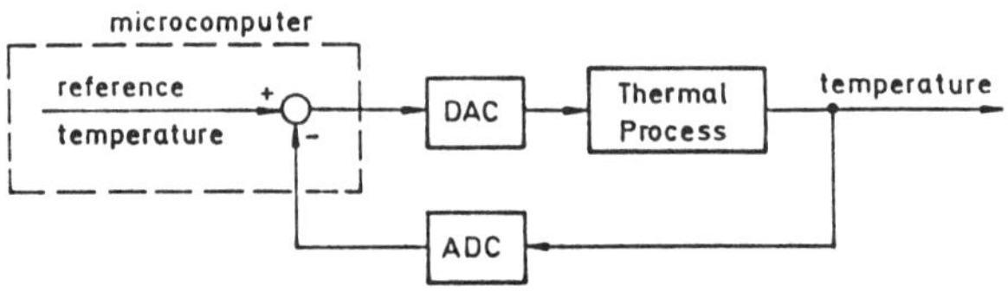

Figure 2.6. Principle of a temperature control system

digital form is again transformed into analog form through a digital-to-analog-converter (DAC) and is made to drive a heater coil in the thermal plant. Fig. 2.7 shows a simple scheme in which the control input to heater coils is derived by on-off principle. If the output temperature is greater than $(T_r + \Delta T)$ the heater is turned off, whereas if the output temperature is less than $(T_r - \Delta T)$ the heater is turned on, where T_r is the reference temperature and ΔT denotes tolerance. The flowchart for this system is shown in Fig. 2.8. Other types of temperature control schemes using microprocessors are discussed in [2,12].

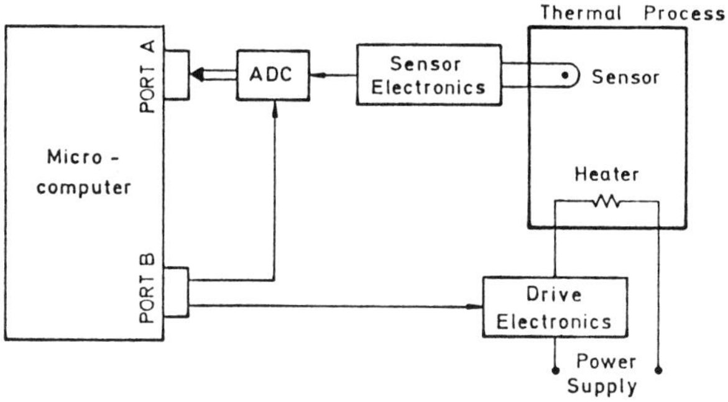

Figure 2.7. Simplified form of a temperature control system

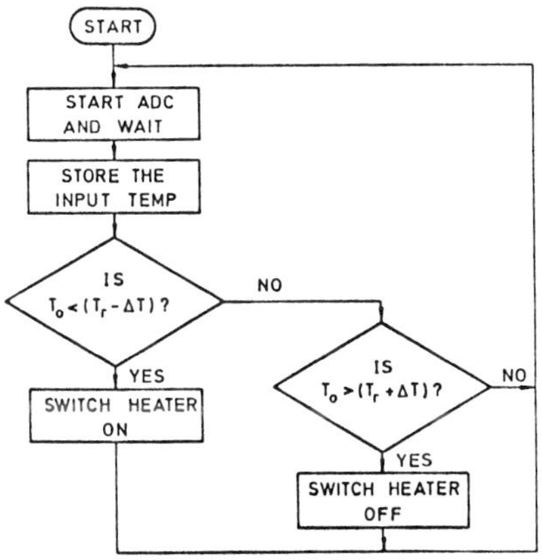

Figure 2.8. Flow chart for temperature control system

III. MOTION CONTROL SYSTEM

3.1. Control theoretic background

The plant to be controlled in a motion control system may be modelled in state space [16] as

$$\left.\begin{aligned} \dot{\underline{x}} &= A\underline{x} + B\underline{u} + E\underline{d} \\ \underline{y} - \underline{y}_r &= C\underline{x} - \underline{y}_r \end{aligned}\right\} \qquad (3.1)$$

where, $\underline{x}$, $\underline{u}$, $\underline{d}$, $\underline{y}$ and $\underline{y}_r$ are the state, control (input), disturbance (input), output and reference signal vectors of length n, m, q, p and p respectively, and A, B, C and E are matrices of dimension (nxn), (nxm), (pxn) and (nxq) respectively.

In the extreme case $\underline{d}$ and $\underline{y}_r$ might have dynamics of their own and may be modelled as systems of exogenous origin [15]. However, for most industrial problems, $\underline{d}$ is a constant vector (step type disturbances) and $\underline{y}_r$ is also constant giving rise to a set point regulation problem.

Mathematically speaking regulation problem with internal stability (RPIS) requires that,

$$\dot{\underline{x}} \to 0, \text{ and}$$

$$\underline{y} - \underline{y}_r \to 0, \text{ as } t \to \infty. \tag{3.2}$$

Assuming $\underline{d}$ and $\underline{y}_r$ constants, differentiation of Eqn.(3.1) yields

$$\dot{\underline{z}} = \hat{A}\ \underline{z} + \hat{B}\ \underline{v} \tag{3.3}$$

where,

$$\underline{z} \triangleq \left[\begin{array}{c} \dot{\underline{x}} \\ \hline \underline{y} - \underline{y}_r \end{array}\right], \quad \underline{v} \triangleq \dot{\underline{u}}, \quad \hat{A} = \left[\begin{array}{c|c} A & 0 \\ \hline C & 0 \end{array}\right] \text{ and } \hat{B} = \left[\begin{array}{c} B \\ \hline 0 \end{array}\right].$$

If the augmented system eigenvalues are made to lie in the open left half plane by state feedback or <u>otherwise</u>, the RPIS is solved.

3.1.1. <u>State feedback</u>

The fundamental theorem of feedback control states that for a system

$$\dot{\underline{x}} = A\ \underline{x} + B\ \underline{u} \tag{3.4}$$

with a state feedback

$$\underline{u} = K\ \underline{x} \tag{3.5}$$

the closed loop system

$$\dot{\underline{x}} = (A + B\ K)\ \underline{x} \tag{3.6}$$

can have any preassigned set of eigenvalues, if and only if, the pair (A,B) is controllable. The theorem gives rise to the famous pole assignment technique of linear system theory.

The formulae for pole assignment [7] are: <u>Real pole</u> (shifting from λ_i to $\hat{\lambda}_i$):

$$k_i = \frac{(\hat{\lambda}_i - \lambda_i)}{(\underline{v}_i^T\ \underline{b}_j)} \tag{3.7}$$

where,

k_i is scalar gain,

$\underline{v}_i^T$ is the eigenvector of A^T corresponding to λ_i, and

$\underline{b}_j$ corresponds to the control input used for shifting the eigenvalue, $j \leq m$.

The poles are shifted one by one, and after each shifting the resulting closed loop system is taken as the base for next shifting. The vector gain, $\underline{k}^T$ is given by,

$$\underline{k}^T = k_i \, \underline{v}_i^T \tag{3.8}$$

and develops into a gain matrix K, when multiple inputs are used for pole shifting.

Pair of complex poles (λ_i, λ_i^* shifted to $\hat{\lambda}_i$, $\hat{\lambda}_i^*$):

$$k_i = \frac{(\hat{\lambda}_i - \lambda_i)(\hat{\lambda}_i^* - \lambda_i)}{(\underline{v}_i^T \, \underline{b}_j)(\lambda_i^* - \lambda_i)} \, . \tag{3.9}$$

The resulting state feedback control is however real and is given by

$$u_j = [k_i \, \underline{v}_i^T + k_i^* \, \underline{v}_i^{*T}]\underline{x} \, . \tag{3.10}$$

For RPIS, the controllability condition of pair ($\hat{A}$, $\hat{B}$) may be broken up into following simpler conditions,

i) pair (A, B) is controllable, and

ii) rank $\left[\begin{array}{c|c} A & B \\ \hline C & 0 \end{array}\right] = (n+p)$. (3.11)

A direct corollary of (ii) above is $m \geq p$, that is the number of inputs m should be greater than or at least equal to the number of outputs p to be regulated.

For asymptotic regulation a weaker condition of stabilisability is sufficient. Stabilisability requires that either

i) pair (A, B) is controllable, or
ii) the uncontrollable modes are stable.

In this case, the control is not achieved in a finite time T, but at $t = \infty$.

The structure of the pole-placement controller comes out as

$$\underline{u} = K_1 \, \underline{x} + K_2 \int (\underline{y} - \underline{y}_r) \, dt \tag{3.12}$$

where, K_1 and K_2 are obtained by partitioning K matrix. It is a combined state feedback and integral-of-output-error control.

3.1.2. Combined observation and control

The problem of incomplete state feedback, or alternatively output feedback, can be taken care of by combined observation and control. Provided the pair (C, A) is observable, the same techniques of controller design by pole placement can be extended to the design of a Luenberger type full order observer [10]. However, if some of the states are available directly as outputs or combination of outputs, a reduced order observer serves the purpose.

A minimal order observer is a dynamical system given by

$$\dot{\hat{\underline{\zeta}}} = D \, \hat{\underline{\zeta}} + F \, \underline{y} + G \, \underline{u} \tag{3.13a}$$

$$\underline{\zeta} = T \, \underline{x} \tag{3.13b}$$

T is a transformation such that $[C' \mid T]^T$ is invertible, where C' consists of the linearly independent rows of C. Thus the order of the observer is given by (n-r), where r is the rank of C, $r \leq p$. For the purpose of observation no difference is made between control and disturbance inputs; B is extended by inclusion of E.

Differentiating Eqn.(3.13b),

$$\dot{\underline{\zeta}} = T(A\,\underline{x} + B\,\underline{u}). \tag{3.14}$$

Subtracting Eqn.(3.14) from (3.13a)

$$\dot{\hat{\underline{\zeta}}} - \dot{\underline{\zeta}} = D(\hat{\underline{\zeta}} - \underline{\zeta}) + (G - TB)\,\underline{u} + (FC' - TA + DT)\,\underline{x}. \tag{3.15}$$

The problem of observability is stated as,

$$\dot{\hat{\underline{\zeta}}} - \dot{\underline{\zeta}} \rightarrow 0 \quad \text{as } t \rightarrow \infty. \tag{3.16}$$

For the error in estimate $(\hat{\underline{\zeta}} - \underline{\zeta})$ to decay,

i) $TA - DT - FC' = 0$,

ii) the eigenvalues of D must lie in the open left half plane with the restriction $\{\lambda_i\}_D \neq \{\lambda_i\}_A$, otherwise F becomes trivial.

iii) $G = TB$; however, not with any arbitrary choice of T. (3.17)

A choice of T is

$$T = \sum_{i=1}^{n} \underline{t}_i\,\underline{v}_i^T \tag{3.18}$$

where $\underline{v}_i$ is as in (3.7).

The eigenvalues of D should be deep in the left half plane for a 'fast' observer. As in the controller design a weaker condition of <u>detectability</u> is applicable here, if finite time observation is not required.

$\underline{t}_i$'s of (3.18) are given by

$$\underline{t}_i = [\lambda_i\, I - D]^{-1}\, F[C'\,\underline{\omega}_i] \tag{3.19}$$

where $\underline{\omega}_i$ is the eigenvector of A corresponding to λ_i. F is by choice.

The structure of the observer comes out same as that of Kalman Filter except that the gain is sub-optimal.

Combined observation and control may give rise to structural instability which cannot be foretold. However by trial and error this difficulty can be circumvented.

3.1.3. Feedforward control

Dynamics of the system is controlled by pole assignment. However for improvement in the transient performance of the system, zeroes also need be assigned. The ageold method of feedforward control, used by the process control engineers, does implicit zero assignment [15]. Particularly for complete disturbance rejection and quick following of command feedforward control from disturbance and reference input signals is very effective [16].

Assuming $\underline{d}$ and $\underline{y}_r$ as vectors of constants for ease of illustrating the technique and that the system has reached steady state, Eqn.(3.1) changes to

$$0 = A\,\bar{\underline{x}} + B\,\bar{\underline{u}} + E\,\underline{d}$$

$$0 = C\,\bar{\underline{x}} - \underline{y}_r \tag{3.20}$$

where $\bar{x}$ and $\bar{u}$ are the steady state values.

By condition (ii) of (3.11) the above equation has a solution which is given by,

$$\begin{bmatrix} \bar{\underline{x}} \\ \bar{\underline{u}} \end{bmatrix} = -[S^T\,S]^{-1}\,[S^T\,H] \begin{bmatrix} \underline{d} \\ \underline{y}_r \end{bmatrix} \tag{3.21}$$

where,

$$S = \left[\begin{array}{c|c} A & B \\ \hline C & O \end{array}\right] \text{ and } H = \left[\begin{array}{c|c} E & O \\ \hline O & -I \end{array}\right].$$

For any change in $\bar{\underline{x}}$ (and $\bar{\underline{u}}$) induced by $\underline{d}$ and/or $\underline{y}_r$,

$$\tilde{\underline{x}} = \underline{x} - \bar{\underline{x}} \text{ and}$$

$$\tilde{\underline{u}} = \underline{u} - \bar{\underline{u}}\,. \tag{3.22}$$

Retaining the same state feedback law as in (3.12),

$$\tilde{\underline{u}} = K_1\,\tilde{\underline{x}} \tag{3.23}$$

leaving out the integral-of-output-error term for the time being.

Substituting (3.21) and (3.22) in (3.23),

$$\underline{u} = K_1\,\underline{x} + [K_1 - I]\,[S^T\,S]^{-1}\,[S^T\,H] \begin{bmatrix} \underline{d} \\ \underline{y}_r \end{bmatrix}. \tag{3.24}$$

It is to be noted that effective feedforward control requires exact knowledge of the system model, that is, A, B, C and E. For this reason, in practice, feedforward control is always used in conjunction with integral feedback control. The total control law thus becomes,

$$\underline{u} = K_1\,\underline{x} + K_2 \int (\underline{y} - \underline{y}_r)\,dt + K_{FF} \begin{bmatrix} \underline{d} \\ \underline{y}_r \end{bmatrix} \tag{3.25}$$

where,

$$K_{FF} = [K_1 - I]\,[S^T\,S]^{-1}\,[S^T\,H].$$

If K_{FF} is partitioned into K_3 and K_4, the feedforward term can be written as,

$$\underline{u}_{FF} = K_3\,\underline{d} + K_4\,\underline{y}_r. \tag{3.26}$$

The structure of the controller is given in Fig. 3.1.

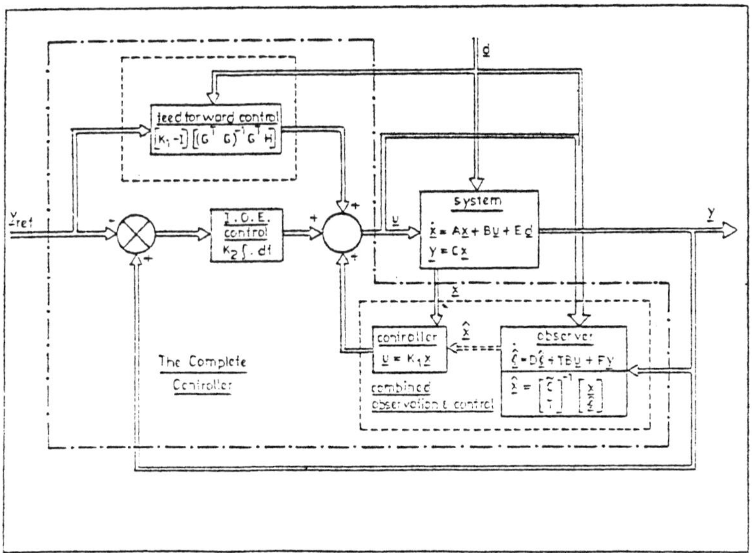

Figure 3.1. Ultimate controller structure consisting of (reconstructed) state feedback, integral of output error (IOE), and feedforward control

3.2. Microprocessor-based Controller

The implementation of the control and observation laws of Sec. 3.1 requires a controller capable of performing multiple tasks of data acquisition, computation of the control input and driving a power device. For motion control system the power device is usually a thyristor device which is amenable to direct digital control. A microcomputer-based controller thus appears to be an ideal solution.

This section describes the hardware and software of an Intel 8085 microprocessor-based controller. The controller does analog data acquisition, computation of the control voltage and direct digital control of a thyristor chopper or a single phase thyristor bridge. In the latter case zero-crossing detection of the on line voltage and linearization of the voltage-phase angle characteristic are also implemented in the controller. The controller besides the basic minimum microcomputer configuration, contains an analog input system (AIS) on a separate PCB, a synchronising (SM) and a direct digital control (DDC) module.

3.2.1. Microcomputer system

The VMC-85 microcomputer kit used for this purpose consists of:

i) 8-bit parallel processor chip 8085,

ii) two EPROM chips 2716 (2Kx8) having 4K bytes of program memory, where monitor program is stored.

iii) Four RAM chips 2114 (1Kx4) having 2K byte of memory to be used to store user programs and data during execution and also for stack operation.

iv) Two programmable peripheral interface (PPI) chips 8255, each each having three 8-bit ports used as input/output (I/O) ports.

v) One programmable timer/counter (PTC) chip 8253 having three counter/timer, and

vi) one keyboard/display controller chip 8279 for interfacing a small keyboard with 24 keys and six 7-segment LED displays used to display address (four digits) and data (two digits) in hexadecimal code.

Decoders (8205) are also used to interface the various chips to the processor.

3.2.2. Analog input system (AIS)

The analog input system consists of an 8-channel analog multiplexer (MUX) and an 8-bit analog-to-digital converter (ADC). The MUX/ADC are accommodated in a single PCB. The diagram of the system is shown in Fig. 3.2. The signals obtained from the transducers are amplified to the value of 0-5 V and filtered to remove ripples. The channel selection bits of MUX and start conversion bit of ADC are provided by the processor. The busy bit of ADC is tested by the processor in a flag-testing mode and data is read when ADC is ready.

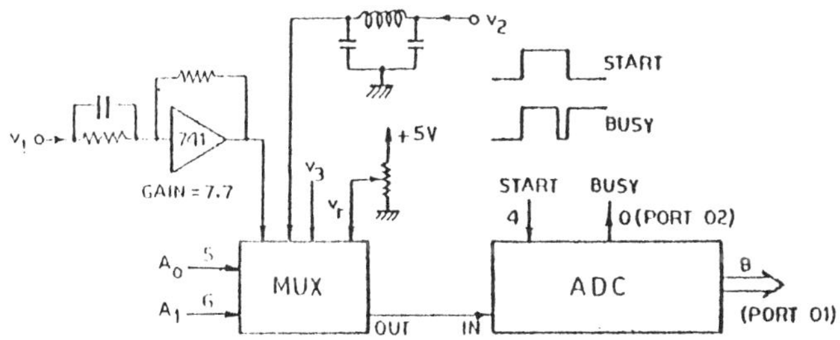

Figure 3.2. Analog input system

3.2.3. Direct digital control (DDC) module

The thyristor chopper is driven by a direct digital control (DDC) module. Two timer/counters are used to control the chopper, which is operated in the pulse-width-modulated (PWM) mode, the chopping frequency remaining constant. This frequency is taken as 200 Hz. The timer 1 is operated in mode II (astable mode), the time period being 5 ms. The clock of both timers is driven by 1.5 MHz signal obtained from 3 MHz clock signal of 8085 processor. High frequency clock is chosen for higher resolution. The gate of timer 1 is held high and the count register is loaded with the value 7500_{10}.

The timer 2 is operated in mode I (monostable mode). The gate of timer 2 is connected to the output of timer 1. The count register is loaded by the CPU in each cycle to get variable time, T_{ON}. The output of timer 2 goes low as it is gated by the output of timer 2. The output of timer 2 goes high as its count register goes to zero. This output is buffered and amplified and then fed to the gates of main and auxiliary thyristors through pulse transformers. The chopping frequency may be altered through software.

3.2.4. Synchronising module (SM)

For the single phase thyristor bridge the thyristors are to be fired in pairs after proper time delay from the zero-crossing of the ac line signal. For this purpose a separate synchronising module is developed to detect zero-crossing of ac line signal. This module consists of a step-down transformer (centre-tapped), zener diodes used as clippers and Schmitt inverters used as buffer to feed two zero-crossing points to the processor. The diagram is shown in Fig. 3.3. The other module used

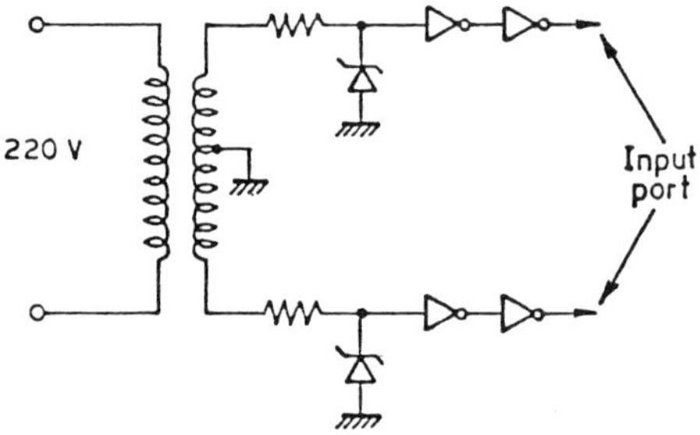

Figure 3.3. Synchronising module

is the DDC module using only buffers and amplifiers. The voltage-time delay relationship being non-linear, the time delay is read from a look-up table stored in memory. The actual time delay is realized by software.

3.2.5. Software for the controller

The two main considerations in the software development are: 1) discretisation of the continuous control and observation equations, sampling interval T being an important parameter, and 2) approximation of the equation coefficients by integer values, as it is much simpler and faster to implement integer arithmetic in the processor. The two choices of T here are 5 ms for the chopper amplifier and 10 ms for the bridge amplifier. For an electromechanical system faster action does not contribute much to the efficacy of the control. Approximation involved due to induction of integral arithmetic does not affect the quality of control to any great extent.

The software developed consists of the main program for computing the control and observation variables; integrations are changed to summations, negative numbers are represented in two's complement and a multiplication routine is used. The software package is sufficiently general and can accomodate similar or more complex control laws without any difficulty. The supporting subroutines are for data acquisition and power line synchronisation. The time delay is implemented through the programmable timer chip in the case of chopper, thus keeping the CPU free. The interrupt feature though not used in this particular case, could be used for abnormal condition monitoring. The program is written in assembly language; though BASIC or PROM-based object modules could be either extremes.

3.3. Study of a Motion Control System

The motion control system studied in the laboratory consists of a separately excited armature voltage controlled dc motor connected to an identical dc generator with lamp load.

The dc machine particulars are:

English Electric make 0.75 KW, 220 V, 4.8 A, 1500 RPM CONT. Measured values of machine parameters and constants are:

$R_a = 2.3\ \Omega$, $L_a = 46.5$ mH, $k = 1.1$ V-s/rad or N-m/A.

The field current was kept constant for both motor and generator. The combined load inertia and friction of the motor-generator set was found out by experiments and the values are:

$J = 0.07$ Kg-m^2, $\nu = 0.002$ N-m-s/rad.

The speed of the set was picked up by a dc tachogenerator, its constant being 33.8 V-s/rad, and low-pas filtered. The motor armature current was measured across 0.1 Ω resistance placed in the circuit, amplified by an OP AMP of gain 10.0 and low-pass filtered. The generator current, which is approximately proportional to load torque, neglecting losses, was measured across a 1.0 Ω resistance placed in the circuit. The complete set-up of the dc motion control system (with the chopper) is shown in Fig. 3.4.

3.3.1. Thyristor amplifiers

If the brain of the motion control system is the programmable controller, the heart of the system is the PWM chopper amplifier or the single phase thyristor bridge. In the absence of GTO thyristors, Jones chopper configuration is used. The chopper has been modelled as a linear gain element in spite of discontinuous conduction. The single phase thyristor bridge has a non-linear input-output relationship, linearisation has been done by means of a look-up table stored in the memory.

3.3.2. Model of DC motor

The electrical port equation of the motor is

$$L_a \, i_a + R_a \, i_a + k \, \omega = v_a \tag{3.26}$$

where ω is the motor speed in rad/s and i_a, v_a are the armature current and voltage respectively.

The mechanical port equation of the motor is,

$$J \, \dot{\omega} + \nu \, \omega + T_L = k \, i_a \tag{3.28}$$

where T_L is the load torque in N-m.

With $[\omega \ i_a]^T$ as state vector, v_a as control input u, T_L as disturbance input d, and ω as output y, the state model of the dc motor is,

$$\dot{\underline{x}} = \begin{bmatrix} -\nu/J & k/J \\ -k/L_a & -R_a/L_a \end{bmatrix} \underline{x} + \begin{bmatrix} 0 \\ 1/L_a \end{bmatrix} u + \begin{bmatrix} -1/J \\ 0 \end{bmatrix} d \tag{3.29}$$

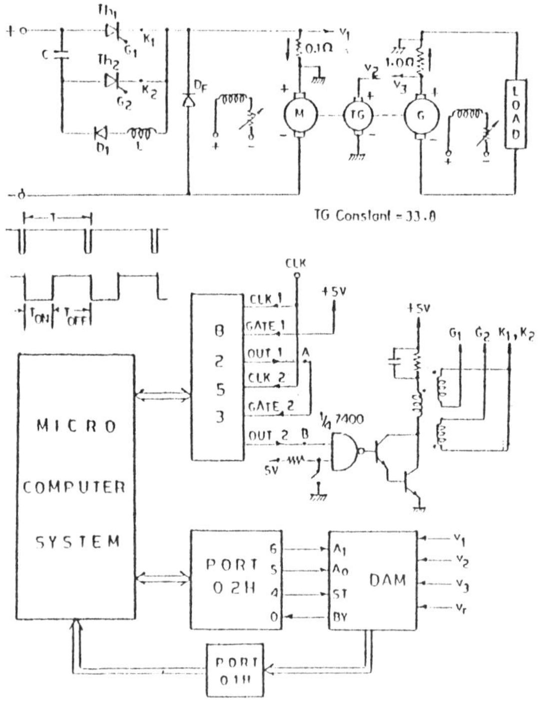

Figure 3.4. Experimental set-up

$$y = [1 \quad 0]\ \underline{x} \tag{3.30}$$

3.3.3. Determination of control and observation laws

For a set point regulation to y_r, the augmented system equation (3.3), after substituting the numerical values, is

$$\dot{\underline{z}} = \begin{bmatrix} -0.0286 & 15.714 & 0.0 \\ -23.656 & -48.387 & 0.0 \\ 1.0 & 0.0 & 0.0 \end{bmatrix} \underline{z} + \begin{bmatrix} 0.0 \\ 21.505 \\ 0.0 \end{bmatrix} v\ . \tag{3.31}$$

The eigenvalues are at $\lambda_1 = 0.0$, $\lambda_2 = -9.617$ and $\lambda_3 = -38.8$. The system passes the controllability tests of Eqn.(3.11). If λ_1 is shifted to $\hat{\lambda}_1 = -20.0$, the corresponding scalar gain k_1 of Eqn.(3.7) is -22.082.

The integral feedback control law:

$$u = [-2.864 \quad -0.93] \begin{bmatrix} x_1 \\ x_2 \end{bmatrix} -22.082 \int (y-y_r)\ dt\ . \tag{3.32}$$

If feedforward control is added, the modified control law is,

$$u = -2.864\ x_1 - 0.93\ x_2 - 22.082 \int (y-y_r)\ dt + 2.89\ d + 3.97\ y_r\ . \tag{3.33}$$

If instead of measuring the armature current directly, it is estimated by a minimal order observer as:

$$\dot{\hat{\zeta}} = -50.0\ \hat{\zeta} - 28.784\ y + 21.505\ u + 1.466\ d$$

$$\hat{x}_2 = \hat{\zeta} + 0.103\ y \tag{3.34}$$

where $d_1 = -50.0$ for a 'fast' observer and $f = -28.784$. For combined observation and control x_2 of Eqn.(3.32) is replaced by $\hat{x}_2$ of (3.34).

The discretised version of (3.34) with sampling time of 5 ms [15], for microprocessor implementation is:

$$\hat{\zeta}(k+1) = 0.7788\ \hat{\zeta}(k) + 0.095\ u(k) + 0.0065\ d(k) - 0.1273\ y(k)$$

$$\hat{x}_2(k) = \hat{\zeta}(k) + 0.103\ y(k). \tag{3.35}$$

Converted into the integral co-efficients, the Eqn.(3.35) becomes

$$\hat{\zeta}(k+1) = (199/256)\ \hat{\zeta}(k) + (24/256)\ u(k) + (3/256)\ v_3(k) - 4\ v_2(k)$$

$$\hat{x}_2(k) = \hat{\zeta}(k) + (7/2)\ v_2(k). \tag{3.36}$$

Taking the chopper gain as 1.2, that is, output voltage to time delay, u is converted to T_{ON}, the control law of Eqn.(3.33) for computation becomes,

$$T_{ON} = -81\ v_1 - v_2 - 3 \sum_{n=0}^{p} (v_1 - v_r) + 112\ v_r + (5/2)\ v_3 \tag{3.37}$$

where the sampling time and measuring constants for the transducers have also been taken into account. The diagram of the controller for the dc motion control system is shown in Fig. 3.5.

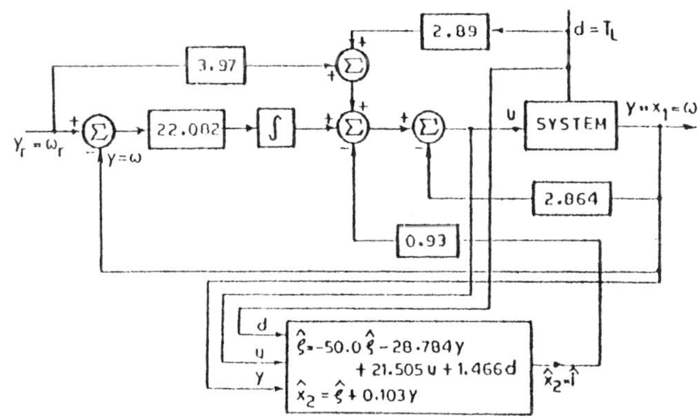

Figure 3.5. The combined controller-observer structure for dc motion control system

3.4. Experiments and Results

The following simulation and laboratory experiments were performed on the dc motion control system. The results are presented in the graphical form in Figs. 3.6 and 3.7.

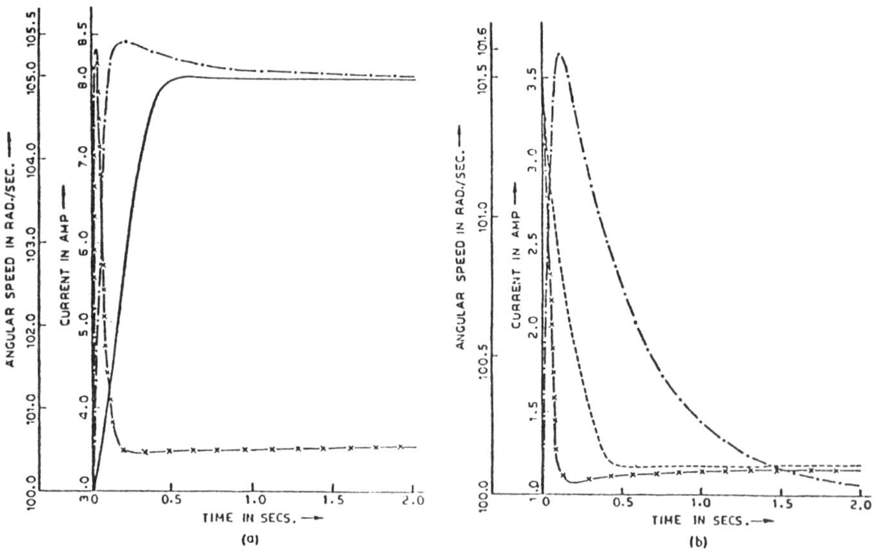

Figure 3.6. Experimental and simulation results for dc thyristor chopper driven motor

(a) Feedback and feedforward control in terms of reference without an observer, the actual states being measured for upward step change in reference speed from 100.0 rad/s to 105.0 rad/s with load torque of 3.65 N.m. remaining constant.

(b) Feedback control with minimal order observer for estimation of motor armature current assumed to be inaccessible for downward step change in load torque from 3.65 N.m. to 1.26 N.m. with the reference speed of 100.0 rad/s remaining constant.

The results are in agreement within limitation of the laboratory set-up. The differences are due to the following reasons: (i) inaccurate modelling of the thyristor amplifier (an accurate model of the chopper amplifier is given in [11]),(ii) the microcomputer, because of its limited word length, introduces a saturation type non-linearity. This is in no way harmful to the practical applications of the dc motion control systems.

Perfect regulation is obtained with the microcomputer controller and the case study proves the points that the microcomputer controller based on the theory outlined is a viable practical proposition.

Comments: The microcomputer-based controller could be time-shared for multi-machine control, e.g. rolling mill motor stands. Though a 16-bit microprocessor would achieve faster control and easier programming, the experience (with Intel 8086) shows that the faster and more powerful processors are not that suitable for the industrial environment of heat and dust.

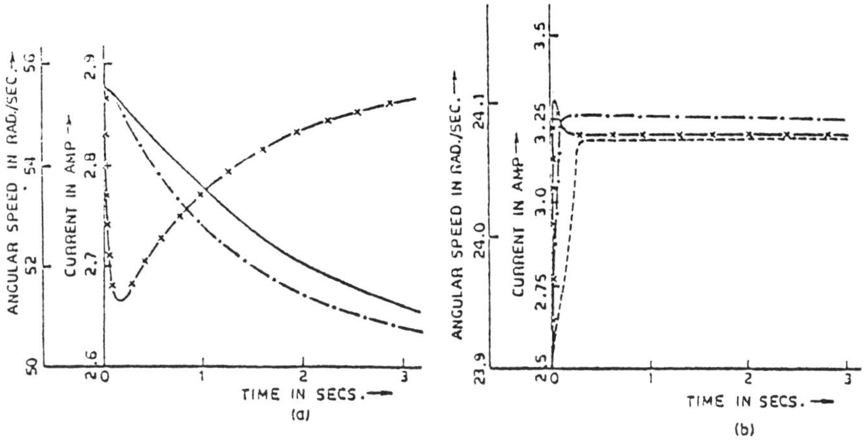

Figure 3.7. Experimental and simulation results for single phase thyristor bridge driven dc motor

(a) Feedback control with minimal order observer for estimation of motor armature current for downward step change in reference speed from 55.5 rad/s to 50.27 rad/s with load torque of 3.08 N.m. remaining constant.

(b) Feedback and feedforward control in terms of load torque (disturbance) without an observer for upward step change in load torque of 2.7 N.m. to 3.47 N.m. with reference speed of 24.09 rad/s remaining constant.

Motor speed { —— Experimental, -·- Simulation }   Armature current { --- Experimental, -x- Simulation }

IV. ON-LINE PARAMETER IDENTIFICATION IN CONTINUOUS DYNAMICAL SYSTEMS

4.1. Introduction

A microprocessor unit for on line identification of the parameters in linear lumped continuous systems based on the so called Poisson moment functional (PMF) method [8,13] will now be described. The PMF method is aimed at mainly averting the difficulties in estimating the effect of derivative terms in continuous models and is considerably immune to zero-mean additive disturbances. A signal f(t), $t \in (0,t_o)$ is treated as a distribution or generalized function and expanded about a time t_o in an exponentially weighted series as

$$f(t) = \sum_{k=0}^{\infty} M_k\{f(t)\} \exp[-\lambda(t-t_o)]\ \delta^{(k)}(t-t_o), \tag{4.1}$$

where $\delta^{(k)}(t-t_o)$ is the k-th generalized time derivative of an impulse distribution occuring at $t = t_o$.

$$M_k\{f(t)\} \triangleq f_k^o = \int_0^{t_o} f(t)\ p_k(t_o-t)\ dt \tag{4.2}$$

$$p_k(t_o) \triangleq p_k^o = \frac{t_o^k}{k!} \exp(-\lambda t_o) \tag{4.3}$$

and λ is a positive real number.

p_k^o is termed as the k-th order Poisson pulse function at t_o and f_k^o is the k-th PMF of f(t) about t_o. f_k^o can be viewed as the output, due to an input f(t), at $t = t_o$, of the (k+1)-th stage of a cascaded filter with identical stages, each element of which has a transfer function $1/(s+\lambda)$. Such a filter chain is known as a Poisson filter chain (PFC).

In order to illustrate the role of PMF transformation in lumped linear models let us consider the following problem.

"Given f(t) and r(t) along with all the related PMF's over an arbitrary but active interval of time (O,T), determine the parameters of a system modelled by

$$(1 + a_{o,1}t) \frac{d^2 f(t)}{dt^2} + (a_{1,o} + a_{1,1}t) \frac{df(t)}{dt} + (a_{2,o} + a_{2,1}t)\, f(t)$$
$$= (b_{2,o} + b_{2,1}t)\, r(t)\text{"} \quad .$$

To set up a procedure to solve this problem, we first take the k-th PMF transformation of the model equation about $t_o \in (O,T)$ as

$$\begin{aligned}
&[f_{k-2}^o - 2\lambda f_{k-1}^o + \lambda^2 f_k^o - (p_{k-1}^o - \lambda p_k^o)\, f^{(o)}(o) - p_k^o\, f^{(1)}(o)] \\
&+ a_{o,1}[t_o\, f_{k-2}^o - \{2\lambda t_o + (k+1)\}\, f_{k-1}^o + \lambda\{\lambda t_o + 2(k+1)\}\, f_k^o \\
&- \lambda^2 (k+1)\, f_{k+1}^o + p_k^o\, f^{(o)}(o)] + a_{1,o}\, [f_{k-1}^o - \lambda\, f_k^o - p_k^o\, f_{(o)}^{(o)}] \\
&+ a_{1,1}[t_o\, f_{k-1}^o - \{\lambda t_o + (k+1)\}\, f_k^o + \lambda(k+1)\, f_{k+1}^o] + a_{2,o}\, f_k^o \\
&+ a_{2,1}\{t_o\, f_k^o - (k+1)\, f_{k+1}^o\} \\
&= b_{2,o}\, r_k^o + b_{2,1}\{t_o\, r_k^o - (k+1)\, r_{k+1}^o\} \quad .
\end{aligned}$$

The above relation may be arranged in the form:

$$[-(f^o_{k-1} - \lambda f^o_k),\ -f^o_k,\ -\{t_o f^o_{k-2} - (2\lambda t_o + \overline{k+1})\, f^o_{k-1}$$

$$+ \lambda(\lambda t_o + 2\,\overline{k+1})\, f^o_k - \lambda^2 (k+1)\, f^o_{k+1}\} \quad ,$$

$$-\{t_o f^o_{k-1} - (\lambda t_o + k+1)\, f^o_k + \lambda(k+1)\, f^o_{k+1}\} \quad ,$$

$$-\{t_o f^o_k - (k+1)\, f^o_{k+1}\},\ r^o_k,\ \{t_o\, r^o_k - (k+1)\, r^o_{k+1}\} \quad ,$$

$$(p^o_{k-1} - \lambda\, p^o_k),\ p^o_k]$$

$$[a_{1,o},\ a_{2,o},\ a_{o,1},\ a_{1,1},\ a_{2,1},\ b_{2,o},\ b_{2,1},\ \theta_{1,o},\ \theta_{2,o}]^T$$

$$= [f^o_{k-2} - 2\lambda f^o_{k-1} + \lambda^2 f^o_k] \quad ,$$

where $\theta_{1,o} = f^{(o)}(o)$ and $\theta_{2,o} = \{(a_{1,o} - a_{o,1})\, f^{(o)}(o) + f^{(1)}(o)\}$.

To get the 9 unknowns (seven essential model parameters and two related to the initial conditions), we require 9 equations. These may be formed by following two methods:

Method 1:

In this method we use PMF's about a single instant of time $t_o \in (0,T)$ with $k = 2,3\ldots,10$. The resulting set of 9 equations may be written in the form

$$\Phi\, U = C\ ,$$

where Φ consists of elements in terms of PMF's of process signals, Poisson pulse functions, the Poisson filter constant λ, and the time instant t_o about which PMF transformations are taken; C consists of elements in terms of PMF's of the process output and λ. The vector U will be obtained as

$$U = \Phi^{-1}\, C.$$

The matrix Φ may be partitioned as

$$\Phi = [\Phi_F \mid \Phi_R \mid \Phi_P]\ ,$$

where the block suffixes F, R and P indicate that the blocks are related to the PMF's of f(t), r(t) and Poisson pulse functions.

In the present case

$$\Phi_P = \begin{bmatrix} (p^o_1 - \lambda\, p^o_2) & p^o_2 \\ (p^o_2 - \lambda\, p^o_3) & p^o_3 \\ \cdot & \cdot \\ \cdot & \cdot \\ \cdot & \cdot \\ \cdot & \cdot \\ (p^o_9 - \lambda\, p^o_{10}) & p^o_{10} \end{bmatrix}, \qquad \Phi_R = \begin{bmatrix} r^o_2 & (t_o r^o_2 - 3r^o_3) \\ r^o_3 & (t_o r^o_3 - 4r^o_4) \\ \cdot & \cdot \\ \cdot & \cdot \\ \cdot & \cdot \\ \cdot & \cdot \\ r^o_{10} & (t_o r^o_{10} - 11 r^o_{11}) \end{bmatrix},$$

and

$$\Phi_F = \begin{bmatrix} -(f_1^o-\lambda f_2^o),-f_2^o,-\{t_o f_o^o-(2\lambda t_o+3)f_1^o+(\lambda^2 t_o+6\lambda)f_2^o-3\lambda^2 f_3^o\},-\{t_o f_1^o-(\lambda t_o+3)f_2^o+3\lambda f_3^o\}, \\ -(t_o f_2^o-3f_3^o) \\ -(f_2^o-\lambda f_3^o),-f_3^o,-\{t_o f_1^o-(2\lambda t_o+4)f_2^o+(\lambda^2 t_o+8\lambda)f_3^o-4\lambda^2 f_4^o\},-\{t_o f_2^o-(\lambda t_o+4)f_3^o+4\lambda f_4^o\}, \\ -(t_o f_3^o-4f_4^o) \\ \vdots \\ -(f_9^o-\lambda f_{10}^o),-f_{10}^o,-\{t_o f_8^o-(2\lambda t_o+11)f_9^o+(\lambda^2 t_o+22\lambda)f_{10}^o-11\lambda^2 f_{11}^o\},-\{t_o f_9^o-(\lambda t_o+11) \\ f_{10}^o+11\lambda f_{11}^o\},-(t_o f_{10}^o-11f_{11}^o) \end{bmatrix} .$$

The 9-vector C is obtained as

$$C = [(f_o^o - 2\lambda f_1^o + \lambda^2 f_2^o),\ (f_1^o - 2\lambda f_2^o + \lambda^2 f_3^o), \ldots\ldots \ldots\ldots (f_8^o - 2\lambda f_9^o + \lambda^2 f_{10}^o)]^T .$$

U is given by

$$U = [a_{1,o},\ a_{2,o},\ a_{o,1},\ a_{1,1},\ a_{2,1},\ b_{2,o},\ b_{2,1},\ \theta_{1,o},\ \theta_{2,o}]^T .$$

<u>Corollary 1</u>

If it is known a priori that the system is time invariant, then the matrix Φ and vectors C and U will be reduced in dimension (since $a_{o,1} = A_{1,1} = a_{2,1} = b_{2,1} = 0$) and become

$$\Phi = \left[\begin{array}{cc|c|cc} -(f_1^o-\lambda f_2^o) & -f_2^o & r_2^o & (p_1^o-\lambda p_2^o) & p_2^o \\ -(f_2^o-\lambda f_3^o) & -f_3^o & r_3^o & (p_2^o-\lambda p_3^o) & p_3^o \\ \vdots & \vdots & \vdots & \vdots & \vdots \\ -(f_5^o-\ f_6^o) & -f_6^o & r_6^o & (p_5^o-\ p_6^o) & p_6^o \end{array}\right]$$

$$= [\Phi_F \,\vdots\, \Phi_R \,\vdots\, \Phi_P] ,$$

$$C = [(f_o^o - 2\lambda f_1^o + \lambda^2 f_2^o),\ (f_1^o - 2\lambda f_2^o + \lambda^2 f_3^o), \ldots,\ (f_4^o - 2\lambda f_5^o + \lambda^2 f_6^o)]^T ,$$

$$U = [a_{1,o},\ a_{2,o},\ b_{2,o},\ \theta_{1,o},\ \theta_{2,o}]^T .$$

<u>Corollary 2</u>

If the initial conditions are further known a priori to be all zero in the above, then it becomes a case of transfer function synthesis. In this situation

$$\Phi = \begin{bmatrix} -(f_1^o - \lambda f_2^o) & -f_2^o & r_2^o \\ -(f_2^o - \lambda f_3^o) & -f_3^o & r_3^o \\ -(f_3^o - \lambda f_4^o) & -f_4^o & r_4^o \end{bmatrix} ,$$

$$C = [(f_o^o - 2\lambda f_1^o + \lambda^2 f_2^o),\ (f_1^o - 2\lambda f_2^o + \lambda^2 f_3^o),\ (f_2^o - 2\lambda f_3^o + \lambda^2 f_4^o)]^T ,$$

$$U = [a_{1,o},\ a_{2,o},\ b_{2,o}]^T .$$

Method 2:

In order to determine the 9 unknowns we can alternatively form 9 simultaneous equations by choosing $k = 2$ at 9 different instants of time, $t_i \in (0,T)$; $i = 1,2,\ldots,9$. We then get

$$\hat{\Phi}_P = \begin{bmatrix} (p_1^1 - \lambda p_2^1) & p_2^1 \\ (p_1^2 - \lambda p_2^2) & p_2^2 \\ \vdots & \vdots \\ (p_1^9 - \lambda p_2^9) & p_2^9 \end{bmatrix} , \qquad \hat{\Phi}_R = \begin{bmatrix} r_2^1 & (t_1 r_2^1 - 3\ r_3^1) \\ r_2^2 & (t_2 r_2^2 - 3\ r_3^2) \\ \vdots & \vdots \\ r_2^9 & (t_9 r_2^9 - 3\ r_3^9) \end{bmatrix} ,$$

and

$$\hat{C} = [(f_o^1 - 2\lambda f_1^1 + \lambda^2 f_2^1),\ (f_o^2 - 2\lambda f_1^2 + \lambda^2 f_2^2),\ \ldots\ldots \\ (f_o^9 - 2\lambda f_1^9 + \lambda^2 f_2^9)]^T .$$

These can be written as

$$\hat{\Phi}\, U = \hat{C} ,$$

$$\hat{\Phi}_F = \begin{bmatrix} -(f_1^1-\lambda f_2^1), -f_2^1, -\{t_1 f_1^1 - (2\lambda t_1+3) f_1^1 + (\lambda^2 t_1+6\lambda) f_2^1 - 3\lambda^2 f_3^1\}, -\{t_1 f_1^1 - (\lambda t_1+3) f_2^1 + 3\lambda f_3^1\}, \\ -(t_1 f_2^1 - 3 f_3^1) \\ -(f_1^2-\lambda f_2^2), -f_2^2, -\{t_2 f_1^2 - (2\lambda t_2+3) f_1^2 + (\lambda^2 t_1+6\lambda) f_2^2 - 3\lambda^2 f_3^2\}, -\{t_2 f_1^2 - (\lambda t_2+3) f_2^2 + 3\lambda f_3^2\}, \\ -(t_2 f_2^2 - 3 f_3^2) \\ \vdots \quad \vdots \quad \vdots \quad \vdots \quad \vdots \\ -(f_1^9-\lambda f_2^9), -f_2^9, -\{t_9 f_1^9 - (2\lambda t_9+3) f_1^9 + (\lambda^2 t_9+6\lambda) f_2^9 - 3\lambda^2 f_3^9\}, -\{t_9 f_1^9 - (\lambda t_9+3) f_2^9 + 3\lambda f_3^9\}, \\ -(t_9 f_2^9 - 3 f_3^9) \end{bmatrix} ,$$

where

$$\hat{\Phi} = [\ \hat{\Phi}_F \mid \hat{\Phi}_R \mid \hat{\Phi}_P] ,$$

and consequently

$$U = \hat{\Phi}^{-1}\ \hat{C} .$$

The time invariant case and the case of transfer function synthesis may be treated in a similar manner as discussed in corollaries of Method 1.

If the initial conditions are known a priori to be zero, the scheme is as shown in Fig. 4.1.

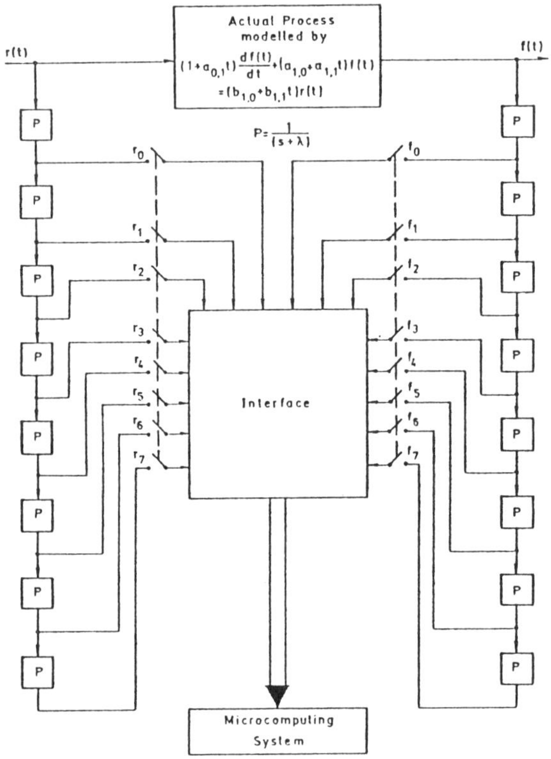

Figure 4.1. PMF method of parameter identification in a first order time-varying model

4.2. Development of Interface Between the Process and Microprocessor

A microprocessor unit dedicated to the task of estimating the parameters of a continuous dynamic process through the PMF method requires an interface for on line operation with the process under identification. We discuss here some details of development of such an interface. The interfacing unit consists of three subsystems:

i) A pair of Poisson filter chains (PFC), one each to generate the PMF's of the process input and the process output.

ii) A multiplexer (to essentially economize hardware) enabling a single A/D converter to handle a number of signals arising out for the PFC.

iii) An A/D converter to convert the analog PMF signals into digital form.

<u>a) Poisson filter chains:</u> Two identical filter chains with a number of stages, each stage having transfer function of the form $-\lambda/(s+\lambda)$, are in-

corporated. Poisson filter element of this type is easily realized by means of an OP Amp (IC 741) in the inverting mode with RC feedback as shown in Fig. 4.2 that can be set to any desired value. The effect of the

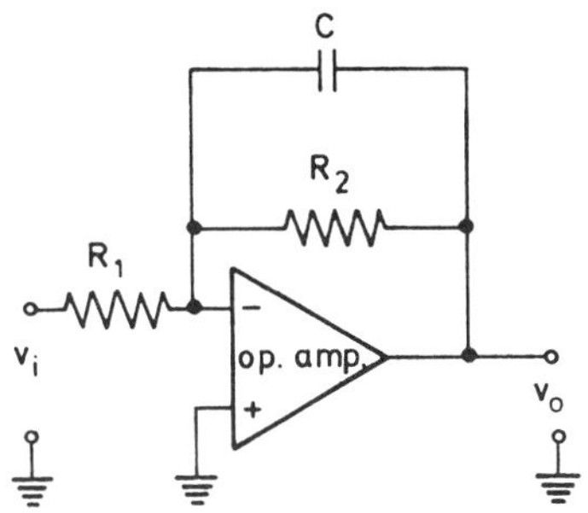

Figure 4.2. Poisson filter element ($P(s) = \frac{-\lambda}{s+\lambda}$)

gain term '$-\lambda$' in the PFC element can be easily included in the PMF formulation. This additional gain term will helpfully strengthen the higher order PMF's.

Since the Poisson filter element is designed for inversion, the PMF's of order 0,1,2,3,... will have signs −,+,−,+,... respectively. This alternation of the signs of the PMF's is taken into consideration in A/D conversion. Use of an A/D converter unit with facility for bipolar operation is suggested, as in Fig. 4.3, instead of unipolar converter with additional compensatory inverters at the required stages of output of the PFC. In the present system each PFC has 3 stages.

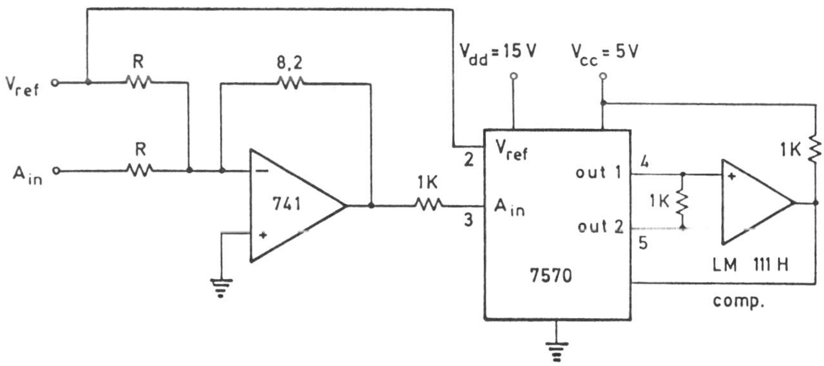

Figure 4.3. Bipolar operation

b) The Multiplexer: An 8 channel analog multiplexer (MP 7501) is used to multiplex the PMF's. f_o, f_1, f_2, r_o, r_1, r_2 are handled through the input lines s_1 to s_6 in 6 channels and the unused channels have been grounded.

The analog inputs are selected by 3 address logic lines (A_2, A_1, A_o). These address and enable lines are connected to port A (in the output mode) of a standard microprocessor kit (SDK 85). The multiplexer address is sequentially incremented so as to access all the PMF's. Connections between PFC and multiplexer, and between the multiplexer and the output port are shown in Fig. 4.4.

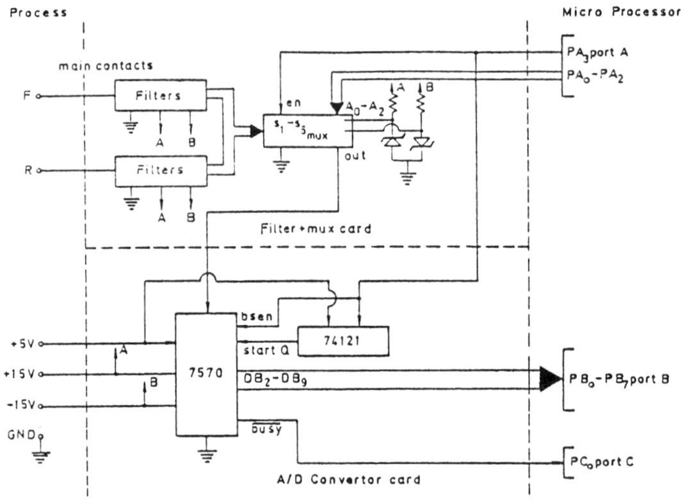

Figure 4.4. The interface

The analog output line of the multiplexer is connected to the analog input of the A/D converter. Since the enable line of the multiplexer is tied to the START line of the 74121 chip which triggers the A/D converter, addressing any input line will automatically initiate the conversion process without any further control signals from the microprocessor.

c) The A/D Converter: An MP 7570 J A/D unit converts the signals from the analog multiplexer into 8 bit digital form. The conversion is initiated by a low to high transition on the START (B) line of 74121 chip, the output of which is connected to the START line of the A/D converter. To start the A/D conversion, a pulse of at least 500 ns duration is required, but the conversion does not start until the start line goes low again. In view of the absence of sample and hold gates here, the START pulse must be made as short as possible without going below the critical width to make sampling synchronous for all practical purposes.

Data must be acquired as fast as possible. This is possible if a tight wait loop is programmed such that a byte is read when it is ready at regular intervals and the conversion then restarted. Alternatively, the BUSY line may be used. The former method has the drawback that, should the A/D converter clock drift slightly due to changes in V_{CC}, the microprocessor may start reading in early, or late. Therefore, the latter method is employed here with the help of the BUSY line. The BUSY line is connected to the last bit of port C and the data bits to port B as shown in Fig. 4.4.

Binary values 08 H to OC H are outputted through port A enabling the multiplexer and addressing the input sequentially. The microprocessor is programmed to input data when the BUSY line goes high, store the value and then increment the multiplexer address.

Fig. 4.5. shows the timing diagram and Fig. 4.6. shows general flow

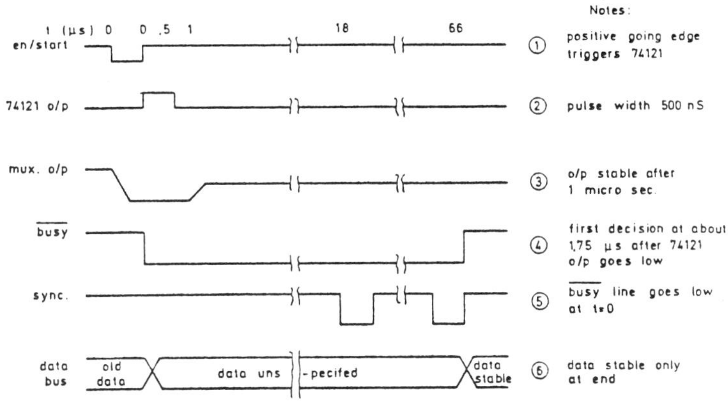

Figure 4.5. Timing diagram

chart for the parameter identification algorithm with the PMF method.

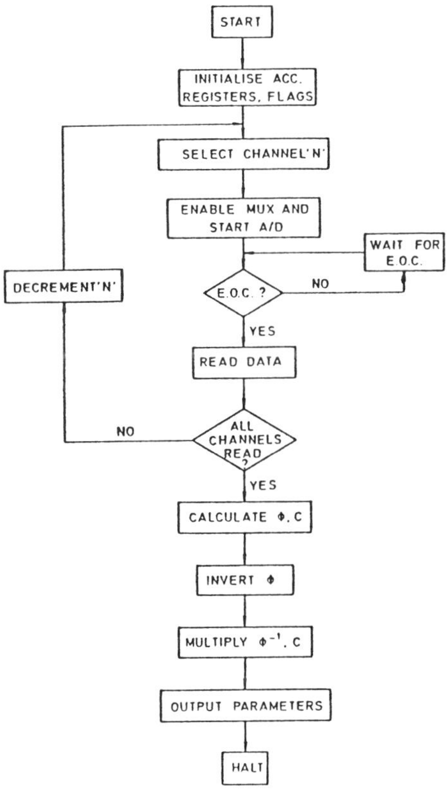

Figure 4.6. Flow chart

4.3. Performance

The unit has been tested with a simulated first order dynamical system whose transfer function is given by 4/(s+4). The dynamical system simulated is completely electrical, i.e. an OP Amp with resistive and capacitive feedback. A positive going unit step is the input to the system. This input and the resulting output are fed to a pair of PFC at $t = 0$. Thus the effect of initial conditions is removed. Consequently in the identification algorithm Φ_p or $\hat{\Phi}_p$ were not inserted. The number of unknowns is 2.

If the PFC is connected arbitrarily some time after the main process went into action, the initial conditions need consideration. At the same time if λ is chosen sufficiently large, and if t_o is also long, the effect of initial conditions will decay very fast in the PMF's measured about that time. Our algorithms therefore, may be written ignoring the initial conditions for all purposes. In the general algorithm, however, the choice of λ is completely free.

Returning to the discussions of our simulation, we choose $\lambda = 3.33$ ($R = 150$ kilo ohm and $C = 2$ microfarad). The parameters of the main process are found to be 4.34 and 4.32 in Method 1 and 3.23 and 2.91 in Method 2. The errors are due to the following:

i) Limited A/D converter resolution: The 8 bit A/D converter finds it difficult to resolve accurately the higher order PMF's. Higher order PMF's are generally weaker than the lower order ones. This generally weakens the determinant of Φ or $\hat{\Phi}$. Consequently, the results become very sensitive to errors in Φ (or $\hat{\Phi}$) and C (or $\hat{C}$).

ii) Lack of truely synchronized samples: The PMF's corresponding to an instant of time are not truely synchronous in the present unit which emplys only one A/D converter for reasons of economy of hardware. The various PMF's corresponding to an instant of time are actually "queued" for a short time which includes the conversion time. The slight delay is considered negligible in view of large time constant of the main process and the consequent slow modes of the signals. When λ is large, this however, cannot be considered negligible. Moreover, when a large number of PMF's are queued, the largest delay experienced by the last PMF for conversion may not be negligible.

4.4. Conclusion and Discussion

An implementation of the PMF method of parameter identification in linear continuous systems is presented. The PMF algorithm may be modified to handle models with unknown time lags [13]. In fact, all models which are linear in the parameters may be directly handled with suitable modification of the basic algorithm. It is interesting to note that the interface in all such cases remains the same in form. It will have to be expanded only in size to handle larger data in more complicated models.

The PMF's of the process signals are found to be considerably immune to zero mean additive noise, if λ is fairly small. In spite of the freedom available to us in the choice of λ, we will be guided by the actual situation in the particular application to choose a suitable value for λ. Large λ will practically eliminate the effects of the initial conditions, but owing to its large band-width in this condition, the PFC will admit more noise than when λ is small. It is possible to use Kalman filtered PMF's [13] if we wish to have optimal filtering.

The PMF's of higher order are generally weak. Consequently, when higher order PMF's are to be used (inevitably in Method 1 and particularly when the model order is high), the matrices Φ and $\hat{\Phi}$ have weakened determinants, making the results more sensitive to errors. In order to overcome this problem, we suggest the use of multiple PFC with various values of λ to generate data for the required number of independent equations, inspite of its increased cost.

V. REFERENCES

[1] AL-ANBUKY, A.H., and SWADI, G.A.: 'Microprocessor position control system', Microprocessors and Microsystems, 3, 1979, 163-167.

[2] AUSLANDER, D.M., and SAGUES, P.: Microprocessors for measurement and control, Osborne/McGraw Hill, Berkeley, 1978.

[3] AUSLANDER, D.M., TAKAHASHI, Y., and TOMIZUKA, M.: 'Direct digital process control: practice and algorithms for microprocessor application', Proc. IEEE, 66, 1978, 199-208.

[4] BARKER, H.A.: 'The microprocessor in control', Proc. IEE, 126, 1979, 77-80.

[5] BASSET, S.: (Ed.) 'Special report on microprocessors/microcomputers', Computer Design, 22, 1983, 173-246.

[6] BOYET, H., and KATZ, R.: Microprocessor Applications, 1, E & L Instruments Inc., Derby, 1979.

[7] DE, N.K., SINHA, S., and CHATTOPADHYAY, A.K.: 'Microcomputer as a programmable controller for state feedback control of a d.c. motor employing thyristor amplifier', Proc. of IEEE Ind. Appl. Soc. Annual Meeting, Chicago, U.S.A., 1984, 586-592.

[8] GANTI PRASADA RAO, SAHA, D.C., RAO, T.M., BHAYA, A., and AGHORAMURTY, K.: 'A microprocessor based system for on-line parameter identification in continuous dynamical systems", IEEE Trans., IE-29 (3), 1982, 197-201.

[9] LIN, W.C.: 'Microprocessor based digital system design fundamentals and the development laboratory for hardware designers and engineering executives', Proc. IEEE, 65, 1977, 1138-1161.

[10] LUENBERGER, D.G.: 'Observers for multivariable systems', IEEE Trans., AC-11 (2), 1966, 190-197.

[11] RAJASEKARAN, K., BHADRA, S.N., and SINHA, S.: 'Analysis of state feedback control of chopper driven separately excited d.c. motor', Elec. Mach. and Electromech., 7 (3), 1982, 155-170.

[12] ROLLS, T., WILLIS, A., and GWINN, R.: 'Design concept for a microprocessor based temperature controller', Microprocessors and Microsystems, 6, 1982, 225-233.

[13] SAHA, D.C., and GANTI PRASADA RAO: Identification of continuous dynamical systems - The Poisson Moment Functional (PMF) Approach, Springer Verlag, Berlin Heidelberg New York Tokyo, 1983.

[14] SCHMIDT, G., and SWIK, R.: 'Microcomputers in automatic control applications - A software point of view', Proc. 7th Triennial World Congress of the IFAC, Helsinki, Finland, June 1978.

[15] SINHA, R.U., DE, N.K., SINHA, S., and CHATTOPADHYAY, A.K.: 'Microprocessor based state feedback control of a d.c. thyristor chopper drive motor', Proc. Nat. Syst. Conf., Bombay, India, 1984, 221-227.

[16] SMITH, H.W., and DAVISON, E.J.: 'Design of industrial regulators - Integral feedback and feedforward control', Proc. IEE, 119 (8), 1972, 1210-1216.

[17] WONHAM, W.M.: 'On pole assignment in multiinput controllable linear systems', IEEE Trans., AC-12 (6), 1967, 660-665.

ACKNOWLEDGMENT

The final preparation of this work took place at the Lehrstuhl für Elektrische Steuerung und Regelung, Ruhr-Universität Bochum, West Germany during the May-June 1985 visit of Prof. Ganti Prasada Rao which was supported by the Alexander von Humboldt Foundation.

CHAPTER 3

SELF-TUNING AND ADAPTIVE CONTROL

D.W.Clarke
Department of Engineering Science
Oxford University
Parks Road
Oxford OX1 3PJ
England

ABSTRACT. Conventional controllers have transfer-functions with fixed parameters which are chosen from knowledge of plant dynamics or by manual tuning experiments. For cases where the dynamics are unknown a self-tuning algorithm can be used; with the advent of microprocessor technology such methods are easily implemented and are increasingly popular. This article describes the background to many current approaches to self-tuning design which are based on a Generalised Minimum Variance strategy. The plant is described as a discrete-time transfer-function with dead-time and with stochastic disturbances; a predictor model is then generated and the unknown parameters of this model are estimated from plant input/output data using a Recursive Least Squares estimator. An outline of the implementation of this self-tuner is given together with a discussion of additional features required to make it work in practice. One important limitation of the GMV approach is the need to know the plant's dead-time; a new method called Generalised Predictive Control is shown to overcome this problem. For simple process-control applications PID regulators are entirely adequate and methods for the self-tuning of their proportional, integral and derivative terms are described - all of which are easy to embed into a microcomputer.

1 INTRODUCTION

The control of manufacturing processes has traditionally been the domain of three-term or PID regulators and relay logic systems. The early use of computers for direct digital control concentrated most of these functions into a large centralised unit where the potential benefits of overall plant optimisation were often not achieved because of the complexity of the system design and its consequential loss of availabilty. Microprocessors have given rise to a change of emphasis towards dedicated single-loop functions linked by a plant-wide network to convenient operator displays. The PID terms are set either remotely

N. K. Sinha (ed.), Microprocessor-Based Control Systems, 35–59.

or by a hand-held unit connected to the microcontroller. The overall design of these distributed control systems can be highly resilient to individual failures as well as allowing for an incremental implementation approach. The basic control algorithms, on the other hand, have hardly changed: the analog function of the PID regulator is simply discretised using fast sampling and some code is added for practical requirements such as integral desaturation (though it is surprising how many implementations fail to take proper account of details such as filtering of the derivative term). This is because it is felt that typical plant engineers are capable of manually tuning a PID loop and that most processes are regulated adequately in this way.

More recently, however, there has been increasing interest in the use of microcomputers for the self-tuning of control loops in which parameter settings are automatically chosen based on direct experimental data. This has been prompted partly by market pressures but also by the recognition that some new applications would be in non-traditional areas (such as biochemical processes) where the manual tuning skill might be absent or where the plants exhibit complex behaviour:

- multiple capacities and dead-time leading to excessive phase-lag for which more advanced laws would give tighter control
- nonlinearities such as in pH regulation where the process gain is a function of the set-point
- variations in dynamics due to catalyst decay or to enzyme poisoning
- interactions between loops arising from the pressure to minimise intermediate holding capactity
- variations in the quality of the feedstock inducing changes in plant behaviour.

In such cases manual tuning is difficult and moreover the closed-loop may become detuned as the plant dynamics vary with time. 'Self-tuning' or 'auto-tuning' (Harris and Billings, 1981) are the terms used to describe the algorithmic approach to determining the parameters of a 'good' fixed controller, whilst 'adaptive' control (Narendra and Monopoli, 1980) implies that the tuning function is sustained so as to provide a consistent closed-loop performance. It is found that the self-tuning of a controller for a plant with fixed dynamics is surprisingly easy but that it is hard to design an adaptive method which can give reliable results over long periods of time.

As a simple example of self-tuning design consider an electrically heated kiln as shown in Fig.1 in which the thermal load is unknown and it is desired to regulate the temperature $y^o(t)$ above ambient to some set-point value w. If $u(t)$ is the control signal to the heater, $Q(t)$ the heater output and α is the heat-transfer coefficient in Newton's law of cooling, the process equations are:

$$c\frac{dy}{dt} = k_1 u - \alpha y, \text{ or:}$$

$$T\frac{dy}{dt} + y = k\ u(t) \tag{1}$$

where the time-constant T and the gain k are unknown. Suppose now that a microcontroller monitors the control u(t) and the measured temperature response y(t) at regular sample-intervals h and that the control between each sample is constant. Then integrating eqn.1 between times a and b gives:

$$T\ [y(b) - y(a)] + \int_a^b y(t)\ dt = k \int_a^b u(t)\ dt$$

and if the sample-interval h is small compared with T this gives:

$$T\ [y(b) - y(a)] + h\ \Sigma_a^b y(ih) = k\ h\ \Sigma_a^b u(ih) \tag{2.1}$$

and if c > b is a later time another equation can be written:

$$T\ [y(c) - y(b)] + h\ \Sigma_b^c y(ih) = k\ h\ \Sigma_b^c u(ih) \tag{2.2}$$

Note that in equations 2.1 and 2.2 all terms except T and k are constructed from data sampled at times t = ih. Hence at time t = c the microcomputer has available two equations in the two unknowns which can be solved to derive estimated values T and k. From these estimates of the plant dynamics the control law parameters can be calculated.

The plant has a Laplace transfer-function k/(1+sT); suppose now that a PI regulator of the form K/(1+1/sTi) is proposed. Noting that the PI function can be written as K(1+sTi)/sTi one easy solution is to choose Ti = T leading to an open-loop transfer-function:

$$G_o(s) = [K(1+sTi)/sTi].[\hat{k}/(1+s\hat{T})] = K\hat{k}/s\hat{T}$$

and the closed-loop transfer-function is $G_o/(1+G_o)$, or:

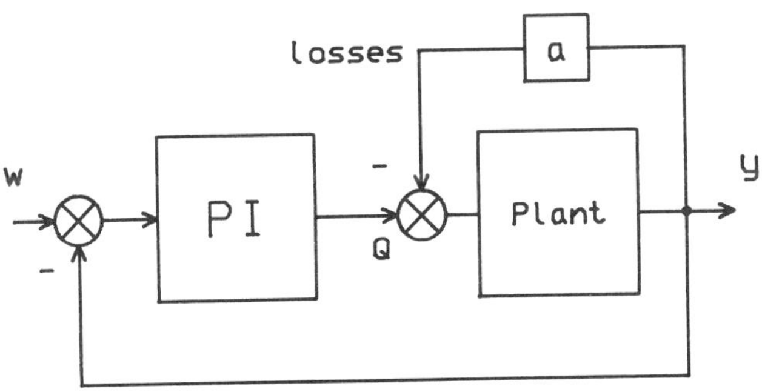

Fig.1 Block diagram of the heater controller

$$G_c(s) = 1 / (1 + s\hat{T}/K\hat{k})$$

The time-constant Tc of the closed-loop is therefore $\hat{T}/K\hat{k}$ and hence can be set to any <u>user-chosen</u> value by the appropriate value of K. This means that <u>despite</u> having no prior knowledge of the thermal load or the heat-transfer coefficients of the furnace, the closed-loop ends up having <u>consistent</u> dynamics Tc which can be specified beforehand by the operator.

The operation of the self-tuner would be as follows:

1. The user chooses the desired closed-loop time constant Tc, the sample-interval h and the three experiment times a,b and c (all these may, of course, have predefined <u>default</u> values).

2. The furnace is switched on at a given control setting u_o and successive measurements of temperature y(t) are made so as to construct equations 2.1 and 2.2.

3. The values $\hat{T}$ and $\hat{k}$ are computed and from them and Tc the controller parameters Ti and K are deduced.

4. The PI controller is then put into automatic mode to bring the temperature to the set-point with the prescribed dynamic behaviour.

The above design is a special case of the self-tuning approach in which the following steps are employed:

1. A candidate input/output structural representation of the plant is proposed together with an analytical design procedure which would produce the best controller coefficients given the correct plant parameters.

2. A parameter estimator using I/O data provides the best-fit model of the plant.

3. The parameter estimates are fed into the design algorithm which accepts them as if they were exact (this is called 'certainty equivalence').

4. The controller coefficients taken from the design process are then used to assert feedback control (this is called an 'explicit' self-tuner; in an 'implicit' self-tuner the controller terms are estimated directly).

Many self-tuners can be developed based on the above principles which differ according to the assumptions made at each stage. For example the model-fitting can be 'one-shot' as in the above example or in an adaptive approach the fitting can be refined recursively using new data at each sample as time evolves so that plant time-variations can be tracked.

Whether a particular self-tuner is successful in practice depends on properly answering many detailed engineering questions: Is the model structure adequate for the class of plants considered? How sensitive is the model-fitting procedure to process disturbances, measurement noise and actuator and plant nonlinearities? Is the control design algorithm robust in coping with specification errors such as incorrectly chosen model-order? Are the user-chosen design parameters (like Tc above) easy for a plant engineer to understand? Is the controlled performance an improvement on that of a simple manually-tuned loop? Can the algorithm be embedded in a microprocessor at a reasonable cost? The development below attempts to answer some of these questions in the context of single-input/single-output loops.

2 GENERAL PLANT AND PREDICTOR MODELS

Often the set-point w(t) of a process-control loop is held constant for long periods so that a locally-linearized plant model which can give a good fit to changes in I/O signals {u(t),y(t)} from their mean values is of the form (Clarke, 1984):

$$y(t) = \frac{B_1(s)}{A_1(s)} u(t-T) + x(t) \tag{3}$$

where T is the dead-time or transport-delay found in many cases. For example it it often claimed that a second-order model with dead-time has a transient response which is a excellent fit to a much more complex plant though this is clearly less true for multi-mode lightly damped flexible systems.

The signal x(t) can represent the modelling errors but we consider it here to be a general disturbance term which affects both the controlled performance and the parameter estimation in a self-tuner - indeed if there is no disturbance the design of a self-tuner is simple. It may include measurement noise but more often x(t) arises from load-disturbances acting on the plant. Hence in particular it will not have zero mean and might be slowly varying with time. Another source of 'noise' is the presence of actuator errors such as stiction which imply that the control transmitted to the plant differs from that calculated by the computer. All these effects will mask the true I/O relationship G(s) and need to be taken into account in the self-tuning design.

As the plant is to be controlled by a microcomputer acting in discrete-time one possibility is to design a continuous controller C(s) and to 'discretize' C(s) by the relationship:

$s \rightarrow (1-z^{-1})/h = \Delta/h$, where z^{-1} is the backward-shift operator

As an example consider $C(s) = K(1+sT) \rightarrow K(1+\Delta T/h)$ so that:

$$u(z^{-1})/e(z^{-1}) = K\,[(1+T/h) - z^{-1}T/h]$$

which gives the implemented controller equation:

$$u(t) = \alpha_0 e(t) + \alpha_1 e(t-1) \; ; \quad \alpha_0 = K(1+T/h), \; \alpha_1 = -KT/h.$$

This is the approach generally adopted with PID discretization where h is generally small compared with T, but for most self-tuning work h is not small and the more general Z-transform theory is used.

Considering the plant to be preceded by the usual Zero-Order-Hold (Franklin and Powell, 1980; Isermann, 1981) the discrete-time model is given by:

$$G(z^{-1}) = \Delta \, Z\{G(s)/s\}$$

where the Z-transform of $G(s)/s$ can be obtained by partial-fraction expansion and term-by-term transformation. If $x(t)$ is stationary it can be shown that with $G(s)$ of order n eqn.3 transforms to the CARMA (Controlled AutoRegressive Moving-Average) model:

$$A(z^{-1})y(t) = B(z^{-1})u(t-k) + C(z^{-1})e(t) \tag{4}$$

or in difference-equation terms:

$$y(t)+a_1 y(t-1)+..+a_n y(t-n) = b_0 u(t-k)+..+b_n u(t-k-n) + e(t)+..+c_n e(t-n)$$

This is the model most often assumed in the self-tuning literature: the time-instants are indexed by the integers; $k = \mathrm{INT}(T/h)+1$ represents the dead-time; $e(t)$ is a 'white-noise' sequence. In practice it is better to assume that the noise is non-stationary and this can be simply done by the model:

$$A(z^{-1})y(t) = B(z^{-1})u(t-k) + x(t) \tag{5}$$

where:

$$x(t) = C(z^{-1})e(t)/\Delta \text{ is Brownian motion.}$$

The representation of eqn.5 will be assumed in general with $x(t)$ interpreted according to context.

A predictive model is the basis of many self-tuning designs: the model forecasts the response at some future time based on currently available data and the control law chooses $u(t)$ so that the predicted output is optimal in some sense. One way is to specify a polynomial:

$$T(z^{-1}) = 1 + t_1 z^{-1} + .. + t_n z^{-n}$$

in which the roots of $T = 0$ will correspond to the dynamic modes in the disturbance response of the closed loop.

Consider now the identity of eqn.6 which is used to obtain polynomials E and F given the plant dynamic parameters {k,A} and the chosen T:

$$T(z^{-1}) = E(z^{-1})A(z^{-1}) + z^{-k}F(z^{-1}) \tag{6}$$

[As an example of the use of eqn.6 take T = C and the plant model:

$$(1 - 0.9z^{-1})\ y(t) = 0.3\ u(t-2) + (1 + 0.5z^{-1})\ e(t)$$

Giving: $(1 + 0.5z^{-1}) = (e_o + e_1 z^{-1})(1 - 0.9z^{-1}) + z^{-2} f_o$

Comparing powers of z^{-1} gives: $e_o = 1$, $e_1 = 1.4$ and $f_o = 1.26$.]

Now multiply the model of eqn.4 by the polynomial $z^2E(z^{-1})$ to get:

$$E(z^{-1})A(z^{-1})y(t+k) = E(z^{-1})B(z^{-1})u(t) + C(z^{-1})E(z^{-1})e(t+k)$$

So: $[T - z^{-k}F]\ y(t+k) = G(z^{-1})\ u(t) + CE\ e(t+k)$, using eqn.6.

i.e. $T\ y(t+k) = F(z^{-1})\ y(t) + G(z^{-1})\ u(t) + CE\ e(t+k)$

Now as T is a user-chosen polynomial we can define filtered signals y'(t) and u'(t) as y'(t) = y(t)/T etc. giving the difference-equation:

$$y'(t) + t_1 y'(t-1) + .. + t_n y'(t-n) = y(t)$$

Hence the future output y(t+k) is given in terms of available data by:

$$y(t+k) = F(z^{-1})\ y'(t) + G(z^{-1})\ u'(t) + CE/T\ e(t+k)$$

It is seen that the prediction has an error term given by CE/Te(t+k) and if T is chosen to equal C the term reduces to Ee(t+k). Inspection of the identity eqn.6 shows that the degree of E is k-1 so that the error is of the form:

$$\tilde{y}(t+k) = e(t+k) + e_1 e(t+k-1) + .. + e_{k-1} e(t+1)$$

This is a moving-average of order k which can be shown to have the minimum possible variance of all such predictors. With the example system the prediction model is:

$$y(t+2) = 1.26y'(t) + 0.3u'(t)+0.426u'(t-1)+e(t+2) + 1.4e(t+1)$$

where: $y'(t) = y(t) - 0.5y'(t-1)$ and $u'(t) = u(t) - 0.5u'(t-1)$.

Consider now the prediction signal p(t) - sometimes denoted $\hat{y}(t+k|t)$ - which depends entirely on data which is known at time t:

$$p(t) = \hat{y}(t+k|t) = F(z^{-1})\ y'(t) + G(z^{-1})\ u'(t) \tag{7.1}$$

and where in the general case the actual future output is:

$$y(t+k) = p(t) + E(z^{-1})\ x'(t+k) \tag{7.2}$$

It is well-known that the difficulty of controlling a dead-time plant is due to the excessive phase-lag which means that the loop-gain needs to be reduced for stability. If the delay is k samples a model which gives an unbiassed prediction of $y(t+k)$ means that exactly the right amount of phase-advance is added into the loop so that the gain can be increased fo tighter control; this is the basis of the Smith Predictor. Eqn.7 represents a more general predictor which can deal with disturbances in the design.

If the general disturbance $x(t)$ drifts so has a Brownian motion model, eqn.7.2 shows that the prediction error also drifts as $x'(t)$ is also Brownian motion unless the polynomial E has a factor Δ. This can be assured by a modified identity:

$$T(z^{-1}) = E'(z^{-1})\Delta A(z^{-1}) + z^{-k} F'(z^{-1}) \tag{8}$$

Now repeating the above derivation with E defined as $E'\Delta$ leads to:

$$y(t+k) = F'(z^{-1})\, y'(t) + E'B\, \Delta u'(t) + E'(z^{-1})\, \Delta x'(t+k)$$

The trick now is to write $F'(z^{-1})$ as $T(z^{-1}) + F(z^{-1})\Delta$ giving:

$$y(t+k) = y(t) + F\, \Delta y'(t) + G\, \Delta u'(t) + E'\, \Delta x'(t+k) \tag{9}$$

where, for example, $\Delta y'(t) = y'(t)-y'(t-1)$: Δ being the differencing operator. Eqn.9 is called an <u>incremental predictor</u> as it provides the prediction $p(t)$ in terms of a 'fixed-point' $y(t)$ and increments of the other signals. In particular the prediction is unbiassed in the steady-state <u>even if</u> the plant parameters are not known exactly; this is of vital importance in applications. A useful benefit of the incremental approach is that it is found naturally to lead to controllers with desirable integrating action and for simple models the resulting structure can be of PID form.

3 PREDICTIVE CONTROL

For simplicity consider the case where $T = 1$ so that the prediction is:

$$p(t) = y(t) + f_o \Delta y(t) + \,..\, + g_o \Delta u(t) + \,..$$

At time t $y(t)$ is measured and $u(t)$ is to be calculated. In a conventional controller $u(t)$ is given by $u(t) = L.[w(t) - y(t)]$ where L is some transfer-function such as PID. A natural extension to this is to choose a controller:

$$u(t) = [\, w(t) - p(t)\,] \,/\, Q(z^{-1}) \tag{10}$$

where $Q(z^{-1})$ is a user-specified transfer-function; this is like Smith predictive control. Note, though, that $u(t)$ appears explicitly in the prediction equation which implies that by appropriate choice of $u(t)$ we can make $p(t)$ and hence $y(t+k)$ have any value we wish. A natural law for 'tight' control is to make $p(t) = w(t)$ which is achieved by:

$$\Delta u(t) = u(t) - u(t-1) = [w(t) - y(t) - f_0 \Delta y(t) - ..]/g_0$$

Note that there is inherent integral action in this control law. When it is in closed-loop the output y(t+k) will differ from the set-point w(t) only by the prediction error which, when T = C, is of minimum variance. Hence this is called a minimum variance controller (Astrom and Wittenmark, 1973) which is useful say in quality control when regulating a variable such as gauge about a fixed value. Fig.2 shows the general case of eqn.10 in which the prediction equations are shown to be equivalent to an advance element with an additive prediction error disturbance. As an additive signal does not affect the stability of the loop the cancellation of the delay by the advance ensures that the gain of the forward-path element 1/Q can be made larger and hence lead to better disturbance rejection. It might be imagined that this control law is unrealizable but in fact the prediction p(t) is calculated from equations 7 or 9 and so is causal.

For a minimum variance design with p(t) = w(t) the control can be written:

$$\Delta u(t) = [w(t) - y(t) - F(z^{-1}) \Delta y(t)] / G(z^{-1})$$

$$\text{where: } G(z^{-1}) = E(z^{-1})B(z^{-1}).$$

Hence the plant zero polynomial $B(z^{-1})$ is cancelled by the algorithm and so if there are zeros outside the stability region the controller will have unstable poles. In practice the position of the plant zeros are not exactly known and then imperfect cancellation leads to control modes which are unstable. Now it turns out that there is a high probability that the B polynomial has these 'nonminimum-phase' roots even if the original continuous-time plant is minimum-phase. Badly positioned discrete-time zeros arise when there is a fractional dead-time (as may occur if the computation of the control takes a significant fraction of

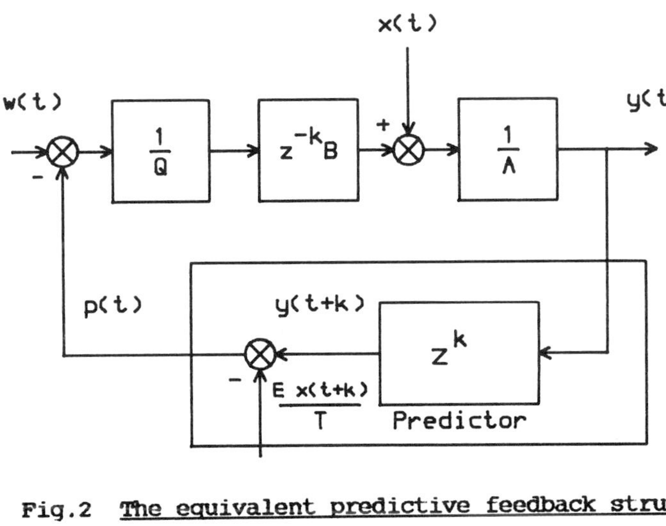

Fig.2 The equivalent predictive feedback structure

the sample interval) or where there are at least two more poles that zeros in G(s) with rapid sampling (small h). Hence minimum variance control has to be used with great care.

Now if instead eqn.10 is used (giving the so-called generalized minimum variance or Clarke-Gawthrop law - Clarke and Gawthrop, 1975) it is readily found by inspection of Fig.2 that the closed-loop characteristic polynomial is:

$$CLCP = B(z^{-1}) + Q(z^{-1})A(z^{-1})$$

Now if Q is written in the form $Q = \lambda Q'(z^{-1})$ it is seen that λ plays the role of a root-locus parameter such that if A is stable then for large enough λ the predictive closed-loop is stable. The GMV label comes from its original derivation where it was shown to minimise:

$$J(u) = E\ \{\ (y(t+k) - w(t))^2 + \lambda'(Q'u(t))^2\ \mid t\ \}$$

This means that λ can be used to trade output against actuation variance; in practice is is found that a significant reduction in the variance of u(t) is accompanied by only a small increase in the variance of y(t). Note that as λ 0 the law reduces to minimum-variance and Fig.2 implies that the forward path gain of 1/Q tends to infinity. The GMV method has proved to be popular because of its flexiblity and there have been several microcomputer implementations.

3.1 Generalized Predictive Control

The GMV law, though an improvement on the original MV design, suffers from two defects. One is that the correct value for λ giving good closed-loop control has often to be found experimentally and the second is that the dead-time k needs to be known. Current self-tuning designs based on the GMV method estimate the F and G parameters provided that k has been prespecified by the user. In practice with some plant the dead-time might not be known and indeed might be time-varying.

The idea behind generalized predictive control or GPC (Clarke et al, 1984) is to predict over a horizon greater than the maximum possible dead-time of the plant. Of course the prediction will depend on both past and future control signals and the method makes suitable assumptions about their values. Consider eqn.8 with T = 1 and k = j:

$$1 = E_j'(z^{-1})\Delta A(z^{-1}) + z^{-j}F_j'(z^{-1}) \qquad (6.1)$$

This Diophantine identity leads to the incremental model:

$$p_j(t) = \hat{y}(t+j|t) = y(t) + F_j\Delta y(t) + G_j'\Delta u(t) + G_j\Delta u(t+j)$$

where the term $G_j'\Delta u(t)$ includes only those controls already transmitted by time t and the term $G_j\Delta u(t+j)$ involves present and future controls.

Suppose the prediction horizon is from j = 1 to j = N (the 'output horizon'). Then the N equations may be written in vector form as:

$$\mathbf{p}(t) = \mathbf{p}_o + \mathbf{G}\,\Delta\mathbf{u}$$

where:

$$\mathbf{p} = \text{col}\,[p_1(t),\ p_2(t),\ ..\ ,p_N(t)]$$

$$\mathbf{p}_o = \text{col}\,[\ \{\ y(t) + F_j \Delta y(t) + G_j' \Delta u(t)\ \}\ ]$$

$$\Delta\mathbf{u} = \text{col}\,[u(t)-u(t-1),\ ..\ ,u(t+N-1)-u(t+N-2)]$$

and

$$\mathbf{G} = \begin{vmatrix} g_o & 0 & \dots & 0 \\ g_1 & g_o & \dots & 0 \\ .. & & & \\ g_{N-1} & & \dots & g_o \end{vmatrix}$$

is a lower-triangular matrix.

Let $\mathbf{w} = \text{col}[w(t+1),\ ..\ ,w(t+N)]$ be a vector of <u>future set-points</u>. Normally in process control w(t+j) is assumed to be constant and equal to w(t) but in some applications such as batch control or robotics future w's will be known. Then if **e** is the vector of predicted future system errors $\mathbf{e} = \mathbf{w} - \mathbf{p}$, giving:

$$\mathbf{e} = \mathbf{w} - \mathbf{p}_o - \mathbf{G}\,\Delta\mathbf{u} \tag{11}$$

This is the key equation of GPC; the problem now is to choose $\Delta\mathbf{u}$ so as to minimise some function of **e** such as sum-of-squares:

$$J(\Delta\mathbf{u}) = \mathbf{e}'\mathbf{e} = \Sigma_1^N\ [\ w(t+j) - \hat{y}(t+j|t)\ ]^2$$

This is achieved by the standard least-squares solution:

$$\Delta\mathbf{u} = [G^T G]^{-1} G^T\ (\mathbf{w} - \mathbf{p}_o) \tag{12}$$

GPC is implemented as a <u>receding-horizon</u> controller for which at each sample t:

1. from the measured data y(t) and old saved data $\Delta y(t-j)$ and $\Delta u(t-j)$ the prediction equations are used to generate $\mathbf{p}_o$;

2. the future set-point sequence **w** is generated leading to the vector $\mathbf{w}-\mathbf{p}_o$;

3. $\Delta\mathbf{u}$ is calculated from eqn.12 where it noted that for a fixed known plant the matrix operations can be precomputed;

4. the first element $\Delta u(t) = u(t)-u(t-1)$ is extracted from $\Delta\mathbf{u}$ and u(t) is transmitted to the plant;

5. the vectors $\Delta u(t-j)$, $\Delta y(t-j)$ are shifted ready for the next sample.

Closer inspection of the method reveals that the control which minimises $\mathbf{e'e}$ is exactly the minimum-variance control described before which, as explained, is unstable for nonminimum-phase plant. So where is the advantage of GPC? It lies in one significant further assumption: that after a 'control horizon' NU which is less than N further control increments $\Delta u(t+j)$ <u>are taken to be zero</u>. This effectively puts <u>infinite cost</u> on future control changes and so stops growing control modes which may arise from the attempted cancellation of B. There is a further advantage: as the trailing elements of $\boldsymbol{\Delta u}$ are zero it can be truncated to a vector of size NU giving the new error equation:

$$\mathbf{e} = \mathbf{w} - \mathbf{p}_o - \mathbf{G}_T\, \boldsymbol{\Delta u}_T$$

where: $\boldsymbol{\Delta u}_T = \text{col}\ [u(t)-u(t-1), \ .. \ , u(t+NU-1)-u(t+NU-2)]$

and $\mathbf{G}_T$ is a matrix of dimension N×NU. The controls which minimise J under the assumed constraint on $\Delta u(t+j)$ are given by eqn.12 but with the truncated $\mathbf{G}$ matrix in place of $\mathbf{G}$. Note in particular that the inversion involves an NU×NU rather than an N×N matrix which is of advantage in adaptive applications where the inversion is to be made at each sample.

An extreme case which has found to be useful for 'simple' process plant is where NU = 1 so the inversion involves just a scalar. For large N the effect of GPC is then to provide that 'mean-level' control which makes the output go to the set-point with the dynamics of the open-loop plant. If it is desired to make the plant respond faster either a set-point prefilter or an increase of NU can be used.

The GPC approach is very flexible and can stabilise open-loop unstable and nonminimum-phase plant (if NU is chosen to be at least equal to the number of lightly damped and unstable poles). The computational demands are not excessive and self-tuned versions have been implemented on LSI 11 and Z80 microcomputers.

4 PARAMETER ESTIMATION

The plant model of eqn.5 can be written as:

$$y(t) = -\, a_1 y(t-1) - \ .. + b_o u(t-k) + \ .. + b_n u(t-k-n) + x(t)$$

and the incremental predictor model of eqn.9 as:

$$y(t) = y(t-k) + f_1 \Delta y'(t-k) + \ .. + g_o \Delta u'(t-k) + x_1(t)$$

Each is a particular example of the general linear model:

$$y(t) = \Sigma_1^n\ \theta_i s_i(t) + e(t) \tag{13}$$

where: $\boldsymbol{\theta} = \{\theta_1, \theta_2, \ .. \ , \theta_n\}'$ is a vector of parameters

and: $\mathbf{s} = \{s_1(t), \ .. \ , s_n(t)\}'$ is a vector of known data.

Suppose at time t there is a vector $\hat{\theta}(t-1)$ available from the previous sample. This can then be used to 'predict the present' giving:

$$\hat{y}(t) = \hat{\theta}'(t-1)s(t) = \Sigma_1^n \hat{\theta}_i(t-1)s_i(t)$$

and this has an associated error $\epsilon(t) = y(t) - \hat{y}(t)$. Now if $\tilde{\theta}$ is the parameter error the prediction error can be written:

$$\epsilon(t) = \tilde{\theta}'(t-1)s(t) + e(t)$$

This equation shows that the prediction error is a combination of parameter error and the disturbance; the job of the parameter estimator is to use $\epsilon(t)$ to improve the estimates under these noisy conditions. This is helped by the existence of data-vectors $s(t)$ which are 'persistently exciting': constant data vectors mean that there is not enough information for the RPE design to work on. In practice it is found that the deliberate injection of test signals (as in the furnace example) is very useful for achieving reliable plant models.

A recursive prediction-error (RPE) algorithm provides a new estimate of the unknown θ using an equation of the form:

$$\hat{\theta}(t) = \hat{\theta}(t-1) + a(t)M(t)s(t)\epsilon(t) \tag{14}$$

where:

a(t) is a scalar 'gain-factor' giving the step length

and: M(t) is a matrix modifying the parameter update direction.

The values of a(t) and M(t) are dependent on the algorithm being used and in general we would like the methods to have the following features:

1. at start-up during the initial tuning phase the estimates should move rapidly as the error $\epsilon(t)$ would be due mainly to parameter errors;

2. during steady-state operation in the adaptive phase the parameter estimates should not move too quickly as $\epsilon(t)$ would be mostly noise;

3. the computational load and the data storage requirements should not increase with time.

There are many RPE methods which differ according to the compromise made between computational complexity and efficiency. The simplest method called LMS (least mean square) simply puts a(t) equal to a small constant μ and M(t) equal to I:

$$\hat{\theta}(t) = \hat{\theta}(t-1) + \mu\, s(t)\, \epsilon(t)$$

This method is used in high-speed applications with sample-rates of the order of kHz for speech processing and communications. Its simplicity is such that it can be readily built into a dedicated microchip, but for adaptive control it has the disadvantage that its convergence rate is low and is moreover data-dependent. Table 1 gives a Fortran code for the LMS method.

4.1 Recursive-least-squares

Recursive-least-squares or RLS is the most popular method in adaptive control as it gives rapid convergence in noise-free conditions and yet is less affected by noise in the steady-state. It has its roots in classical 'off-line' statistical estimation theory where a function:

$$S(\hat{\theta}) = \Sigma_1^N \epsilon^2(j,\hat{\theta})$$

is to be minimised w.r.t. $\hat{\theta}$ given data from t = 1 up to t = N. The solution is like that given in eqn.12 for the GPC law:

$$\hat{\theta} = [\mathbf{X}^T\mathbf{X}]^{-1}\,\mathbf{X}^T\,\mathbf{y}$$

where $\mathbf{X}$ is an N×n matrix of data and $\mathbf{y}$ is an n-vector of 'output' data. If the error sequence {e(t)} is white-noise with common variance it is readily shown that this gives the estimate with minimum parameter error which reduces to 0 (consistent) as N increases. However, this direct formulation has two drawbacks: the estimate tends to a constant vector which is unacceptable for adaptive control and the computational demands increase with N (as the corresponding dimension of $\mathbf{X}$ increases).

To track slowly varying parameters a modified cost S_1 is proposed with a 'fading memory':

$$S_1(\hat{\theta}) = \Sigma_1^N \beta^{N-j}\epsilon^2(j,\hat{\theta})$$

where the weighting β is called a forgetting factor and is a quantity less than 1.0. Its use implies that 'old' data at time t-j in the past is discounted more than recent data. A useful measure here is the 'asymptotic sample length' or ASL which is defined to be $1/(1-\beta)$ and which is interpreted as the data stretch which is of significance in the current estimates. For example if the plant varies 'slowly' a value for

Table 1 Fortran code of the LMS algorithm

```
      SUBROUTINE LMS (Y,S,THETA,N,RMU)
C     CODING OF THE LMS ALGORITHM
      DIMENSION S(1), THETA (1)
      PREDY = 0.0
      DO 1 I = 1,N
1     PREDY = PREDY + THETA(I)*S(I)                  ! ŷ
      PREDER = Y - PREDY                             ! ε
      DO 2 I = 1,N
2     THETA(I) = THETA(I) + RMU * S(I) * PREDER      ! θ̂
      RETURN
```

ASL could be 1000 samples (or $\beta = 0.999$) whilst rapid variations can be tracked with an ASL of say 20 (or $\beta = 0.95$).

To be useful for on-line work RLS needs to be embedded into the general RPE method outlined above and it can be shown that this is achieved by the following equations:

$$\epsilon(t) = y(t) - \hat{\boldsymbol{\theta}}'(t-1)\mathbf{s}(t)$$

$$\hat{\boldsymbol{\theta}}(t) = \hat{\boldsymbol{\theta}}(t-1) + \mathbf{k}(t)\,\epsilon(t) \tag{15a}$$

$$\mathbf{k}(t) = \mathbf{P}(t-1)\mathbf{s}(t) \,/\, [\beta + \mathbf{s}'(t)\mathbf{P}(t-1)\mathbf{s}(t)] \tag{15b}$$

$$\mathbf{P}(t) = [\mathbf{I} - \mathbf{k}(t)\mathbf{s}'(t)]\,\mathbf{P}(t-1)/\beta \tag{15c}$$

Equations 15a and 15b show how the matrix $\mathbf{M}$ is made to vary with time depending on the matrix $\mathbf{P}$. To start the algorithm $\mathbf{P}(0)$ needs to be specified and it can be shown that if $\beta = 1$ and $\mathbf{P}(0) = \gamma\mathbf{I}$ where γ is large then the above recursions give exactly the same estimates as the offline RLS solution. In general the algorithm operates at each sample instance t as follows:

1. the new data $y(t)$ and $\mathbf{s}(t)$ are acquired and the prediction error ϵ is computed from the old parameter estimates;

2. eqn.15b is used to update the n×1 'Kalman gain' vector $\mathbf{k}(t)$;

3. eqn.15a provides the new parameters $\hat{\boldsymbol{\theta}}(t)$;

4. eqn.15c prepares $\mathbf{P}(t)$ ready for the next sample.

In offline RLS the matrix $\mathbf{V} = [\mathbf{X}^T\mathbf{X}]^{-1}$ is proportional to the covariance matrix of the parameter errors so for example its diagonal terms give individual variances. It can be shown that the matrix $\mathbf{P}(t)$ is equal to $\mathbf{V}$ under certain conditions such as $\beta = 1$ and so could be used as a diagnostic to indicate the confidence in the model. For example if there are too many parameters in the model (assumed n greater than the plant n) $\mathbf{P}$ will be singular and a test (say giving the condition number) may indicate this.

Consider the use of an algorithm such as RLS in a self-tuner. Even at the relatively long sample interval of 1 minute there might be many tens of thousands of samples while the tuner is unattended (Clarke, 1980). Hence there must be careful attention paid to the numerical robustness of the estimator, particularly when implemented on a microcomputer with a relatively short word-length. (Note that in most cases the dynamic range of parameters would not be known a priori so that floating-point rather than scaled fixed-point working is recommended.) The problem with eqn.15c is that if $\mathbf{P}$ becomes skew-symmetric because of round-off there is numerical instability in the equations as written. The solution is to use a <u>factored</u> (Bierman, 1977) form for $\mathbf{P}$ as it is known that $\mathbf{P}$ should always be positive-definite. One common form is the UDU' version where $\mathbf{P}$ is written as:

$$\mathbf{P}(t) = \mathbf{U}(t)\,\mathbf{D}(t)\,\mathbf{U}^T(t)$$

and the updates are re-written in terms of the upper-triangular matrix **U** which has units down its diagonal and of the diagonal matrix **D**. Lack of space precludes development of the theory but Table 2 gives the Fortran coding. As written it requires storage allocation for the n-vectors S, D, GK and THETA and for the vector U of dimension n(n-1)/2; with n = 10 this implies a total requirement for 85 real elements. We note from eqn.15 that if $|\mathbf{P}|$ tends to 0 then so does $|\mathbf{k}|$ which means that the parameter estimates will not change despite large values of ϵ. The role of β is to prevent this happening so that the 'adaptive gain' remains nonzero - but consider what happens when the plant is under quiescent conditions with constant input and output. For the case when the data vector $\mathbf{s}(t)$ consists of incremental values the elements of $\mathbf{s}$ are zero and eqn.15c shows that:

$$\mathbf{P}(t) = \mathbf{P}(t-1)/\beta \rightarrow \infty, \text{ if } \beta < 1.$$

This is the so-called phenomenon of 'blow-up': under weakly exciting conditions $|\mathbf{P}(t)|$ grows, leading to an excessive growth in the adaptive gain. This does not affect the estimated parameters until a burst of data such as a set-point change or a noise-spike arrives, whereupon there is a rapid and often unacceptable variation in the estimates.

Blow-up can be prevented in many ways. The simplest are to limit $\mathbf{P}(t)$ by the initial $\mathbf{P}(0)$ or to periodically switch $\mathbf{P}(t)$ to $\mathbf{P}(0)/\delta$ where δ is some prior chosen number such as 10. A useful alternative which seems to work well in practice is to use a variable forgetting-factor (Fortescue et al, 1981) which attains the value of 1 during quiescent conditions yet falls to the minimum value when the data is exciting. One method considers the sum-of-squares of the prediction errors of the updated model:

$$S_2(t) = \Sigma_1^t \epsilon_1(t)^2 = \Sigma\,[y(t) - \hat{\boldsymbol{\theta}}'(t)\mathbf{s}(t)]^2$$

which is affected by the value of β. Keeping S_2 constant at a chosen value S_0 roughly 'balances' the information lost by the estimator by that gained with the new data and can be achieved by choosing:

$$\beta(t) = 1 - [1 - \mathbf{k}'(t)\mathbf{s}(t)]\,\epsilon(t)^2 / S_0$$

We see that if the prediction error ϵ is small the forgetting-factor β is near 1 whereas if the adaptive gain $\mathbf{k}$ is small and there is a large value of ϵ the value of β is reduced to well below 1. A minimum possible value of β can be prespecified in terms of the expected rate of parameter changes.

5 SELF-TUNING

So far we have discussed certain control laws and RPE estimation methods. To obtain a self-tuner all that needs be done is to couple the two algorithms together, as shown in Fig.3.

Table 2 Fortran coding of the UDU' RLS estimator

```
          SUBROUTINE UDU (S,D,U,GK,THETA,NPAR,PERR,BETA)
C
C         Input:  S data vector; D,U old factors, THETA old parameters
C                 NPAR number of parameters, PERR prediction error, BETA
C         Output: D, U, THETA all updated.
C
          DIMENSION GK(1), U(1), D(1), THETA(1), S(1)
C
          FJ = S(1)
          VJ = D(1)*FJ
          ALPHAJ = BETA + VJ*FJ
          D(1) = D(1)/ALPHAJ
          GK(1) = VJ
          IF (NPAR .EQ. 1) GOTO 1000
          KF = 0
          KU = 0
          DO 30  J = 2,NPAR
           FJ = S(J)
           J1 = J - 1
           DO 10  I = 1,J1
             KF = KF + 1
             FJ = FJ + S(I)*U(KF)
10           CONTINUE
           VJ = FJ*D(J)
           ALFLST = ALPHAJ
           ALPHAJ = ALFLST + VJ*FJ
           D(J) = D(J) * (ALFLST/ALPHAJ) / BETA    ! update D
           GK(J) = VJ
           PJ = - FJ/ALFLST
           J1 = J - 1
            DO 20  I = 1,J1
             KU = KU + 1
             W = U(KU) + GK(I)*PJ
             GK(I) = GK(I) + U(KU)*VJ
             U(KU) = W                             ! update U
20          CONTINUE
30         CONTINUE
C
1000      CONTINUE
C            UPDATE PARAMETER ESTIMATES
          DO 40 I  = 1,NPAR
40        THETA(I) = THETA(I) + GK(I) * PERR / ALPHAJ  ! eqn.15a
          RETURN
```

The simplest implicit self-tuner is the minimum-variance design which for $T = C$ and for the CARMA system of eqn.4 it can be deduced from eqn.7.1 that:

$$C(z^{-1})\ p(t) = F(z^{-1})\ y(t) + G(z^{-1})\ u(t)$$

and:
$$y(t+k) = p(t) + E(z^{-1})\ e(t+k)$$

The control law for $p(t) = \hat{y}(t+k|t) = w(t) = 0$ is given by:

$$F(z^{-1})\ y(t) + G(z^{-1})\ u(t) = 0$$

or:
$$u(t) = -\ [f_0 y(t) + f_1 y(t-1) + \ldots + g_1 u(t-1) + \ldots] \ /\ g_0 \tag{16}$$

When this control is asserted $p(t)$ is put to 0 so the closed-loop is:

$$y(t+k) = E(z^{-1})\ e(t+k)$$

This implies that the controlled output has a net disturbance which is simply the prediction errors, provided of course that there are no constraints on the control signal and that the loop is stable. As the prediction errors are of minimal variance so is the regulation of the plant output.

Defining $H(z^{-1})$ to be $z[1-C(z^{-1})]$ the prediction at time t-k is:

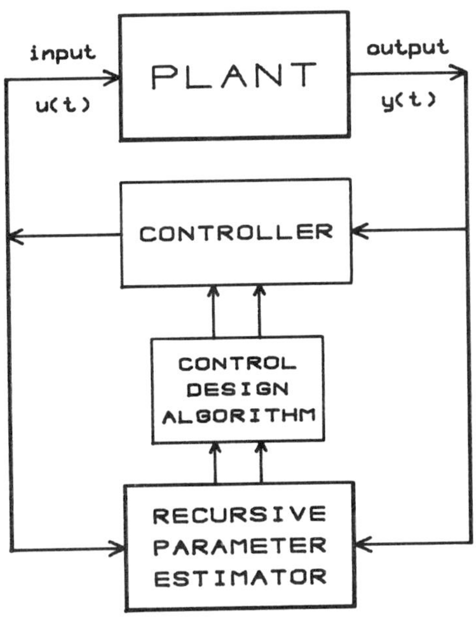

Fig.3 Structure of a self-tuning controller

$$p(t-k) = F(z^{-1})y(t-k) + G(z^{-1})u(t-k) + H(z^{-1})p(t-k-1)$$

But the control law has put p(t-k-1), p(t-k-2), .. to 0, so:

$$y(t) = F(z^{-1})\ y(t-k) + G(z^{-1})\ u(t-k) + E(z^{-1})\ e(t)$$

or: $$y(t) = f_0 y(t-k) + f_1 y(t-k-1) + .. + g_0 u(t-k) + .. + \epsilon(t)$$

This is seen to be a linear model as in eqn.13 so RLS can be used with vectors:

$$\boldsymbol{\theta} = \text{col}\ [f_0,\ f_1,\ \ldots\ ,\ g_0,\ g_1\ ..]$$

$$\mathbf{s} = \text{col}\ [y(t-k),\ y(t-k-1),\ ..\ ,\ u(t-k),\ u(t-k-1),\ ..]$$

However closer inspection of the control law of eqn.16 shows that the same control u(t) is produced if all parameters are multiplied by the same arbitrary constant α. Whilst this might not be important, in practice α could attain very large or very small values and moreover the parameter estimation does not have a unique solution. One way to overcome this minor difficulty is to prespecify a value $\bar{g}$ for the first G parameter g_0. Hence the RLS equations now use data:

$$\epsilon(t) = y(t) - \bar{g}u(t-k) - \hat{\boldsymbol{\theta}}'(t-1)\mathbf{s}(t)$$

where: $$\hat{\boldsymbol{\theta}} = \text{col}\ [\hat{f}_0(t-1),\ ..\ ,\ \hat{g}_1(t-1),\ ..]$$

and: $$\mathbf{s} = \text{col}\ [y(t-k),\ ..\ ,\ u(t-k-1),\ ..]$$

The algorithm of eqn.15 is then used to update the estimate of the unknown $\boldsymbol{\theta}$ at each sample and the control asserted is then:

$$u(t) = -\ [\hat{f}_0 y(t) + .. + \hat{g}_1 u(t-1) + ..]\ /\ \bar{g}$$

This gives a unique fixed point for the parameter estimates (given constant plant dynamics) and it can be shown that if $0.5b_0 < \bar{g} < \infty$ the algorithm will converge so as to give the correct control action. The only effect of choosing a value of $\bar{g}$ much larger than the true value is to reduce the speed of convergence which in practice is found to be rapid.

Note that the method does not involve an intermediate control design calculation as the controller parameters are estimated directly: this is called an 'implicit' or 'direct' self-tuner. It has the advantage that convergence and robustness proofs are more readily generated as it avoids the potential singularities of some explicit designs. Table 3 gives an outline Fortran coding for the minimum-variance method.

Although the control u(t) is based on data stretching an amount MAX(NU,NY) samples into the past the parameter estimation is based on data as far back as NU+K and NY+K. This should be remembered when dimensioning arrays in a program, particularly when the dead-time k is many samples (our practice is to limit the number of parameters to 10 as there is usually not much point in estimating more). Note also that for a 'bumpless transfer' in the estimator the arrays should be filled with real plant data before starting the tuning. This can be done by running the data acquisition for a few samples when the plant is under manual or PI control. If (unlike the example which uses positional data) incremental data is used in the estimation, initialisation simply fills in the appropriate vectors with zeros.

The clipping of the control U by UMAX and UMIN is to overcome the usual problem of actuator limits. If this necessary precaution is not taken and an excessive control U is calculated the saved data for subsequent parameter estimation would not correspond to that transmitted

Table 3 Fortran coding of a minimum-variance self-tuner

```
      SUBROUTINE MOVE (FROM, TO, LENGTH)          ! Move
      DIMENSION FROM(1), TO(1)                    ! data
      DO 1 I = 1,LENGTH                           ! from FROM
1     TO(I) = FROM(I)                             ! to TO
      RETURN
C
      SUBROUTINE SAVE (DATA,ARRAY,LENGTH)         ! Save data
      DIMENSION ARRAY(1)                          ! in array
      I = LENGTH                                  ! but first
1     ARRAY(I) = ARRAY(I-1)                       ! move all
      I = I-1                                     ! elements
      IF ( I .GT. 1) GOTO 1                       ! along 1.
      ARRAY(1) = DATA                             ! Data in first
      RETURN
C
      REAL FUNCTION SCAPRO (A1, A2, LENGTH)       ! Scalar
      DIMENSION A1(1), A2(1)                      ! product
      SCAPRO = 0.0                                ! of two
      DO 1 I = 1,LENGTH                           ! vectors
1     SCAPRO = SCAPRO + A1(I) * A2(I)             ! A1 and A2
      RETURN
C
      < Main part of the coding: acquire Y = y(t) >
      YBAR = Y - GBAR*USAVE(K)                    ! y-gu(t-k)
      CALL MOVE (YSAVE(K),S,NY)                   ! old y(t-k)..
      CALL MOVE (USAVE(K+1),S(NY+1),NU)           ! old u(t-k-1)..
      NPAR = NY + NU                              ! n = dim θ
      PERR = YBAR - SCAPRO(THETA,S,NPAR)          ! ε(t)
      CALL UDU (S,D,U,GK,THETA,NPAR,PERR,BETA)! see Table 2
      U = - (SCAPRO(THETA,YSAVE,NY) +             ! control
     +       SCAPRO(THETA(NY+1),USAVE,NU))/GBAR
      IF (U .GT. UMAX) U = UMAX                   ! assert control
      IF (U .LT. UMIN) U = UMIN                   ! limits
      CALL SAVE (U,USAVE,NU+K)                    ! save u(t)
      CALL SAVE (Y,YSAVE,NY+K)                    ! save y(t)
      < Transmit u(t) = U to the plant and wait for h >
```

to the plant. This would make the estimator think that the plant's gain had fallen so the following control gains would be pointlessly increased.

6 IMPLEMENTATION AND USE

For process control applications a self-tuner can be mounted on a 8-bit microcomputer giving typical sample intervals of 0.1 sec to 1 minute (Clarke and Gawthrop, 1981). With high-performance robotics where sample rates of the order of 1 kHz might be required the microprocessor would need to be 16 or 32 bit with a floating-point coprocessor and even then care would need to be taken in the coding by, for example, using a hierarchy of priorities for the control, control design and estimation phases. In the process field there are several implementation options (Wittenmark and Astrom, 1984):

- o a stand-alone controller (e.g. a personal microcomputer or a purpose-built microcontroller) for applications to important loops (Tuffs and Clarke, 1985; Duff et al, 1982);
- o embedded in a DDC package in a larger computer for tuning several loops in turn on operator request (Fjeld and Wilhelm, 1981);
- o as a node in a distributed control system where the self-tuning capability is either built-in or provided by some central resource (Halme et al, 1981).

A designer of a self-tuning system has first to decide on the basic algorithm: at present PID, GMV and pole-placement are the most commonly chosen, though the new GPC method seems to be better for general use. Given this choice and the method of implementation the next key design problem is that of the operator's interface. With standard PID regulators this design has been exhaustively explored but self-tuning adds the further dimension of the parameter estimation. It is unlikely that the plant operator would do anything more that enable and inhibit the estimator (possibly in conjunction with a standard probing test-signal) but the commissioning engineer would need to have access to the tuner's data-base to set up constants such as P(0) and {NU,NY,K} etc. This can be done either by a standard terminal or by a hand-held unit which is connected to the microcomputer via a private serial link.

For use simply as a self-tuner in the presence of an engineer there is no difficulty in obtaining good performance as he would be able to monitor its operation and allow the injection of properly exciting test-signals. For plants which do not vary much with time the estimator can then be switched off and provided the basic controller is robust then a fixed-parameter law would then be adequate. However if tight control is required (as with Smith Predictors or MV designs) recall that this is accompanied by sensitivity to changes in dead-time. If the method can cope with such changes (e.g. GPC) it is possible to use the algorithms described above for adaptive control provided that the estimator is properly 'jacketted'. The jacket tries to ensure that estimation only takes place when the I/O data is providing information which can improve the model. Estimation can be inhibited under the following conditions:

1. if the change in control signal $\Delta u(t)$ is too small to affect the process (e.g. with an actuator with dead-band);

2. if the output $y(t)$ is near to the set-point where the measurements are affected by noise and quantization;

3. if the prediction error ϵ is too large, indicating that the output has been affected by a big load-disturbance.

It is not necessary to couple the estimated model to the control design algorithm at all times. Indeed in some cases it might be useful to allow the engineer to enable this transfer and provide some measure such as the model's predictions or prediction error variance to guide him in this operation. This is most valuable for process plant where the set-point is constant for long periods and there is little natural disturbance to provide good models. Full adaptive control in such cases might only be possible if there is an additive test-signal which actively perturbs the plant.

7 PID SELF-TUNERS

There are many ways of designing a self-tuner for PID operation but these can be placed into two main categories: model-based (as in the self-tuners described above) or based on 'expert-system' and pattern recognition principles. A typical model-based design will specify that the plant can be described by the second-order model with dead-time (Gawthrop, 1982; Radke and Isermann, 1984):

$$y(t) + a_1 y(t-1) + a_2 y(t-2) = b_0 u(t-k) + b_1 u(t-k-1)$$

In most current designs the value of k needs to be prespecified and the estimator then provides values for $\{a_1, a_2, b_0, b_1\}$. Now a PID regulator can be implemented by the discrete-time equation:

$$u(t) = u(t-1) + p_0 e(t) + p_1 e(t-1) + p_2 e(t-2)$$

Hence the forward-path of controller+plant can be wriiten as:

$$y(t) = z^{-k} B(z^{-1})/A(z^{-1}) \;.\; P(z^{-1})/\Delta$$

and one possibility is to cancel the plant's poles by the controller:

$$P(z^{-1}) = p_o A(z^{-1}), \text{ giving } y(t) = z^{-k} B(z^{-1}) p_o / \Delta$$

The coefficient p_o can now be chosen based, say, on phase-margin considerations where the user specifies a margin of, say, 60°. This can be achieved analytically or numerically from the relationship $z \rightarrow \exp(j\omega h)$. It is seen, for example, that the the effective PID gain using this approach reduces as the dead-time k increases (Banyasz et al, 1985).

The 'expert system' methods are developments of the classical Ziegler-Nichols' (1942) tuning rules in which the loop is made to oscillate under proportional feedback and the period of oscillation together with the critical gain are noted. The PID gain, derivative and integral settings are then deduced from these experimentally found values. The manual method is inconvenient for automation and instead two approaches are used: Astrom and Hagglund (1984) use a relay controller and describing-function ideas whereas Bristol (1977), Kraus and Myron (1984) track error responses to steps in load-disturbance or set-points and adjust PID settings according to measured and desired damping factors and overshoots. In both cases the tuning does not depend on a mathematical model, though the user has to have a rough idea of timescales to set the tuner up and the designs (like the original method) imply use on typical sluggish process plant. It is found that the tuning is generally very good and often gives better control than a manual approach, though the adaptive case as usual needs to be employed with care so as to avoid being fooled by disturbances which differ from those assumed in the designs.

8 CONCLUSIONS

Self-tuning has matured over the last few years from a theoretical subject to one which has become a serious industrial tool. Practical experience is growing rapidly and it appears that the methods will become a standard part of a control engineer's armoury. Microprocessors have played a crucial role in achieving acceptance of self-tuning for they allow the computationally demanding algorithms to be implemented cheaply. Self-tuners do not remove the need for the control engineer's skill but allow him to think more positively about the real control needs of the plant rather than simply hoping that classical fixed PID laws will solve all his problems.

REFERENCES

[1] Astrom,K.J. and Hagglund,T.; 'Automatic tuning of simple regulators with specifications on phase and amplitude margins'. Automatica, 20, No.5, pp.645-651, (1984)

[2] Astrom,K.J. and Wittenmark,B.; 'On self-tuning regulators'. Automatica, 9, No.2, pp.185-199, (1973)

[3] Banyasz,Cs., Hetthessy,J. and Keviczky,L.; 'An adaptive PID regulator for MBCC's'. IFAC symposium on Identification, York, (1985)
[4] Bierman,G.J.; Factorization methods for discrete system estimation. Academic Press, (1977)
[5] Breddermann,R.; 'Realisation and application of a self-tuning on-off controller'. Symposium on Adaptive Systems, Bochum, (1980)
[6] Bristol,E.H.; 'Pattern recognition: an alternative to parameter identification in adaptive control'. Automatica, **13**, pp.197-202, (1977)
[7] Clarke,D.W.; 'Some implementation considerations of self-tuning controllers'. in Numerical techniques for stochastic systems ed. by F.Archetti and M.Cugiani, North-Holland, (1980)
[8] Clarke,D.W.; 'Self-tuning control of nonminimum-phase systems'. Automatica, **20**, No.5, pp.501-517, (1984)
[9] Clarke,D.W. and Gawthrop,P.J.; 'Self-tuning controller'. Proc.IEE, **122**, No.9, pp.929-934, (1975)
[10] Clarke,D.W. and Gawthrop,P.J.; 'Implementation and application of microprocessor-based self-tuners'. Automatica, **17**, No.1, pp.233-244, (1981)
[11] Clarke,D.W., Mohtadi,C. and Tuffs,P.S.; 'Generalized predictive control. Part 1: the basic algorithm'. OUEL report 1555/84, (1984)
[12] De Keyser, R.M.C. and van Cauwenberghe, A.R.; 'Microcomputer-controlled servo system based on self-adaptive long-range prediction'. MECO Conference, Athens, (1983)
[13] Dexter,A.L.; 'Self-tuning control algorithm for single-chip microcomputer implementation'. Proc.IEE, **130**, Pt.D, No.5, pp.255-260, (1983)
[14] Duff,J., Dwolatzky,B., Sanoff,S.P. and Wellstead,P.E.; 'Hardware and software for self-tuning control'. IEEE Conference on Applications of Adaptive and Multivariable Control, Hull, (1982)
[15] Fjeld,M. and Wilhelm,R.G.; 'Self-tuning regulators - the software way'. Control Engineering, October, pp.99-102, (1981)
[16] Fortescue,T.R., Kershenbaum,L.S. and Ydstie,B.E.; 'Implementation of self-tuning regulators with variable forgetting factors'. Automatica, **17**, No.6, (1981)
[17] Franklin,G.F. and Powell,J.D.; Digital control of dynamic systems. Addison-Wesley, (1980)
[18] Gawthrop,P.J.; 'Self-tuning PI and PID controllers'. IEEE Conference on Applications of Adaptive and Mutivariable Control, Hull, (1982)
[19] Glattfelder,A.H., Huguenin,F. and Schaufelberger,W.; 'Microcomputer based self-tuning and self-selecting controllers'. Automatica, **16**, pp.1-8, (1980)
[20] Halme,A., Ahava,O., Karjalainen,T., Torvikoski,T. and Savolainen,V.; 'Implementing and testing of

some advanced control schemes in a microprocessors based process instrumentation system'. IFAC Congress, Kyoto, (1981)

[21] Harris,C.J. and Billings,S.A. (editors); Self-tuning and adaptive control: theory and applications. Peter Perigrinus, (1981)

[22] Isermann,R.; Digital control systems. Springer-Verlag, (1981)

[23] Kalman,R.E.; 'Design of a self-optimising control system'. Trans.ASME, 80, pp.468-478, (1958)

[24] Kraus,T.W. and Myron,T.J.; 'Self-tuning PID controller uses pattern recognition approach'. Control Engineering, June, pp.106-111, (1984)

[25] McDermott,P.E. and Mellichamp,D.A.; 'An auto-pole-placement self-tuning controller'. Int.J.Control, 40, No.6, pp.1131-1147, (1984)

[26] Narendra,K.S. and Monopoli,R.V.; Applications of adaptive control. Academic Press, (1980)

[27] Nishikawa,Y., Sannomiya,N., Ohta,T., Tanaka,H. and Tanaka,K.; 'A method for auto-tuning of PID-control parameters'. IFAC 8th World Congress, Kyoto, (1981)

[28] Radke,F. and Isermann,R.; 'A parameter-adaptive PID-controller with stepwise parameter optimization'. IFAC 9th World Congress, Budapest, Hungary, (1984)

[29] Schmid,C. and Unbehauen,H.; 'Identification and CAD of adaptive systems using the KEDDC package'. IFAC Symposium on Identification and System Parameter Estimation, Darmstadt, (1979)

[30] Tayler,C.; 'Patterns of development in self-organising control'. Process Engineering, pp.34-39, February, (1985)

[31] Tuffs,P.S. and Clarke,D.W.; 'FAUST: a software package for self-tuning control'. IEE Conference: Control '85, Cambridge, (1985)

[32] Wittenmark,B. and Astrom,K.J.; 'Practical issues in the implementation of adaptive control'. Automatica, 20, No.5, pp.595-605, (1984)

[33] Ziegler,J.G. and Nichols,N.B.; 'Optimum settings for automatic controllers'. Trans.ASME, 64, pp.759-768, (1942)

CHAPTER 4

ADAPTIVE CONTROL OF SYNCHRONOUS MACHINE EXCITATION

O.P. Malik
Department of Electrical Engineering
The University of Calgary
Calgary, Alberta, T2N 1N4
Canada

ABSTRACT. Adaptive control techniques with particular emphasis on the characteristics that make them suitable for application as synchronous machine stabilizers are discussed. The control algorithms described use least-squares identification with different control strategies. Microcomputer implementation of two algorithms and results of experimental tests are also described. Use of a varying forgetting factor with the identification routine and a self-searching control strategy is introduced to improve the overall flexibility of the controller when applied to varying operating conditions encountered in power systems.

1. INTRODUCTION

Synchronous machine excitation control and its role in improving power system stability has been an important topic of investigation since the early 1940's. Although the high gain continuously acting voltage regulators help improve the dynamic limits of power networks, they could also introduce negative damping, particularly in large weakly coupled systems, and consequently make the system unstable. To overcome this problem, supplementary stabilizing signals were introduced into the excitation system.

The purpose of the supplementary stabilizing signal is to enhance system damping by producing a torque in phase with the speed. It is usually derived by processing any one of a number of possible signals, i.e. speed, acceleration, power, frequency, through a suitable circuit called the power system stabilizer (PSS), to obtain the desired phase relationship. The output of the PSS, i.e. the stabilizing signal, is introduced into the excitation system at the input to the AVR/exciter alongwith the voltage error.

Power system stabilizer operates through the exciter, generator and the transmission system. The characteristics of the plant are non-linear. For example, the gain of the plant increases with generator loading and ac system strength [1]. Also, the phase lag of the plant increases as the ac system becomes stronger. Thus controller

N. K. Sinha (ed.), Microprocessor-Based Control Systems, 61–79.

parameters which are optimum for one set of operating conditions may not be optimum for another set of operating conditions.

In the conventional fixed parameter controllers, the gains and other parameters must be chosen by a trial and error method, and will not ideally suit the entire spectrum of operation. In general, it would be desirable to track the operating conditions and to compute, in real time, control that would provide optimum performance. Adaptive regulators provide a satisfactory solution to meet such a requirement.

2. ADAPTIVE CONTROL

Adaptive control can be described as the changing of controller parameters based on the changes in system operating conditions. Whenever an adaptive controller detects changes in system operating conditions, it responds by determining a new set of control parameters. Therefore, the application of adaptive control strategy to power system control is attractive because the effective system response changes with load level and system configuration.

Although many aspects of power system control problem are different from the problems for which the adaptive control algorithms were originally developed, a number of studies [2-5] indicate that the benefits of adaptive control can still be realized in power systems. Fixed parameter controllers are always a compromise between the best settings for light and heavy load conditions. Adaptive technique assumes that the controller parameters are optimal for the operating conditions, and thus the system stability is enhanced.

The basic functions of an adaptive controller can be generalized as [6]:

(i) identification of unknown parameters, or measurement of a performance index,

(ii) decision of the control strategy,

(iii) on-line modification of controller parameters.

Depending upon how these functions are synthesized, different algorithms for adaptive controllers are obtained. Certain algorithms exhibit characteristics which influence their applicability in power systems. Characteristics of a few adaptive control algorithms are discussed in this Chapter from the point of view of their applicability to power systems. Generator excitation control through an adaptive stabilizer loop is used as an example to illustrate the effect of adaptive control.

2.1 Algorithms with Implicit Identification

A very useful adaptive control technique is to specify a desired performance and measure the actual performance against this performance. Such controllers are commonly known as the Model Reference Adaptive Controllers. A reference model representing the desired behaviour of the closed-loop system is driven by the same input as the controlled system. The regulator parameters are adjusted depending upon the error between the system output and the reference model output as shown in Fig. 1.

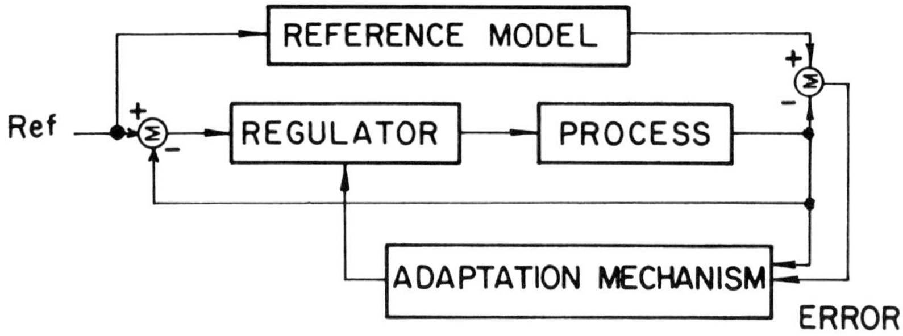

Figure 1. Model reference adaptive control scheme.

The algorithm may be described as below:

(i) Choose a model representing the desired performance.

(ii) Compute the error between the outputs of the reference model and the actual system.

(iii) Based on the correlation between this error and the system states, update the feedback gains of the system.

Let the behaviour of the plant and of the system model be given by the nth order state equations

$$\dot{y}_p(t) = A_p(t)y_p(t) + B_p(t)u(t) \tag{1}$$

$$\dot{y}_m(t) = A_m(t)y_m(t) + B_m(t)u(t) \tag{2}$$

where

y is the output state vector

u is the input control vector

A,B are matrices,

and the subscripts p,m refer to plant and model respectively.

An error function can be formed by writing

$$e(t) = y_m(t) - y_p(t) \tag{3}$$

The aim of the control is to force this error function to zero. This can be done by choosing a Liapunov function [7] or Popov's 'hyperstability' theory [2].

For a satisfactory application of this technique, the designer must have a good idea of the performance that the system is capable of achieving within the control constraints. Otherwise the control limitation occurs and the system response will be substantially different to that of the reference model. Also the actual system responds immediately to unpredictable external disturbances but the reference model cannot do so. This performance difference will be interpreted as a system error and an undesirable adaption of system gains will occur.

2.2 Algorithms with Explicit Identification

Another class of adaptive controllers performs an explicit identification of system parameters or system transfer function as shown in Fig. 2.

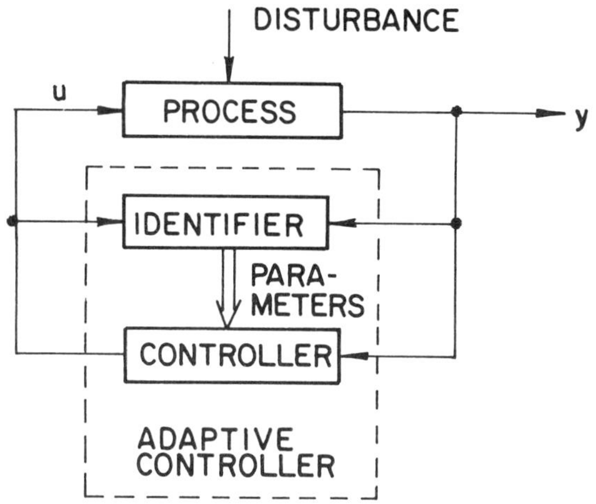

Figure 2 Adaptive control scheme with explicit identification.

A number of identification algorithms have been developed using the discrete domain mathematics. Sampled data design techniques are used to compute control in the following way:

(i) Select a sampling frequency, f, about ten times the normal frequency of oscillation.

(ii) Each sampling interval, T (=1/f), update the system parameters. Least squares or extended least squares technique, in recursive form, is usually used to identify the transfer function of the system.

(iii) Use the updated estimates to compute the controller parameters based on the control strategy chosen.

2.2.1 Identification Recursive identification algorithms can track time varying parameters in real-time. Let the plant output, y(t), at time t be given in terms of the past measurements in difference equation form as

$$y(t) = -a_1 y(t-1) - \dots -a_n y(t-n) + b_1 u(t-1) + \dots + b_m u(t-m) + e(t) \tag{4}$$

where u is the input
t = kT, k = 1, 2, ...,
and e(t) is a set of random variables with zero mean and finite

variance.

Defining the backward shift operator, q^{-1}, as

$$y(t-1) = q^{-1}y(t) \tag{5}$$

eqn. (4) can be written as

$$A(q^{-1})y(t) = B(q^{-1})u(t) + e(t) \tag{6}$$

where the polynomials A and B are given by

$$A(q^{-1}) = 1 + a_1 q^{-1} + \dots + a_n q^{-n} \tag{7}$$

$$B(q^{-1}) = b_1 q^{-1} + \dots + b_m q^{-m} \tag{8}$$

If the noise is uncorrelated, a prediction model of the plant of eqn. (6) can be given as

$$\hat{y}(t) = [1 - \hat{A}\ (q^{-1})]y(t) + \hat{B}\ (q^{-1})u(t) \tag{9}$$

where ^ indicates the estimated value.

Using the recursive least squares algorithm, the estimate $\hat{\theta}(t)$ of the system parameters, θ, at time t can be obtained as [8]

$$\hat{\theta}(t) = \hat{\theta}(t-1) + K(t)\eta(t) \tag{10}$$

where $[\theta] = [a_1\ \dots\ a_n\ b_1\ \dots\ b_m]^T$

$\eta(t) = y(t) - h^T(t)\hat{\theta}(t-1)$ is the prediction error,

and $h(t) = [-y(t-1)\ \dots\ -y(t-n)u(t-1)\ \dots\ u(t-m)]^T$

The correction or gain vector K(t) is given by

$$K(t) = P(t-1)h(t)[w(t) + h^T(t)P(t-1)h(t)]^{-1} \tag{11}$$

The covariance of the estimation error, P(t), can be determined recursively as

$$P(t) = \{I - P(t-1)h(t)h^T(t)[w(t) + h^T(t)P(t-1)h(t)]^{-1}\} \times P(t-1)/w(t) \tag{12}$$

where w(t) is a weighting factor used to discount old data when the parameters vary with time. It helps improve the tracking property of the identification routine. Weighting factor may be in the exponential form, or it may be fixed as

$$w(t) = \lambda \qquad 0 < \lambda < 1$$

To start the algorithm, the following initial conditions may be used:

$$\hat{\theta}(0) = 0 \quad \text{or very small numbers, and}$$
$$P(0) = pI$$

where p is a large number, and
I is an identity matrix.

The recursive least squares algorithm yields biased estimates in the presence of correlated noise. One way of avoiding this problem is to use an extended recursive least squares method [9]. In this method, the prediction model incorporates the noise dynamics alongwith the system transfer function. In power systems, the system dynamics are nonlinear, and the noise is attributed to random load fluctuations, unpredictable system disturbances and measurement noise. Provided these disturbances are uncorrelated, the recursive least squares algorithm yields unbiased estimates of the parameters.

To correctly identify the system there must always be some variation in the system output. For adaptive excitation control problem this variation may be caused by continual changes in load demand. However, if that variation does not register on the measurement system, the estimated B parameters will be very small and this might cause the controller to fall asleep. To overcome this problem, a barely perceptible amount of uncorrelated noise signal may be superimposed on the control signal fed into the exciter.

Four control strategies based on the explicit system identification are discussed in the following sections.

2.2.2 Minimum Variance Controller The combination of minimum variance control [10] and recursive least squares identification has been shown to possess desirable properties [11] and has been tried in the excitation control problem [5, 12]. A minimum variance controller first predicts the next measurement for zero control, and then chooses the control value so that the predicted output error is zero.

This form of adaptive controller is very easy to compute. For a single input single output plant, let the process be given by the transfer function

$$y(t) = \frac{B}{A} u(t)$$

or

$$y(t) = -a_1 y(t-1) - a_2 y(t-2) + b_1 u(t-1) + b_2 u(t-2) \tag{13}$$

Then the minimum variance control is given simply as [11]

$$u(t) = \frac{1}{b_1} [a_1 y(t) + a_2 y(t-1) - b_2 u(t-1)] \tag{14}$$

This form of controller may be implemented using comparatively slow microcomputer with a sampling interval of 50 ms or less.

One unfortunate characteristic of this control algorithm is the possibility that the control computation may become unstable [10] particularly under non-minimum phase conditions. In addition, cancellation of large errors in one sample period is unpractical due to the control limits on the excitation system. If the control signal is limited, the resulting control can give rise to poor damping.

2.2.3 Optimal Linear Quadratic Controller The concepts behind this class of regulators is to minimize a linear quadratic performance index of the form

$$J = \sum_{1}^{t} [y^2(t) + r\, u^2(t-1)] \tag{15}$$

For $r \neq 0$, the optimal control minimizes the output error subject to the requirement that the control effort be small. Thus it places a penalty on the control to achieve the best possible control action. For $r \to 0$, the penalty on control is reduced and in the limit gives minimum variance performance. For $r \to \infty$, control action is taken only to stabilize an unstable system. Thus with a proper choice of r, excessive saturation in the controller can be avoided.

Let the system be described in the state space form as

$$x(t) = Fx(t-1) + gu(t-1) \quad , \quad x(0) \text{ known} \tag{16}$$

and

$$y(t) = h^T x(t) \tag{17}$$

where F is the system matrix, and g and h are n x 1 dimension vectors.

With the condition that F, g are controllable, and F, h are observable, the optimal control can be found as the state feedback [13]

$$u(t) = K(t)\, x(t) \tag{18}$$

where K(t) is the steady-state solution of the recursive equations [14]

$$K(t) = -[g^T S(t-1) g - r]^{-1}\, g^T S(t-1) F \tag{19}$$

$$S(t) = [F + gK(t-1)]^T\, S(t-1) F + h^T h \tag{20}$$

To obtain the states x(t) in eqn. (16) from the measurements of input and output, an observer is required. For on-line computations, the observer design can be simplified using an observable canonical form [13] by which the states can be defined in terms of the past and present measurements. For example,

$$y(t) = x_n(t) = x_{n-1}(t-1) - a_1 y(t-1) + b_1 u(t-1) \tag{21}$$

and similar expressions hold for the states x_1 to x_{n-1}.

System parameters are identified every sampling interval using a recursive identification technique and the state vector is computed using the observer design given in eqn. (21). The control is then computed from eqn. (18).

To obtain steady-state solution of eqns. (19) and (20) requires infinite number of iterations, which is not practical for real-time applications. In practice, a limited number of iterations of these two equations gives satisfactory results. The computations can be also simplified since the matrix S is symmetric.

This controller has the advantage that, if the parameter estimates are true, the controller will always be stable. It also has a single parameter, r, to describe the best control action. Moreover, the deterioration of performance with parameter error is low. However, these desirable characteristics are obtained at the cost of a significantly heavy computational burden.

2.2.4 Pole-Assigned Controller This type of controller is similar to the model reference controller in that the desired response is prespecified by specifying the desired system closed-loop poles, but the update of the controller parameters is based on explicit system identification. Whereas in the minimum variance case all the poles are shifted towards the origin, the pole assigned controller has the freedom to place them at other locations. This permits a trade-off between the performance and the control effort.

For the system given by eqn. (13), the control is computed from [15]

$$u(t) = -\frac{G(q^{-1})}{F(q^{-1})}\, y(t) \tag{22}$$

where the polynomials G and F are given by

$$F(q^{-1}) = 1 + f_1 q^{-1} + \ldots + f_{m-1} q^{-m+1} \tag{23}$$

$$G(q^{-1}) = 1 + g_1 q^{-1} + \ldots + g_{n-1} q^{-n+1} \tag{24}$$

Choosing the closed-loop poles as roots of the polynomial T given by

$$T(q^{-1}) = 1 + t_1 q^{-1} + \ldots + t_r q^{-r} \tag{25}$$

the control parameters f_i and g_i are computed from

$$A(q^{-1})F(q^{-1}) + B(q^{-1})G(q^{-1}) = T(q^{-1}) \tag{26}$$

where $r \leq n + m - 1$

For $n = m$, solution of eqn. (25) can be written as

$$Z = M^{-1}L \tag{28}$$

where M is a (m+n-1) x (m+n-1) matrix having as its elements the estimated values $\hat{A}$ and $\hat{B}$ of the A and B parameters [16], and its inverse exists provided that the polynomials $\hat{A}$ and $\hat{B}$ have no common factors [17].

$$[Z] = [f_1 \; \dots \; f_{m-1} g_0 \; \dots \; g_{n-1}]^T \tag{29}$$

$$[L] = [(t_1 - a_1) \; \dots \; (t_n - a_n) \; t_{n+1} \; \dots \; t_{m+n-1}]^T \tag{30}$$

If the elements of L are very small, the coefficients of F and G are small. Then the roots of F are close to zero and a stable controller is achieved. If the poles are moved a small distance from their open-loop locations, only a small amount of control is required. Hence, even if the control is bound limited, it will still provide better damping than the minimum variance controller. Because the control is not directly computed from the inverse of the polynomial B as in the minimum variance algorithm, the possibility of achieving a stable controller is higher for a zero mean or non-minimum phase system. However, there is no specific guideline on the choice of T and a poor choice might lead to unstable control computations.

2.2.5 Pole-Shift Controller The problems associated with the pole-assigned controller can be overcome by adding a constraint to eqn. (26) such that the closed-loop poles are obtained by radially shifting the open-loop poles towards the origin. Thus T becomes

$$T(q^{-1}) = \hat{A}(\alpha q^{-1}) = 1 + \alpha \hat{a}_1 q^{-1} + \dots + \alpha^n \hat{a}_n q^{-n} \tag{31}$$

where α is close to but less than one. Then 'L' in eqn. (29) becomes

$$L = [\hat{a}_1(\alpha - 1) \; \hat{a}_2(\alpha^2 - 1) \; \dots \; \hat{a}_n(\alpha^n - 1) \; 0 \; \dots \; 0]^T \tag{32}$$

With α close to one, elements of 'L' are close to zero, elements of polynomials F and G are small and a stable controller is achieved. So, for a stable plant, a properly identified pole-shift controller always yields a stable controller provided that the polynomials $\hat{A}$ and $\hat{B}$ have no common factors. This approach has the advantage of having a single parameter, α, trade-off for best and stable control.

The adaptive controller is thus achieved by computing on-line the closed-loop characteristic eqn. (31) using the system parameters identified by the recursive relationships of eqns. (10) to (12). The control, u, is computed as

$$u(t) = [-y(t), \; \dots, \; -y(t-n-1), \; -u(t-1), \; \dots, \; -u(t-m-1)][Z]$$

or $$u(t) = x^T(t)Z \tag{32}$$

based on controller parameters calculated from eqn. (29). For high order systems, the solution can be computationally expensive. But because of the special form of the matrix [M] in eqn. (28), this computation can be simplified considerably [18].

3. ADAPTIVE CONTROLLERS FOR POWER SYSTEMS

Desirable properties of a good power system adaptive controller may be described as:

(i) the controller should not saturate under transient conditions [19],

(ii) the controller should guarantee stability under non-minimum phase conditions [11],

(iii) as an 'a priori' decision cannot be made on the parameters, preselection of the required performance be avoided [15],

(iv) trade-off between control effort and control action should preferably be made by adjusting a single parameter [16].

Characteristics associated with the minimum variance controller indicate that it does not satisfy the first two properties, but the fourth property is not a constraint because there are no parameters to be preselected. The model reference and the pole-assigned controllers are not suitable from the point of view of the third property and they do not possess the fourth property.

The linear quadratic and pole-shift controllers are not constrained by the first three properties and they possess the desirable fourth property. They therefore seem to be more suitable for application in power systems.

Simulation studies have been performed on a power system consisting of a generating unit connected to a fixed voltage bus through a long double-circuit transmission line. The machine was simulated by a set of non-linear differential equations based on Park's equations. Incorporating a simple governor and an AVR gave a tenth order non-linear model of the system.

Electrical power output was used as the input to the adaptive stabilizer shown in Fig. 3. The power signal was processed through a washout circuit to eliminate the dc component and sampled every 100 ms. The system between the control input u and the output P was identified as a third order system with m=n=3. A forgetting factor of 0.95 was used to discount old data.

Results for a representative study of a short-circuit applied at the middle of one of the two parallel lines and the loss of the faulted line 0.4s after the inception of the short-circuit are shown in Fig. 4. Load angle responses for three different algorithms, i.e. the linear quadratic, the pole shifting and a conventional stabilizer designed according to Ref. [20] show that the adaptive stabilizers perform much better than the conventional stabilizer.

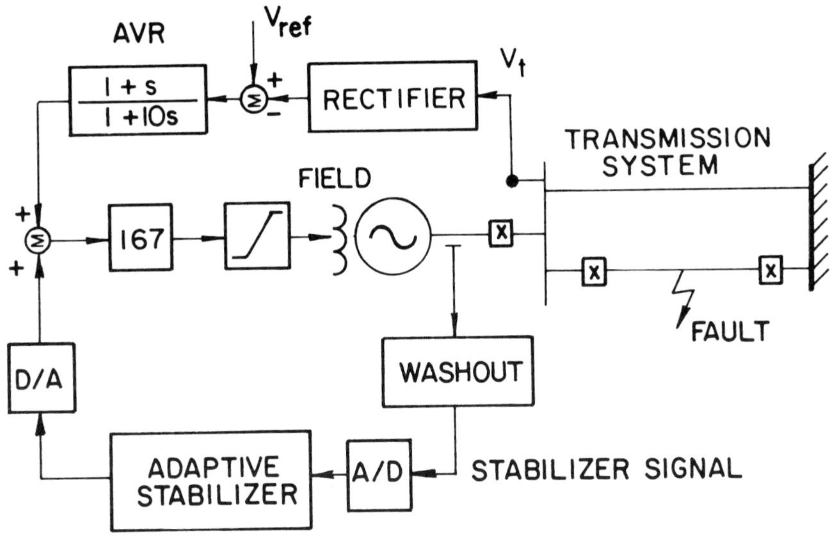

Figure 3 Power system model with excitation control scheme and adaptive stabilizer

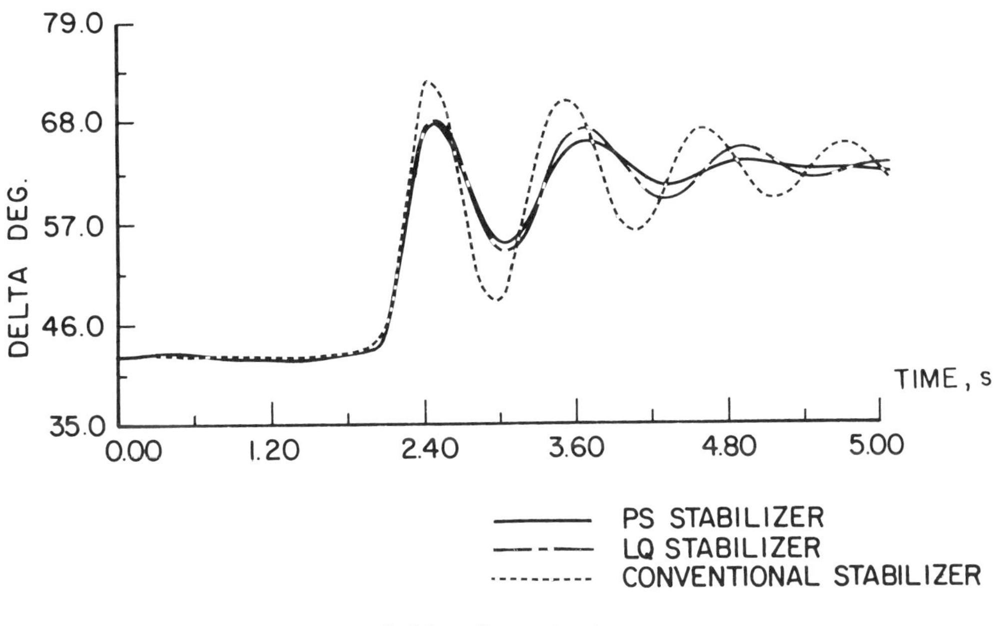

Figure 4 Short-circuit and line loss test

4. MICROCOMPUTER IMPLEMENTATION OF ADAPTIVE CONTROLLER

The linear quadratic and the pole-shifting algorithms described above have been implemented on a 16 bit microcomputer as a stand-alone adaptive controller.

4.1 Hardware Requirements

An Intel single board computer iSBC 86/12 with 5 MHz clock was used for implementation. It has 32 kilobytes of dynamic RAM (data memory) and can also have up to 16 kilobytes of program memory (ROM). It also has a serial RS232 interface for communications. The input/output was done through two multibus compatible I/O ports each consisting of an Intel 8205 address decoder, 8287 octal bus transceivers and 8282 octal latches. These ports are capable of handling either 16 bit or 8 bit data in parallel. The output of the computer was converted into an analog signal using a 12 bit D/A converter and added to AVR output.

The input signal to the adaptive controller could be either the power or the error in speed. The speed signal was obtained by a speed decoder circuit which counts the time period of one shaft revolution through a 1 MHz clock and the signal was digital in format.

The power signal was obtained using a Hall-effect wattmeter. The signal was amplified and processed through a washout filter to obtain the error in electrical power. The power signal was fed into a data acquisition system controlled by a PDP 11/03 computer acting as a front end I/O processor for the SBC 86/12. In this experimental setup, the PDP 11/03 was used simply because of its easy accessibility. Ideally the controller will have its own data acquisition system.

4.2 Software Structure

The software development was done on an Intel development system. The software contains one main program and 31 sub-routines. The controller is based on the recursive least-squares identification algorithm used in conjunction with either the linear quadratic or pole-shifting control algorithm. Either of the two control algorithms can be selected by the user.

The software is divided into foreground and background tasks. The function of the foreground task is to interface with the operator console. The background task is interrupt driven and its function is to perform identification and control update. The program is interactive and thus the operator can examine and/or change the parameters in the foreground mode. The foreground task is not so time constrained and hence comparatively lengthy jobs such as floating point to and from ASCII string conversion are done in the foreground mode.

The foreground task is interrupted every 100 ms and once the interruption occurs, the computer switches to background mode. In the background mode, the computer performs the required task before returning to the foreground mode.

The recursive least squares parameter identification requires a large dynamic range of numbers. This is handled by using floating point arithmetic. The use of assembler for arithmetic and input/output routines, and PL/M language for control algorithm gives a good combination of speed and ease of programming.

In any on-line control, computation time is a major factor. Using a 5 MHz clock, the recursive least squares identification routine for n=m=3 takes about 20.7 ms, the pole-shifting controller identity solving takes about 3.5 ms and 5 iterations of linear quadratic controller Riccati equation solving routine takes about 29.7 ms. Combined with an overhead of 1.8 ms for functions such as I/O, control computation, etc., the adaptive controller based on the pole-shifting algorithm takes 30 ms and that using the linear quadratic algorithm takes 52.2 ms.

5. EXPERIMENTAL RESULTS

The adaptive controller designed utilizing the Intel SBC 86/12 was tested as a power system stabilizer. The experimental setup was as shown in Fig. 3. The power system is modelled by a 3 kW microalternator driven by a dc motor and is connected to a large system bus through a pair of long transmission lines each modelled by five π sections equivalent to a length of 250 km of a 500 kV line.

An extensive series of tests has been conducted in the laboratory. However, only an illustrative set of test results is given in Figs. 5 and 6. In the studies described here, a fixed forgetting factor of 0.9 is used. Results for a sudden opening of one of the two parallel transmission lines with $p\delta$ as the input to the adaptive stabilizer is shown in Fig. 5 for the linear quadratic and the pole-shifting controllers.

System performance with a short-circuit applied at the midpoint of one of the two parallel transmission lines with successful reclosure after 32 cycles is shown in Fig. 6 using power as the stabilizing signal. These results demonstrate that the adaptive stabilizer provides good damping and brings the system back to steady-state quickly.

6. ADAPTIVE CONTROLLER WITH VARIABLE FORGETTING FACTOR AND SELF-SEARCHING POLE-SHIFT

The most suitable values of the forgetting factor, λ, in the identification algorithm and of the pole-shift factor, α, in the pole-shifting control algorithm are related to the operating conditions of the system. To maintain the controller performance in an optimum state, it is desirable that both λ and α vary with the operating conditions.

6.1 Variable Forgetting Factor

The best value of the forgetting factor depends upon the operating conditions. If this value is chosen appropriate for the fault conditions, and the system operates under steady-state for a long time, the matrix P in eqn. (12) will blow up. Variable forgetting factor can be used to

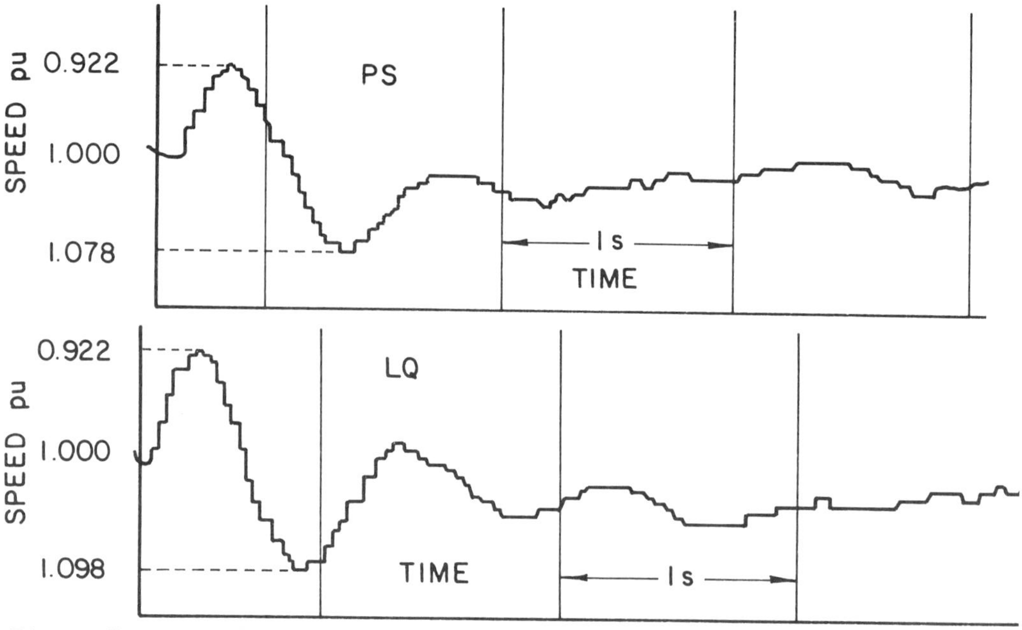

Figure 5. One of the two parallel transmission lines opened suddenly

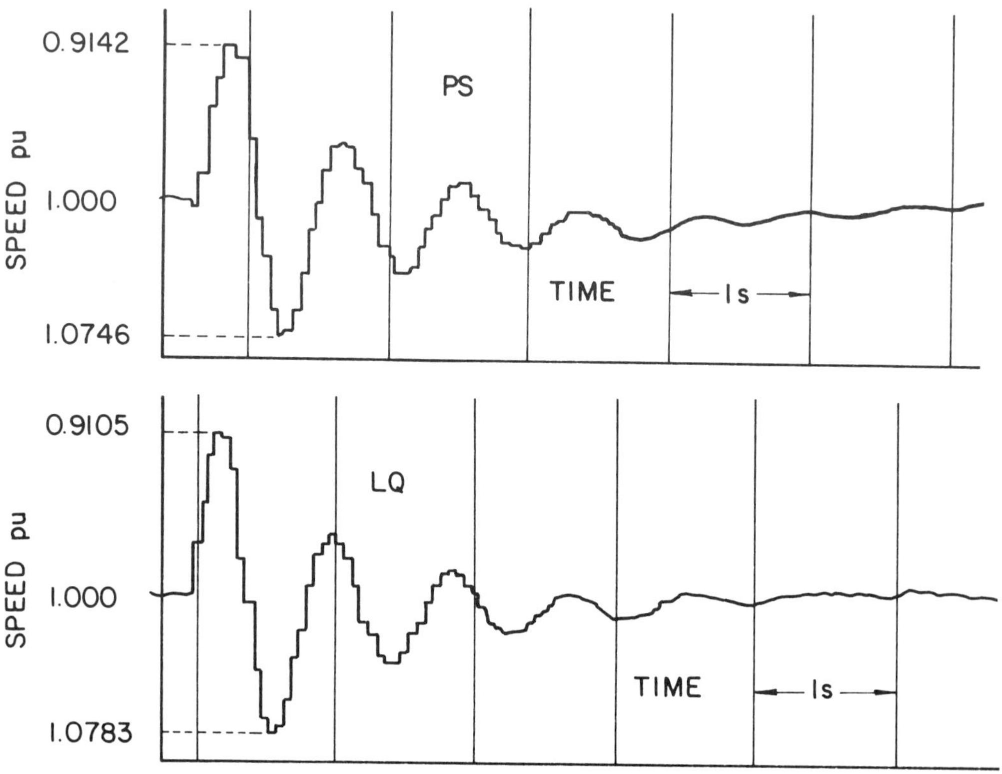

Figure 6. Short-circuit with successful reclosure

overcome this problem.

In the variable forgetting factor algorithm, the weighting factor, w(t), is determined at each sampling instant by

$$w(t) = 1 - [1 - h^T(t-1)\cdot K(t)]\, \eta^2(t)/\Sigma_0 \tag{33}$$

Under steady-state operation, η will be very small or equal to zero. This forces w(t) near or equal to 1 which prevents the P matrix from blowing up. Upon the occurrence of a disturbance, w(t) decreases as η increases. This improves the tracking property of the identifier. For large disturbances, w(t) decreases and fast parameter tracking results.

The Σ_0 in eqn. (33) is a preselected constant which can be expressed as

$$\Sigma_0 = \sigma^2 N_0 \tag{34}$$

where σ^2 is the expected measurement noise variance based on the knowledge of the process. N_0 controls the speed of adaptation as it corresponds to a nominal asymptotic memory length. For power system stabilizer application, a recommended value of Σ_0 is 0.8.

6.2 Self-Searching Pole-Shift

The fixed value of the pole-shift factor in the control algorithm described in Section 2.2.5 is chosen to achieve the best compromise between small and large system disturbances. A value of α close to zero will be ideal for system operation under dynamic state but it may not provide proper damping during transient conditions after a large system fault. To determine the appropriate value of α according to system operating conditions, the design of a self-searching pole-shifting algorithm is developed using the following basic principles:

(i) theoretically, as the closed-loop poles are shifted closer to the origin of the unit circle in the z-domain, the system becomes more stable,

(ii) practically, as the poles are shifted towards the origin, more control effort is required. The control is bound limited and if these limits are exceeded unsatisfactory control action results. In the worst case, the system may become unstable.

To obtain the ideal closed-loop system response under varying conditions, and in view of the above-mentioned principles, the pole-shifting factor must be variable. In the self-searching pole-shifting technique, the algorithm for calculating $\alpha(t)$ is formulated as below.

Let the control be constrained as

$$u_{min} \leq u \leq u_{max}$$

To meet principle (ii), a sensitivity function can be computed from eqn. (32) as

$$\frac{\partial u}{\partial \alpha} = x^T \frac{\partial Z}{\partial \alpha} \tag{35}$$

Substituting eqn. (28) in eqn. (35) gives

$$\frac{\partial u}{\partial \alpha} = x^T M^{-1} \frac{\partial L}{\partial \alpha} = -x^T M^{-1} [a_1 \;\; 2a_2\alpha_1 \;\; \ldots \;\; na_n\alpha^{n-1}] \tag{36}$$

Defining a control margin as

$$\Delta u = \begin{cases} u_{max} - u & u \geq 0 \\ u - u_{min} & u < 0 \end{cases}$$

the α modification factor is given by

$$\Delta\alpha = K(\frac{\partial u}{\partial \alpha})^{-1} \Delta u = -K \, (|\frac{\partial u}{\partial \alpha}|)^{-1} \Delta u \tag{37}$$

where K is a positive constant chosen to avoid excessive variation in α.
The pole-shift factor $\alpha(t)$ is given by

$$\alpha(t) = \alpha(t-1) + \Delta\alpha \tag{38}$$

This calculation may be repeated every sampling period until the control u is within the limits. In applying this algorithm to a synchronous machine excitation control it has been found that in general only one iteration is necessary. On start up, α can be assigned any value between 0 and 1. When the system is operating in the steady-state, shifting the closed-loop poles even to the origin of the unit circle will result in only a small control effort.

6.3 Results

Response of the power system described in Section 3 to a 50% step change in input torque is shown in Fig. 7 for the cases of no stabilizer, a conventional stabilizer and the pole-shift stabilizer with variable w and α. The variations in w and α, and the output of the adaptive stabilizer are shown in Figs. 8 and 9 respectively.

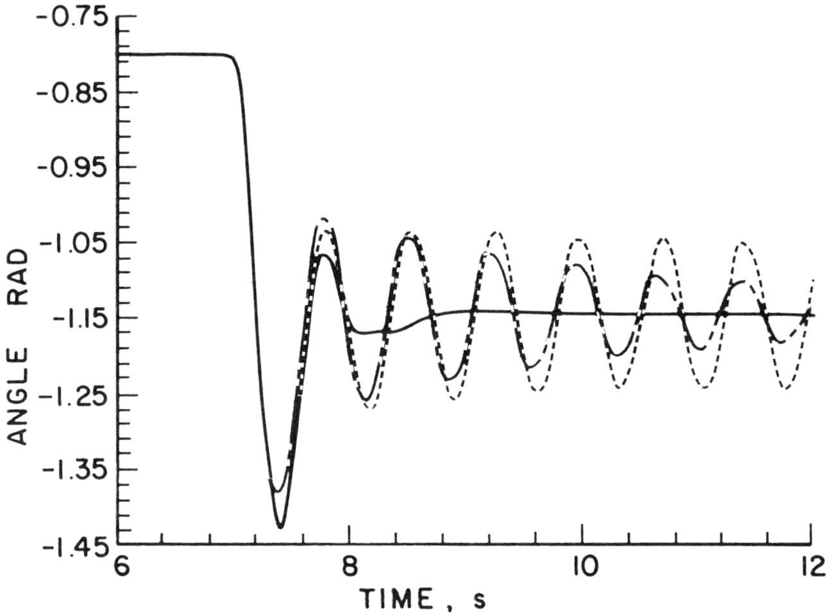

Figure 7. Response to a 50% step increase in torque
···· No stabilizer
—·— Fixed w, α stabilizer
——— Variable w, α stabilizer

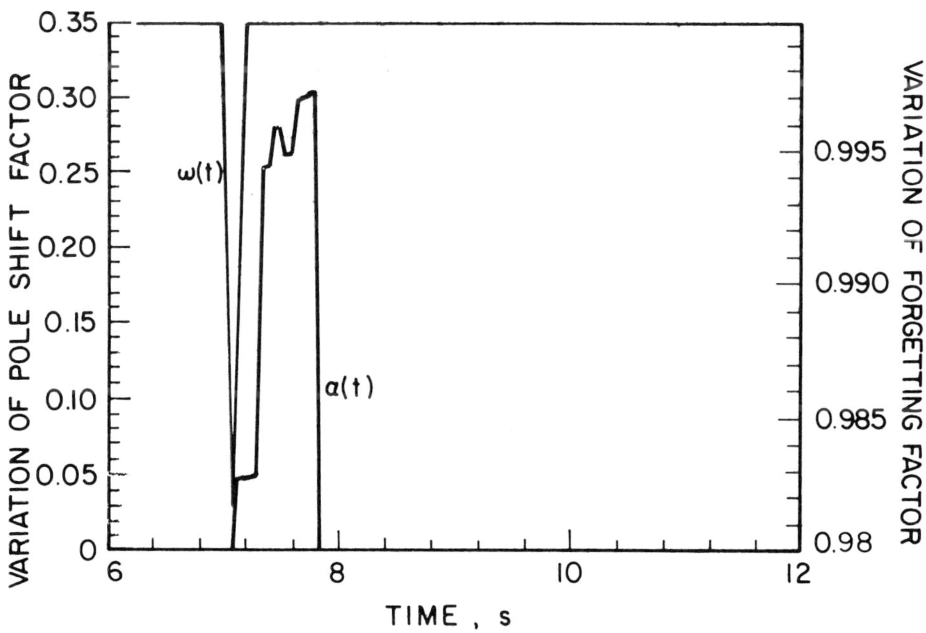

Figure 8. Variation of forgetting, w, and pole-shift, α, factors

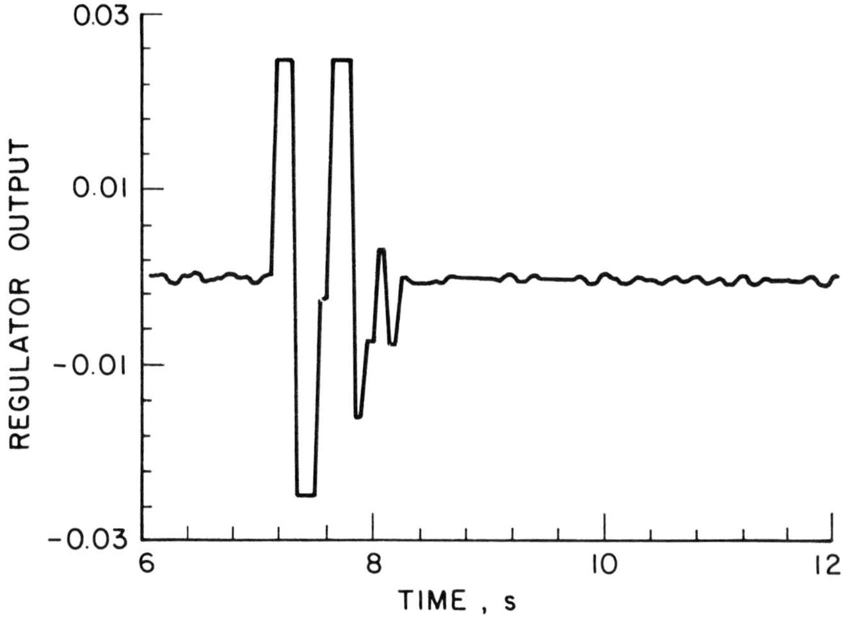

Figure 9. Output of adaptive stabilizer

7. CONCLUDING REMARKS

An overall view of the applicability of an adaptive control scheme to improve power system stability has been presented. An important characteristic of an adaptive stabilizer is that it continually tunes itself to the current operating conditions. It thus improves both the dynamic and transient stability. Two adaptive control algorithms have been implemented on a 16 bit microcomputer. Real-time operation of the adaptive controller as a stabilizer for a generating unit has demonstrated the viability of the microcomputer implementation of such controllers.

8. REFERENCES

1. E.V. Larsen and D.A. Swann: 'Applying power system stabilizers, Part I: General concepts', *IEEE Trans. on Power Apparatus and Systems*, Vol. PAS-100, pp. 3017-3024, 1981.
2. E. Irving, J.P. Basset, C. Charcossey and J.P. Nonville: 'A method to improve the power network 'steady state stability' and to reduce the unit stress. The multivariable adaptive control of generators', *Proceedings IFAC Symposium on Automatic Control and Protection of Electric Power Systems*, Melbourne, 1977, pp. 60-69.
3. O.P. Malik, G.S. Hope and R. Ramanujam: 'Real-time model reference adaptive control of synchronous machine excitation', *IEEE Power Engg. Society Winter Meeting*, New York, 1978, Paper #A78 297-4

4. G. Ledwich: 'Adaptive Excitation Control', *Proc. IEE,* Vol. 126, pp. 249-253, 1979.
5. M.A.H. Sheirah, O.P. Malik and G.S. Hope: 'A self-tuning automatic voltage regulator', *Electric Power System Research,* Vol. 2, pp. 199-213, 1979.
6. B. Wittenmark: 'Stochastic adaptive control methods: A survey', *International J. of Control,* Vol. 21, pp. 705-730, 1975.
7. H.R. Outhred and F.J. Evans: 'A model reference adaptive controller for turbo-alternators in large power systems', *Proc. 4th Power Systems Computations Conference,* Grenoble, 1972, Paper #3.1/9.
8. G.C. Goodwin and R.L. Payne : *Dynamic System Identification, experiment, design and data analysis*, Academic Press, New York, 1977.
9. V. Panuska: 'An adaptive recursive least squares identification algorithm', *Proc. 8th IEEE Symposium on Adaptive Processes,* Pennsylvania State University, 1969.
10. K.J. Åström: *Introduction to stochastic control theory,* Academic press, New York, 1970.
11. K.J. Åström and B. Wittenmark: 'On self-tuning regulators', *Automatica,* Vol. 9, pp. 185-199, 1973.
12. J. Kanniah, O.P. Malik and G.S. Hope: 'Excitation control of synchronous generators using adaptive regulators, Parts I and II, *IEEE Trans. on Power Apparatus and Systems,* Vol. PAS-103(5), pp. 897-910, 1984.
13. T. Kailath: *Linear systems*, Prentice-Hall, 1980.
14. B.D.O. Anderson and J.B. Moore: *Linear optimal control*, Wiley, 1972.
15. P.E. Wellstead, D. Prager and P. Zanker: 'Pole assigned self-tuning regulator', *Proc. IEE,* Vol. 126, pp. 781-787, 1979.
16. A. Ghosh, G. Ledwich, O.P. Malik and G.S. Hope: 'Power system stabilizer based on adaptive control techniques', *IEEE Trans. on Power Apparatus and Systems,* Vol. PAS-103(8), pp. 1983-1989, 1984.
17. L.E. Dickson: *First course in the theory of equations,* Wiley, New York, 1922.
18. A. Ghosh: 'Adaptive stabilizers for generator excitation', Ph.D. Thesis, University of Calgary, 1983.
19. E.V. Larsen and D.A. Swann: 'Applying power system stabilizers Part III: Practical considerations, *IEEE Trans. on Power Apparatus and Systems,* Vol. PAS-100, pp. 3034-3041, 1981.
20. F.P. deMello, L.N. Hannett and J.M. Undrill: 'Practical approaches to supplementary stabilizing from accelerating power', *IEEE Trans. on Power Apparatus and Systems,* Vol. PAS-97, pp. 1515-1522, 1978.

CHAPTER 5

MICROPROCESSOR-BASED SYSTEM IDENTIFICATION BY SIGNAL COMPRESSION METHOD

N. Aoshima
Institute of Applied Physics
University of Tsukuba
Sakura, Ibaraki 305
Japan

ABSTRACT. Measurements of impulse response, frequency response, step response and Bode diagram of linear systems by signal expansion and compression techniques are explained. This method uses special computer generated test signal, and all signal processings are performed by a microprocessor based measuring system. The S/N ratio can be improved immensely by several computer techniques. So this method is useful for measurements in noisy conditions. Since the test signal has flat power spectrum and constantly zero phase, equivalently, the frequency response function and the Bode diagram can be obtained directly. Applications for the measurement of heat conduction system, acoustic system and electrical second order system are described. When an impulse response of a wave propagation system is concerned, a direct wave and reflected waves can be separately analyzed. This method is also applicable for at least some kinds of nonlinear systems.

1. INTRODUCTION

One input one output linear system can be characterized completely by its impulse response or frequency response function. These functions can be determined from analyzing the input and output signals. Progresses in computer techniques make these measurements rather easy task. From the Fourier transformations of the input and output signals, the complex frequency response function is obtained by just division. (Fig. 1) In practice, there are some problems in this procedure. If the Fourier transform of the input signal $X(j\omega)$ becomes zero, division is impossible. An impulse signal is ideal as an input signal because it contains all frequency components, but practically it cannot be realized. So short duration finite amplitude pulse signal is

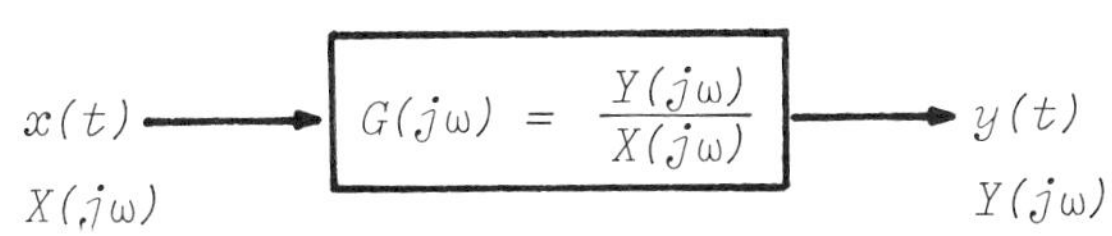

Figure 1. One input one output linear system.

N. K. Sinha (ed.), Microprocessor-Based Control Systems, 81–103.

often used, but such signal has rather small signal power and tend to be masked by noise. The signal compression method discussed here is an answer to these problems.

The signal compression method was derived and named by this author and applied first for sound measurement.[1,2,3] This method is also useful for a measurement of impulse response, frequency response, impedance function, step response and Bode diagram of any linear system even in very noisy conditions. Additionally, it can be applied for analyzing some kinds of nonlinear systems.

2. PRINCIPLE OF THE SIGNAL COMPRESSION METHOD

2.1. Signal Expansion and Compression

A test signal for a measurement of frequency response is desired to have flat power spectrum in wide frequency range like Fig. 2. To generate time domain signal whose power spectrum is like Fig. 2, an inverse Fourier transformation is used. Next frequency domain function is defined first.

$$X(n) + jY(n) = 60 \exp\{-(\frac{n - 900}{800})^{12}\} , \quad 0 \le n \le 2047$$

$$X(n) + jY(n) = X(4096 - n) - jY(4096 - n) , \quad 2049 \le n \le 4095$$

$$X(2048) + jY(2048) = 0 \tag{1}$$

This is shown in Fig. 3. Since an inverse Fourier transformation is made by FFT, Fig. 3 is symmetrical and the center point corresponds to the maximum frequency. For the application in which low frequency response is important, the spectrum can extend to zero frequency. Smooth cutoff shape is necessary to suppress long transient of time domain signal. The inverse Fourier transformed signal of Fig. 3 by FFT is shown in Fig. 4(a). This is a cyclic signal. Shifted and magnified figure is shown in Fig. 4(b). Since phase angle of Eq. (1) is constantly zero, all frequency components concentrate to the time origin and a sharp pulse appears in Fig. 4.

Then, another frequency function is defined as

$$X(n) + jY(n) = 60 \exp\{-(\frac{n - 900}{800})^{12}\} \exp(\frac{12n^2}{10000} j) , \quad 0 \le n \le 2047$$

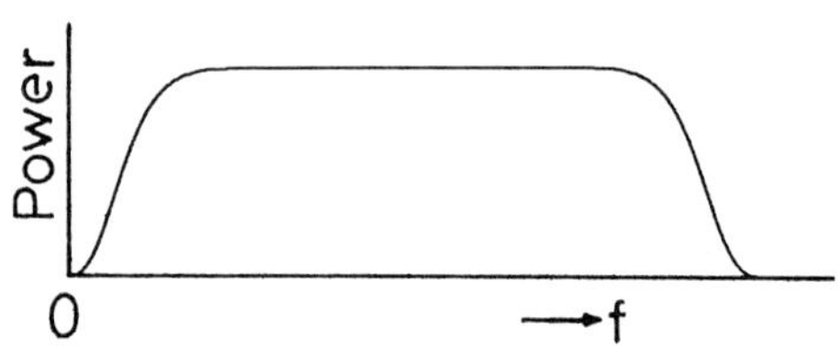

Figure 2. Desired test signal power spectrum.

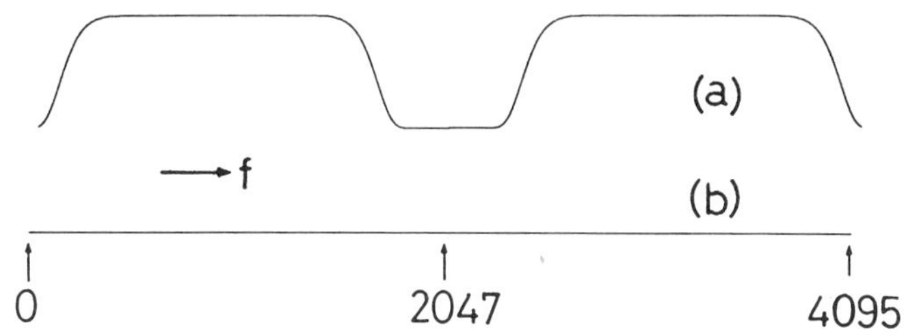

Figure 3. Assumed frequency domain function as Eq. (1). (a) Real part. (b) Imaginary part.

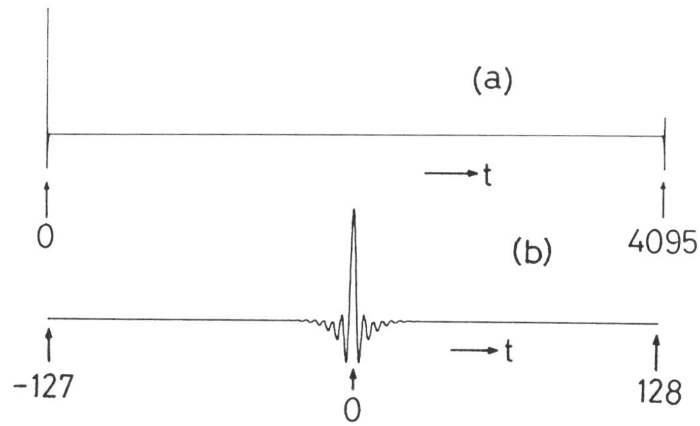

Figure 4. Inverse Fourier transformed function of Fig. 3. (a) Transformed time domain signal. (b) Shifted and magnified figure.

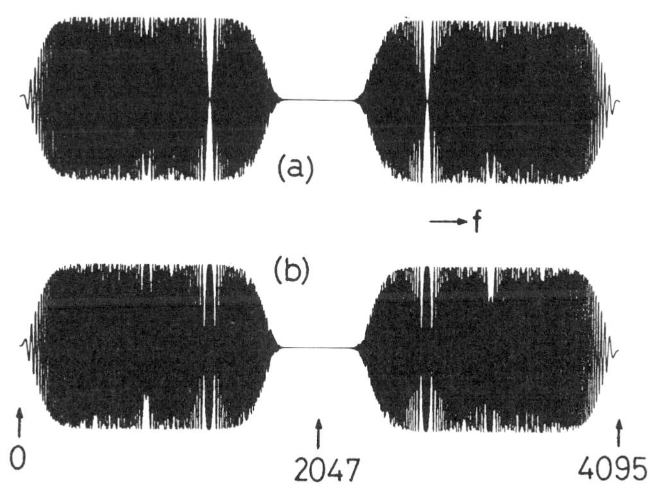

Figure 5. Another assumed frequency domain function as Eq. (2). (a) Real part. (b) Imaginary part.

Figure 6. Inverse Fourier transformed function of Fig. 5.

$$X(n) + jY(n) = X(4096 - n) - jY(4096 - n)\ , \quad 2049 \le n \le 4095$$

$$X(2048) + jY(2048) = 0 \tag{2}$$

This is shown in Fig. 5. Equation (2) can be considered as derived from Eq. (1) by applying phase shift filter $H(jn) = \exp(12n^2j/10000)$. The magnitude of this filter is unity for all frequencies, so Eq. (2) has the same flat power spectrum as Eq. (1). The delay time by this filter is given by

$$-\frac{d}{dn}\{\angle H(jn)\} = \frac{-24n}{10000}\ , \tag{3}$$

so each frequency component of Eq. (2) has different delay time. This makes the inverse Fourier transformed time signal of Eq. (2) spread in time axis as shown in Fig. 6.

Now pulse signal of Fig. 4 is expanded to long signal of Fig. 6 by Fourier transformation (FFT), phase shift filter ($H(jn)$) and inverse Fourier transformation (IFFT). The reverse process is also possible. Long signal of Fig. 6 can be compressed to pulse signal of Fig. 4 by FFT, inverse phase shift filter ($H^{-1}(jn) = \exp(-12n^2j/10000)$) and IFFT. These operations are all linear, so they can be interchanged with other linear operations.

The process applying an input signal to a linear system and obtaining an output signal can be thought as a linear operation. The signal compression method can be summarized as Fig. 7. The pulse signal of Fig. 4 is expanded to the long signal of Fig. 6 and supplied to a linear system as a test input signal. A response of this linear system is then compressed. Since linear operations are interchangeable,

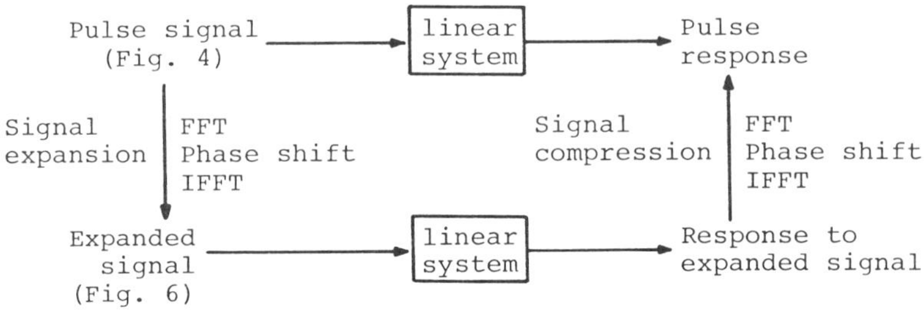

Figure 7. Principle of the signal compression method.

this compressed signal is identical to an output signal when pulse signal is supplied directly to the linear system. By using expanded long signal, larger signal power can be supplied to the system under investigation than by conventional pulse method. This is advantageous to overcome external noise.

2.2. Improvement of S/N Ratio

The expanded test signal is supplied by a computer via D/A converter to the object system and the output signal is taken into the same computer via A/D converter. If the object system is time invariant, time domain averaging is possible, since precise relation is maintained between the input and output signals. If responses are averaged N times, S/N ratio is supposed to increase $20\log\sqrt{N}$ dB assuming external noise is zero mean random and independent to the excitation.

If an object system has wide band characteristics and does not contain pure delay or sharp resonance, the output signal can generally be compressed to a pulse-like signal. In this case, S/N ratio can be improved by picking up the concentrated pulse in the computer memory and making other data zero. This comes from the fact that the signal component has special phase characteristics and it can be compressed by the phase shift filter but the noise component spreads widely in time axis even after the phase shift filter is applied. So by picking up the narrow time interval which contains the signal pulse, large part of noise component is eliminated. This process is named as "signal cut out". Suppose total data length is 4096 points and n points are "cut out", the noise power is reduced to $n/4096$, so the S/N improvement is $10\log(4096/n)$ dB. By this technique, S/N ratio can be improved more than 15 dB without difficulty.

When a wave propagation system is analyzed, several wave packets are often separately obtained. In such situation, "cut out" is useful to select particular wave ray not only to improve S/N ratio.

2.3. Microprocessor based Processing System

Several kinds of microprocessor based processing system are now being used in our laboratory. Two kinds of them will be explained here.

One is based on commercially available low cost personal computer which uses Z80A as a CPU. Signal I/O modules are attached externally and programs are all written by machine language. An FFT of large data number is most important in processing procedure. This is performed by double precision (16 bit) fixed point calculation. We tried 8 bit calculation first, but truncation error was too much. The maximum data number is 4096 points, and the FFT of 4096 complex numbers is performed in 53 seconds. Conversion rate of A/D and D/A is limited up to 25 kHz by software control and 300 kHz by the external DMA equipment.

The second system uses 68000 as a CPU and is constructed by ourselves from IC chips. Programs are contained in ROMs, and 128 k bytes DRAMs are used as a data area. This system can process maximum data length of 32768, and the averaging is possible up to 65536 times. Computation time of complex FFT is 2.8 seconds for 4096 points and 27.7 seconds for 32768 points. Maximum sampling rate is about 300 kHz which can be made faster by using faster A/D converter. The large data capacity of this system is useful to measure long duration wide band impulse response.

3. EXAMPLE OF THE MEASURING PROCEDURE

As an example of the signal compression method, a measurement of heat conduction system like Fig. 8 will be explained. Copper wire (0.5 mm diameter) is heated by an electric heater, which produces time varying temperature signal in it. The temperature is measured by a thermocouple (TC) at d cm apart from the heater. The test signal of the signal compression method is supplied to the heater as a current signal.

Since the time constant of this system is rather long, the test signal is generated from next equation instead of Eq. (2).

$$X(n) + jY(n) = 60 \exp\{-(\frac{n}{170})^{12}\} \exp(\frac{-12n^2}{1000}j) , \quad 0 \le n \le 2047$$

$$X(n) + jY(n) = X(4096 - n) - jY(4096 - n) , \quad 2049 \le n \le 4095$$

$$X(2048) + jY(2048) = 0 \tag{4}$$

The power spectrum of this signal is shown in Fig. 9. It is concentrated in low frequency region and extended to lowest frequency. The inverse Fourier transformed test signal is shown in Fig. 10(a), and the compressed pulse signal is in Fig. 10(b). As Fig. 10(a) shows, the low frequency component is supplied first, in contrast to the previously defined test signal of Fig. 6. This is useful when the system under investigation has large response in low frequency range.

The dc bias is added to the test signal as Fig. 11(a), which shows the current waveform actually supplied to the heater. The output signal of the thermocouple placed at $d = 4$ cm is shown in Fig. 11(b). The clock frequency of the A/D and D/A converter was 4 Hz, so data aquisition time was 1024 seconds. Time domain averaging was not applied in this example. The acquired data are stored in 16 bit form,

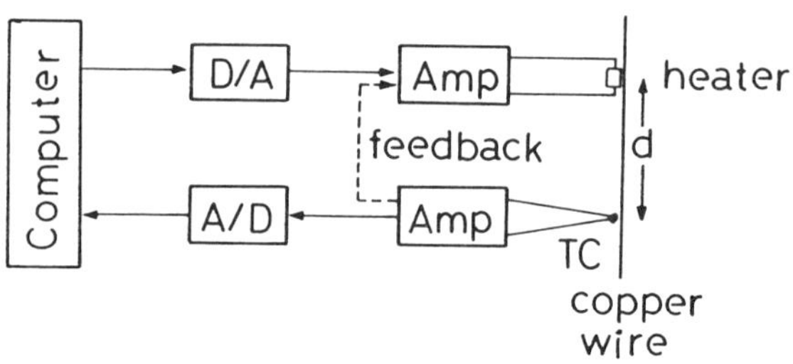

Figure 8. A heat conduction system.

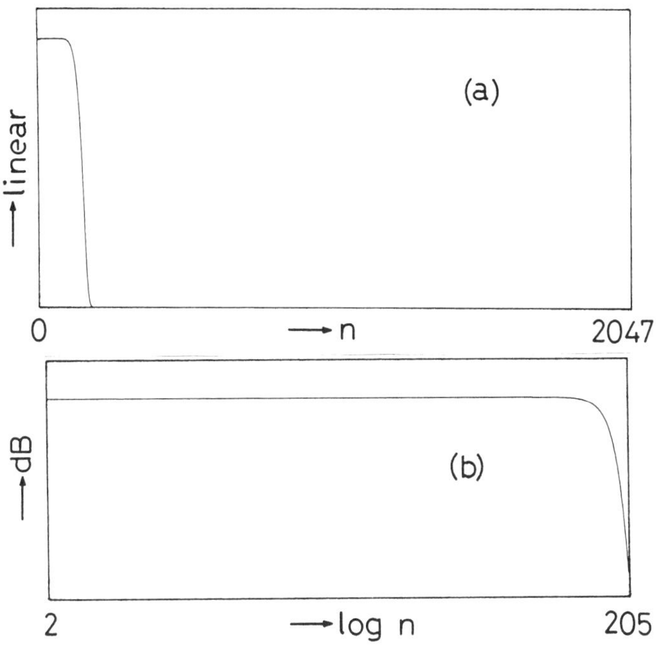

Figure 9. Power spectrum of another test signal defined as Eq. (4). (a) Full range spectrum in linear-linear scale. (b) Lower two decades in log-log scale. Actual frequency value depends on the clock frequency of the D/A converter.

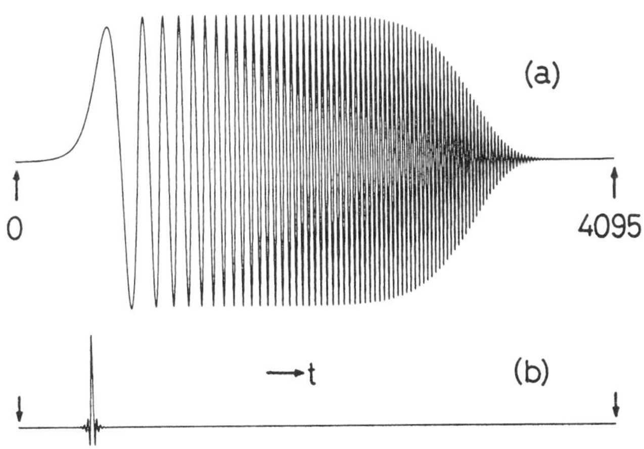

Figure 10. (a) The test signal for the heat conduction measurement, obtained by the inverse Fourier transformation of Eq. (4). (b) Compressed signal of (a).

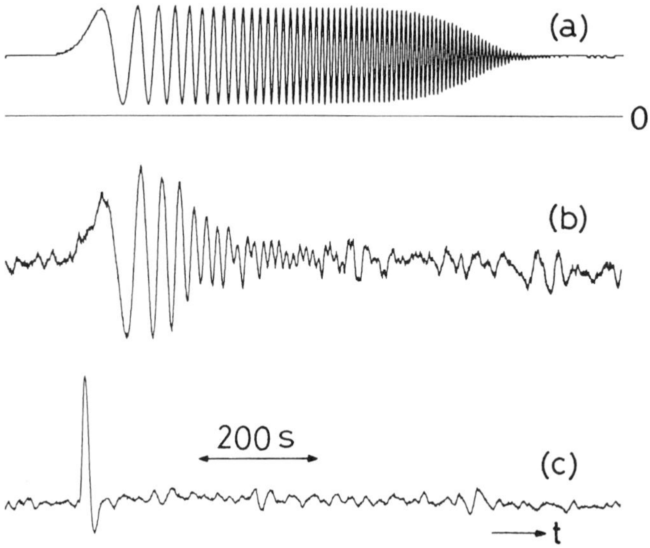

Figure 11. (a) Supplied current to the electric heater.
(b) Amplified voltage signal from the thermocouple.
(c) Compressed signal of (b).

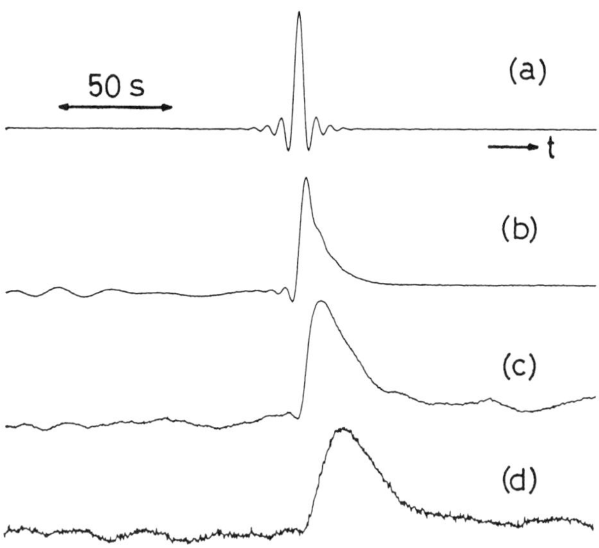

Figure 12. Compressed impulse of the input test signal (a), and compressed impulse response (b) - (d) in magnified form (open loop). (b) $d = 0$ cm. (c) $d = 4$ cm. (d) $d = 8$ cm.

though the A/D conversion is 8 bit length. This makes it possible to obtain more precise data than 8 bit resolution by dither effect of external noise when the averaging is performed. The waveform of Fig. 11(b) is Fourier transformed, applied the compression filter given by the next equation and then inverse Fourier transformed.

$$H^{-1}(jn) = \exp(12n^2j/1000) \tag{5}$$

The result is shown in Fig. 11(c). This is an impulse response, if the compressed input signal of Fig. 10(b) can be considered as an impulse. The signal component of Fig. 11(c) concentrates rather narrow region, and widely spread noise components become relatively small compared with the concentrated signal pulse. Though the relation between current and temperature is thought as nonlinear, the nonlinear component can be eliminated as explained in the later section.

The magnified impulse response varying d is shown in Fig. 12. In this measurement, the feedback loop of broken line in Fig. 8 is not present. When the feedback exists, compressed impulse responses of Fig. 13 are obtained varying feedback gain. Since the input signal has flat power spectrum and constantly zero phase after the compression, the amplitude and phase of the response become directly the amplitude and phase of the frequency response function. Thus the Bode diagram

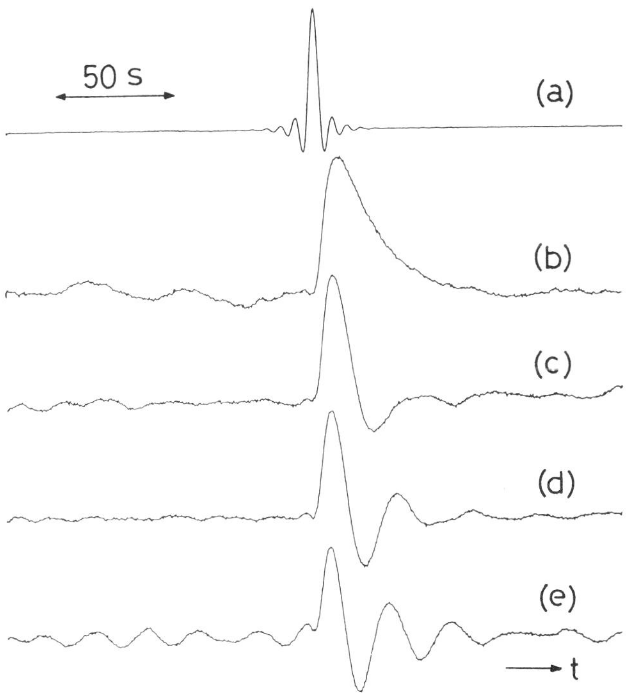

Figure 13. (a) Compressed input test signal. (b) - (e) Impulse response at d = 4 cm, when the feedback loop exists. Relative values of the feedback gain are 0.0 (b), 0.4 (c), 0.7 (d) and 1.0 (e).

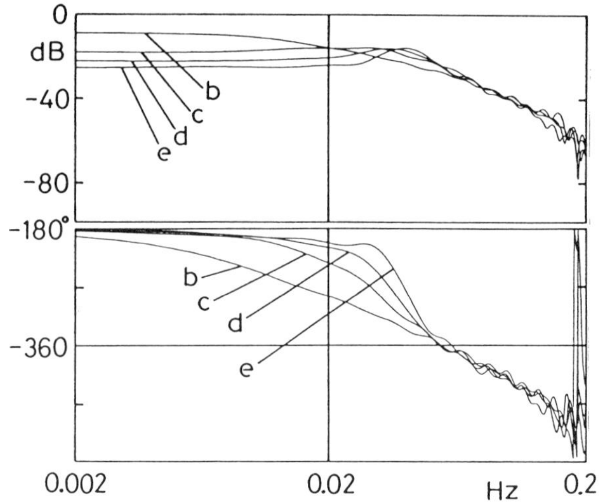

Figure 14. Bode diagram obtained from (b) - (e) of Fig. 13.

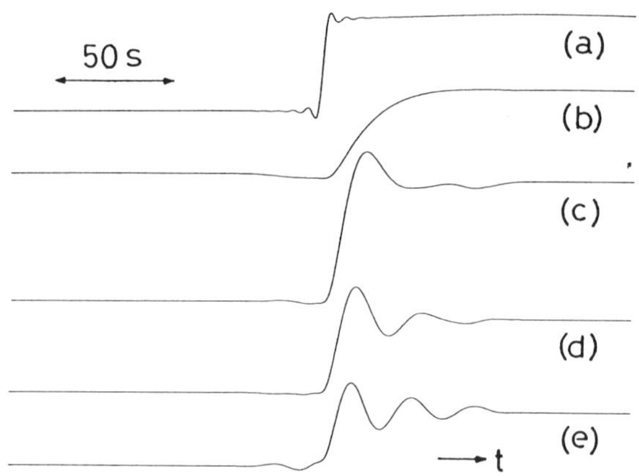

Figure 15. Step response calculated from Fig. 13 (a) - (e).

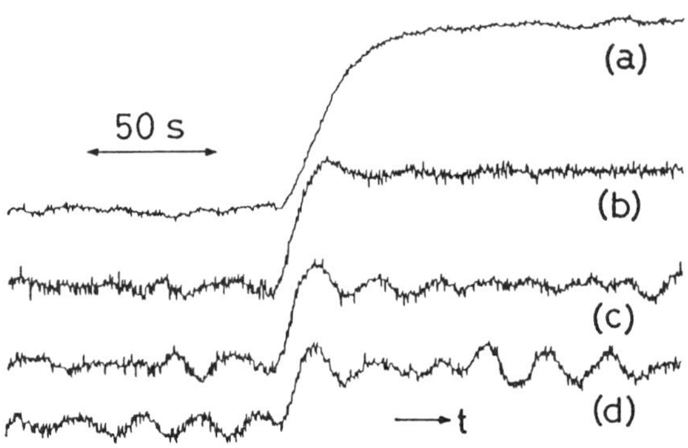

Figure 16. Directly measured step response. Relative values of the feedback gain are 0.0 (a), 0.4 (b), 0.7 (c) and 1.0 (d).

Fig. 14 is obtained from the impulse response of Fig. 13. In doing so, the "signal cut out" was applied and about 94 % of the total data length was made zero in computer memory. This is estimated as 12 dB S/N improvement. The step response is obtained as Fig. 15 by integrating the impulse response of Fig. 13. For comparison, directly measured (using step signal as the heater input current) step response is shown in Fig. 16. The superiority of the signal compression method is obvious.

4. APPLICATIONS

4.1. Measurement of Acoustic System

The impulse response is useful for analyzing sound propagation system because direct sound and reflected sound can be separated in time record. A simple acoustic system Fig. 17 is considered. A loudspeaker is placed near one end of circular iron pipe (both ends are open) and sound field inside the pipe is picked up by a small (1/4 inch diameter) condenser microphone at the center position. The test signal of Fig. 6 is supplied to the loudspeaker. Since acoustic transducers cannot respond to very low frequencies, the signal power spectrum like Fig. 2 is more suitable than Fig. 9. The microphone output signal is shown in Fig. 18. This is 16 times averaged waveform. The clock frequency of the A/D and D/A converter is 20 kHz, and band pass filters (710 - 9000 Hz) are used both after D/A and before A/D conversion. The waveform of Fig. 18 is compressed to Fig. 19, which is shown in magnified form in Fig. 20. Many times reflected sound waves are clearly observed. Phase inversion occurs successively by the open

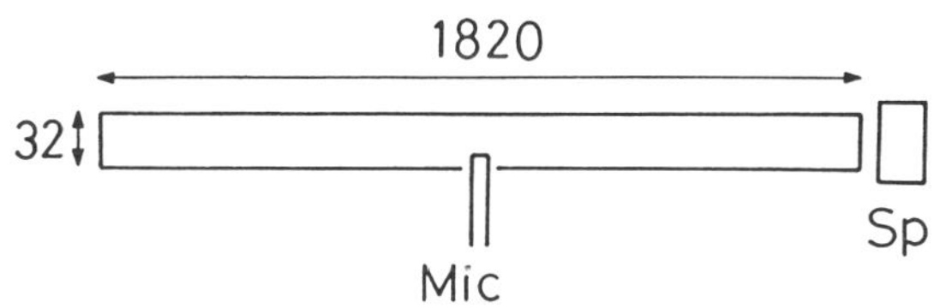

Figure 17. A simple acoustic system consisted of a pipe, a loudspeaker and a microphone. (unit : mm)

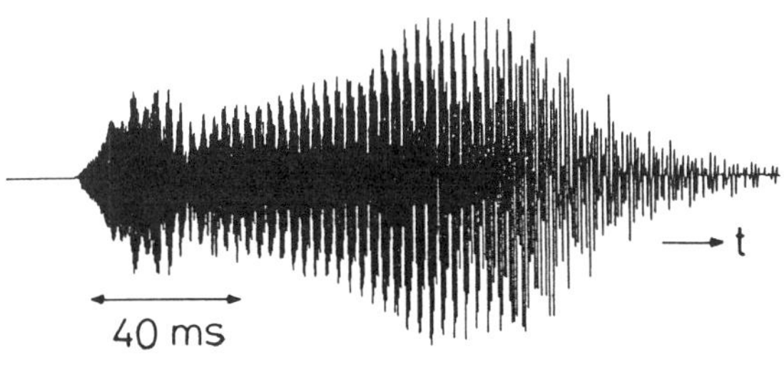

Figure 18. The microphone output waveform of the simple acoustic system of Fig. 17. (16 times averaged)

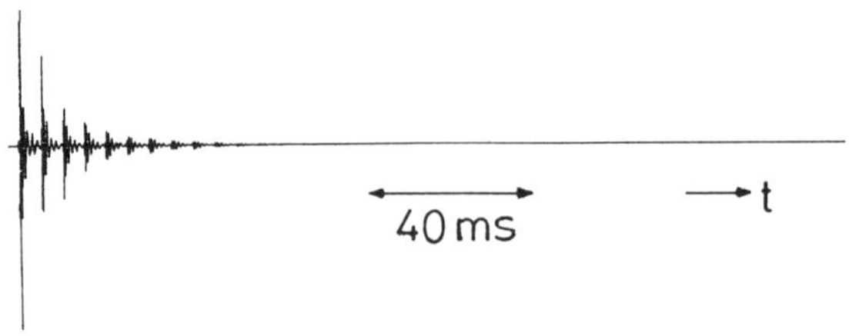

Figure 19. Compressed impulse response of the acoustic system.

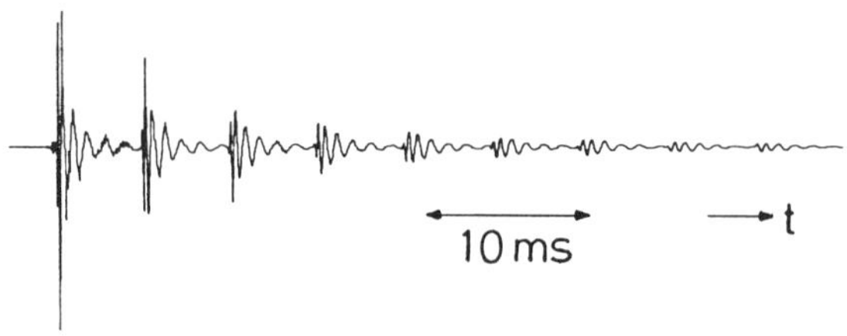

Figure 20. Magnified figure of Fig. 19.

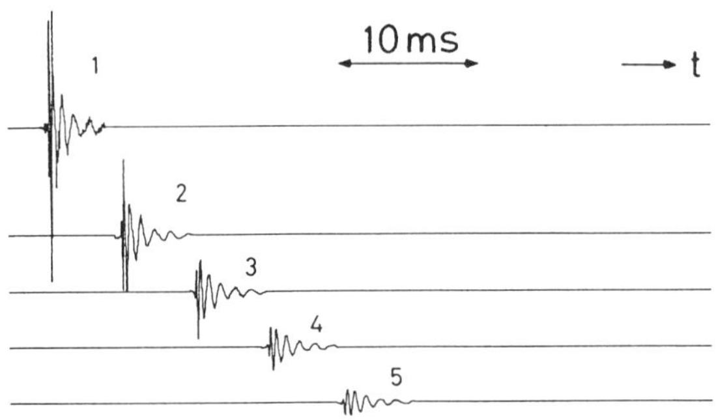

Figure 21. "Cut out" of each reflected sound.

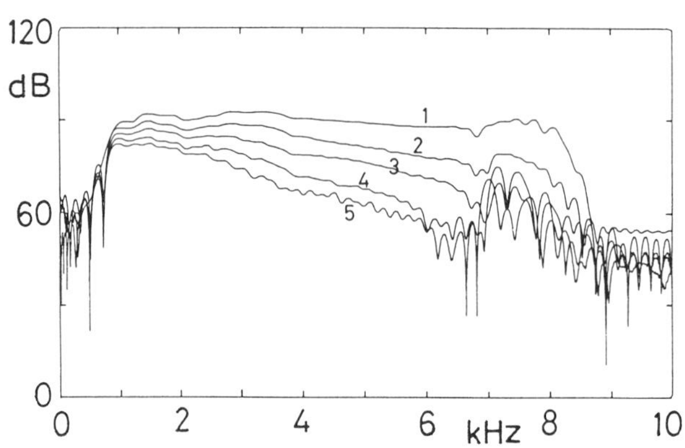

Figure 22. Power spectrum of each reflected sound.

end reflection.

Then "cut out" of each sound is performed as shown in Fig. 21 and power spectrum of each sound is computed as Fig. 22. From Fig. 22, it is observed that as reflections occur many times, high frequency components decrease more. This is explained by considering sound radiation from open ends, which is more effective for higher frequencies. In frequency region higher than about 6.5 kHz, different feature is observed. This is explained by the cross mode in circular cross sectional duct whose cutoff frequency is calculated theoretically as 6.6 kHz.

Next example is a measurement of impulse response of guitar strings. Strings of electric guitar are excited by sound signal, and the string vibration is obtained by a magnetic pickup of the guitar as an electric signal. The impulse response of this acoustically excited mechanical vibration system is measured by the signal compression method and the result is shown in Fig. 23(a). Two same strings (0.032 inch diameter) were used and they were tuned to 95 and 125 Hz. Data size was 16384 points, sampling frequency was 1 kHz, averaging was 16 times and band pass filter of 56 - 450 Hz was used. Power spectrum is calculated as Fig. 23(b). This measurement was performed by 68000 microprocessor system described before.

In room acoustics, the reverberation decay curve is defined as an ensemble average of squared response after termination of input stationary sound. For a stationary input signal of arbitrary spectrum $\Phi(f)$, the reverberation response $R(t)$ is derived by this author as [4]

$$R(t) = \int_{\infty}^{\infty} |G(f,t)|^2 \, \Phi(f) df \qquad (6)$$

$$G(f,t) = \int_{t}^{\infty} h(\tau) \exp(-2\pi j f \tau) d\tau \qquad (7)$$

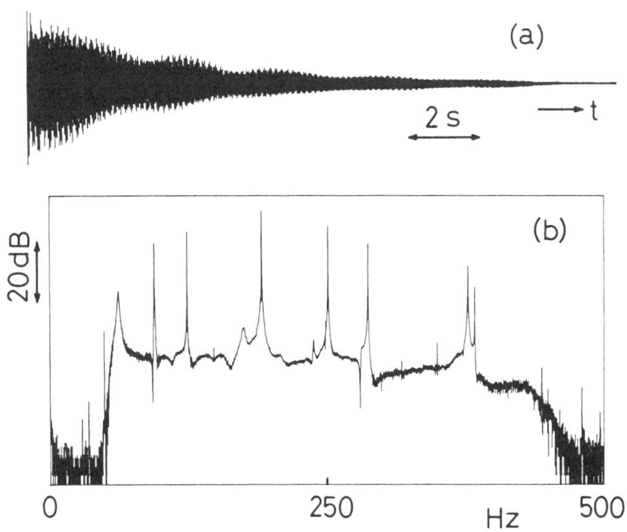

Figure 23. Impulse response (a) and power spectrum (b) of guitar strings excited by sound signal.

where $h(\tau)$ is the impulse response of any interesting system. The reverberation responses calculated from the impulse response of Fig. 23(a) by Eqs. (6) and (7) are shown in Fig. 24. The spectra $\Phi(f)$'s are assumed as successive 10 Hz rectangular band, and $G(f,t)$ is calculated for 16384 points of f and 128 points of t. This calculation was performed by the same 68000 microprocessor system.

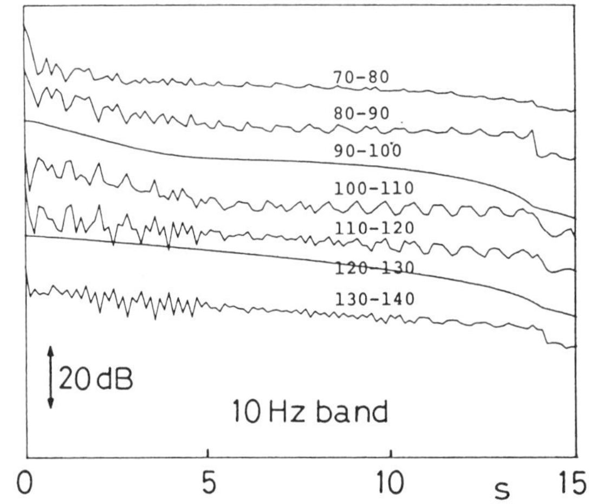

Figure 24. Reverberation response calculated from Fig. 23(a).

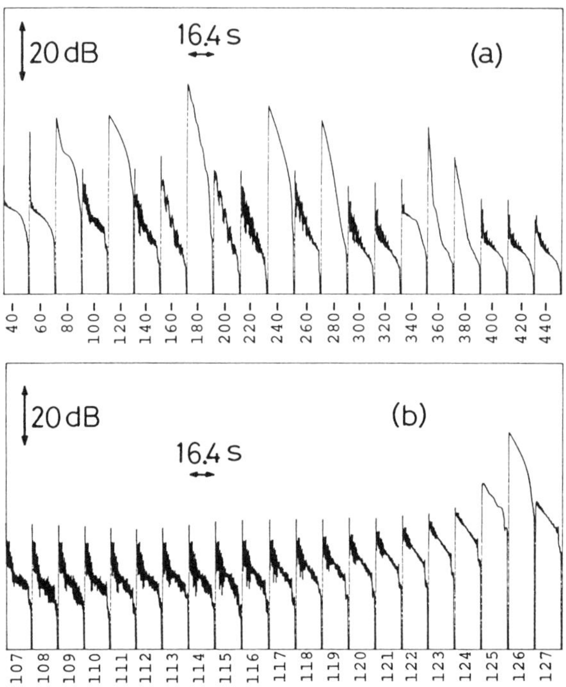

Figure 25. Reverberation response for 20 Hz (a) and 1 Hz (b) band stationary input signal.

These reverberation curves show that only when $\Phi(f)$ contains resonant frequency, the corresponding curve becomes smooth. Relative strength of each response can be observed in Fig. 25. These curves represent reverberation curves compressed in time axis and arranged for successively varying frequency band of 20 Hz (a) and 1 Hz (b). These figures show new aspects of resonant systems.

4.2. Analog Simulation of Second Order System

An approximate second order electronic circuit Fig. 26 is considered. Operational amplifiers A1 and A2 consist approximate integrators, and A3 and A4 are inverting amplifiers. Potentiometers VR1, VR2 and VR3 give coefficients k_1,k_2 and k_3 respectively. The feedback path from A3 to A1 is removed first and the test signal of the signal compression method is applied at "a". From the output signal at "c", the frequency response of the integrator A1 is obtained as Fig. 27. Since this is an inverting integrator, the phase shift is $-90^o - 180^o = -270^o$. The output signal from "e" gives frequency response of Fig. 28, when k_3 is set to zero. This is a response of double integrator and one inverter, so ideally the amplitude should be -40 dB/dec slope and the phase should be constantly $-270^o \times 2 - 180^o = -720^o = 0^o$. In low frequency region, the phase shift is not enough. In high frequency region, the signal power and the S/N ratio become small.

Then the closed loop characteristics are measured. The output signal is obtained from "d" varying loop gain coefficient k_2 and damping coefficient k_3. The frequency responses when k_2 is varied ($k_3 \equiv 0$) are shown in Fig. 29. The damping increases as k_2 becomes smaller. In case of true second order system, the damping should be constant when k_2 is changed. The reason is, as Fig. 28 shows, the phase characteristics are different from ideal second order system. When k_3 is varied ($k_2 \equiv 1.0$), the frequency response is obtained as Fig. 30.

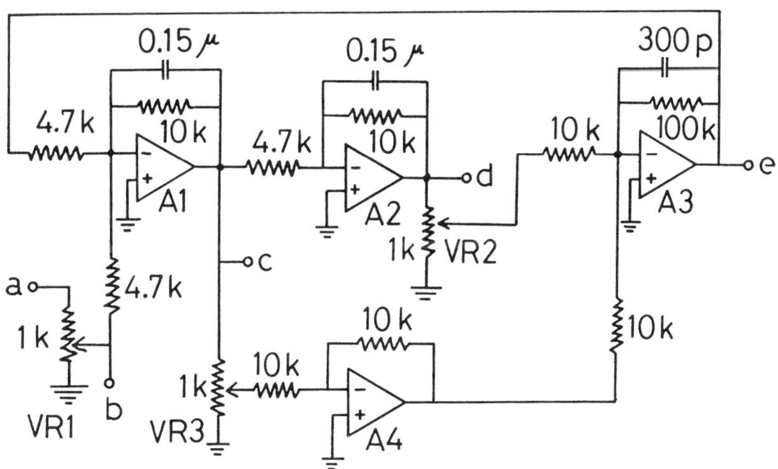

Figure 26. Analog simulation of second order system.

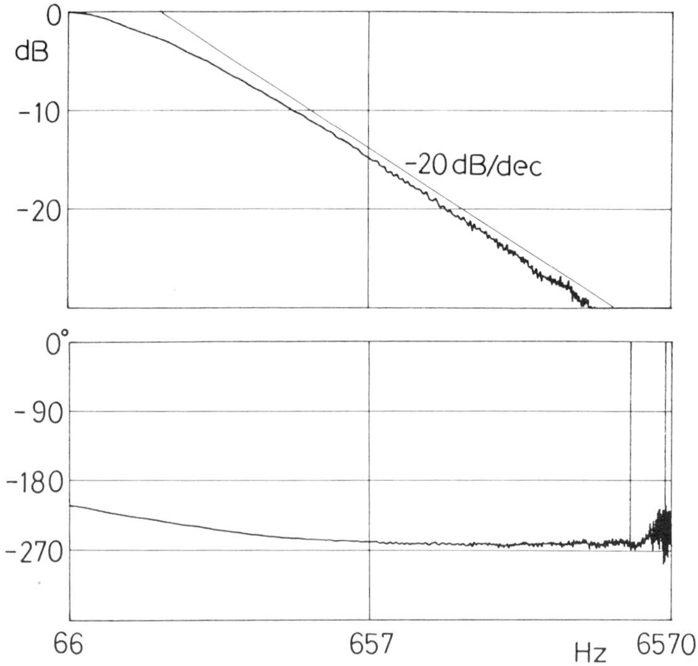

Figure 27. Frequency response of the integrator A1. The input signal is applied at "a" and the output signal is taken from "c". (open loop)

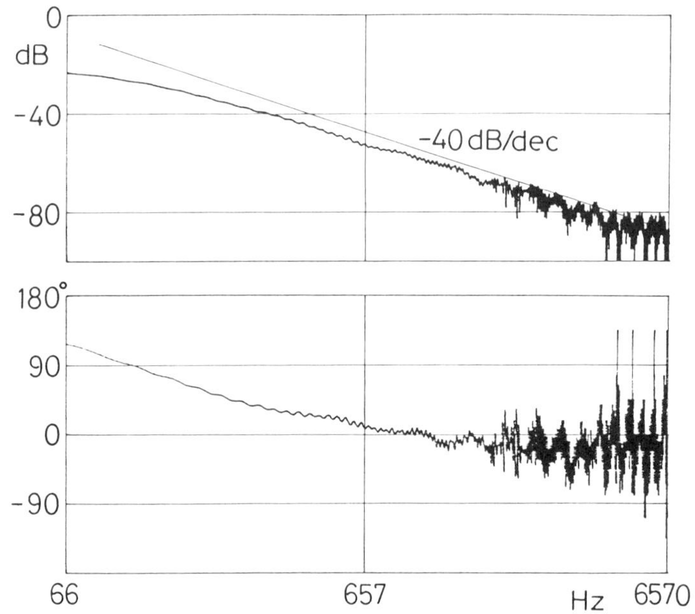

Figure 28. Frequency response of A1, A2 and A3. The input signal is applied at "a" and the output signal is taken from "e". (open loop)

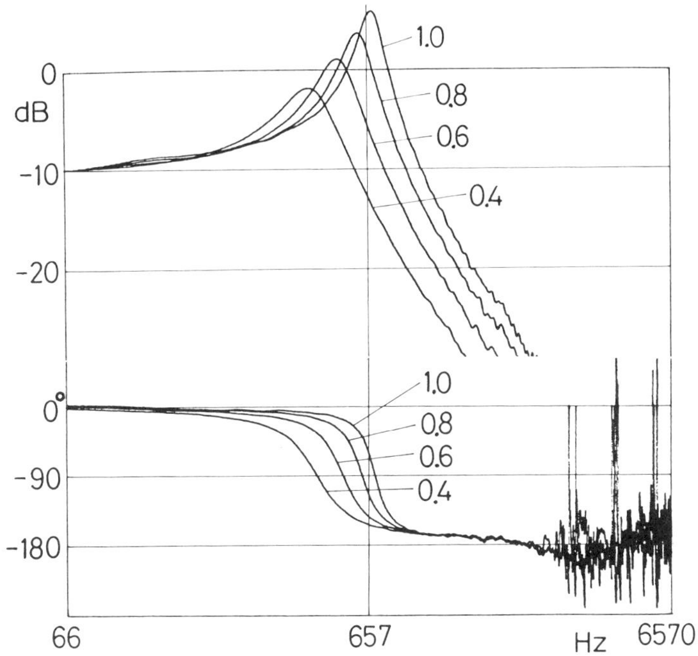

Figure 29. Frequency response of the second order system varying the loop gain k_2. (closed loop)

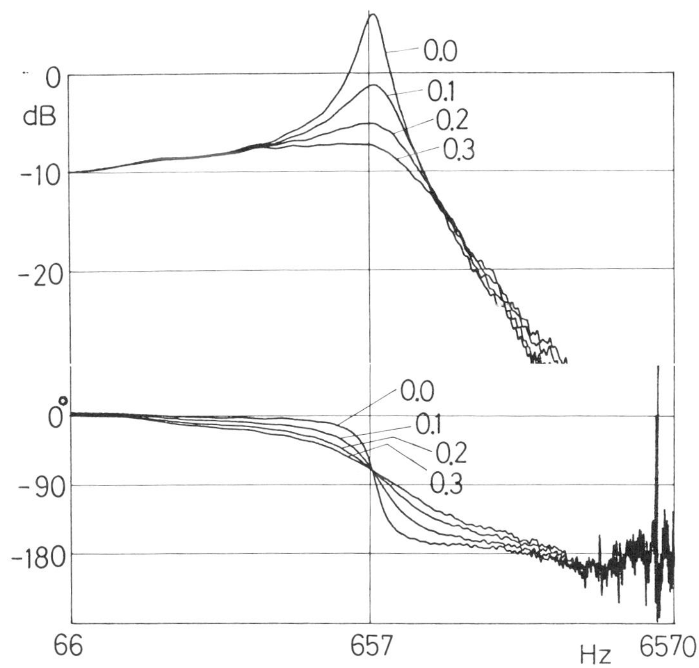

Figure 30. Frequency response of the second order system varying the damping coefficient k_3. (closed loop)

By the similar procedure, the frequency response functions and the impedance functions of various physical systems can be measured by the signal compression method. The microprocessor based system is most useful for the field and industrial applications of such measurements.

4.3. Measurement of Nonlinear System

The signal compression method was developed first for the measurement of linear systems but it was found to be useful for analyzing at least some kinds of nonlinear systems [5]. The nonlinear system considered here is a combination of a linear element and a power law device such as a squarer.

The first example is an electronic squaring module. The test signal of the signal compression method is applied to a bandpass filter (224 - 2240 Hz) and then to the squarer. The input and the output signal of the squarer and their power spectra are shown in Fig. 31. The squared signal happens to be negative because of the digital logic used in the microprocessor system. The input signal Fig. 31(a) is compressed to Fig. 32(a). The output signal Fig. 31(b) is compressed by the following three kinds of filters.

$$H_1^{-1}(jn) = \exp(-j\alpha n^2) \tag{8}$$

$$H_2^{-1}(jn) = \exp\{-j(\alpha/2)n^2\} \tag{9}$$

$$H_3^{-1}(jn) = \exp\{-j(\alpha/3)n^2\} \tag{10}$$

It is assumed that the test signal has been generated by the next expanding filter.

$$H(jn) = \exp(j\alpha n^2) \tag{11}$$

The filters given by Eqs. (8) - (10) will be called first, second and third order compression filter. The compressed waveforms of Fig. 31(b) by these filters are shown in Fig. 32(b) - (d). Expanding the input signal as a Fourier series, the squared output signal is thought to have frequency doubled component and low frequency component. Additionary the linear component is also present by an unavoidable unbalance of the squarer. It has been shown theoretically that ;

(1) If 1st order filter is applied, the linear component concentrates as a pulse, and the frequency doubled component appears at the right side of the pulse, and the low frequency component appears at the left side.

(2) If 2nd order filter is applied, the frequency doubled component concentrates as a pulse, and the linear component and low frequency component appear both at the left side of the pulse.

(3) If 3rd order filter is applied, the frequency tripled component concentrates.

Figures 32(b) - (d) assure these arguments, recalling the cyclic nature of these waveforms. These figures are not in the same scale

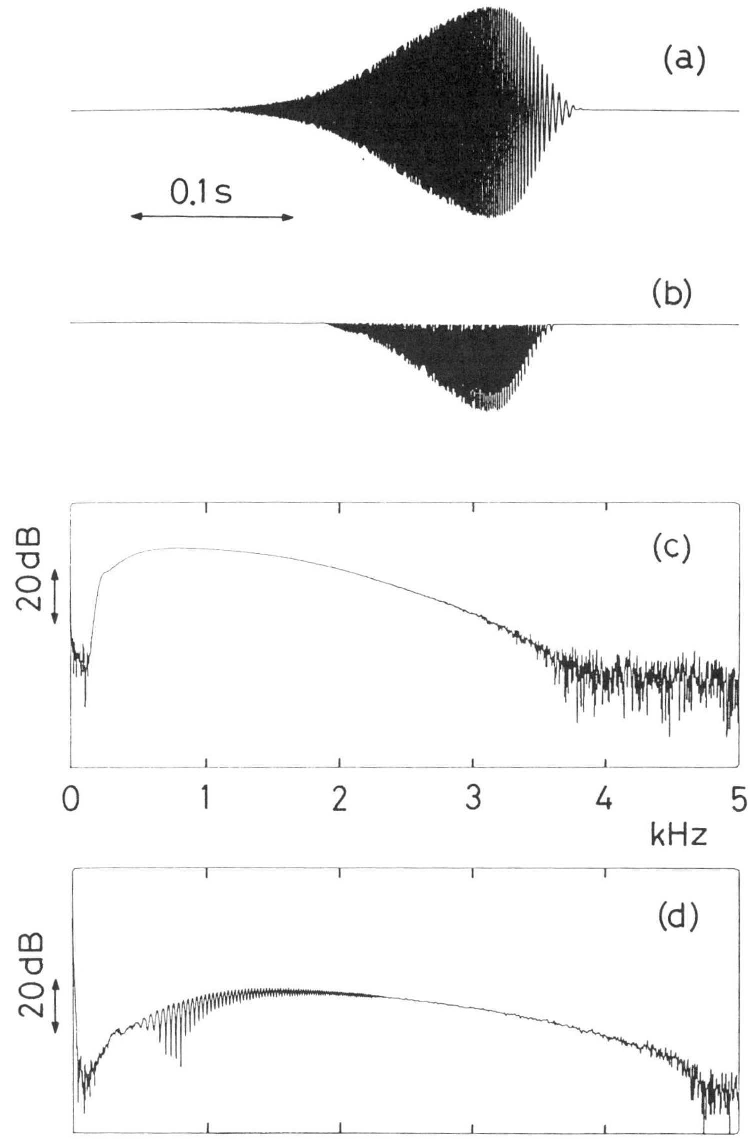

Figure 31. Input (a) and output (b) signal of the electronic squarer. (c) Power spectrum of (a). (d) Power spectrum of (b).

because scale changes are made automatically in the FFT calculation. But by comparing the linear and squared components of Fig. 32(b) and (c), the squared component is seen to be much larger than the linear one. The frequency tripled component is very small as Fig. 32(d) shows.

Next example is a numerical nonlinear model. The test signal of the signal compression method was applied to a bandpass filter. The output signal $x(t)$ was acquired by the microprocessor system and the

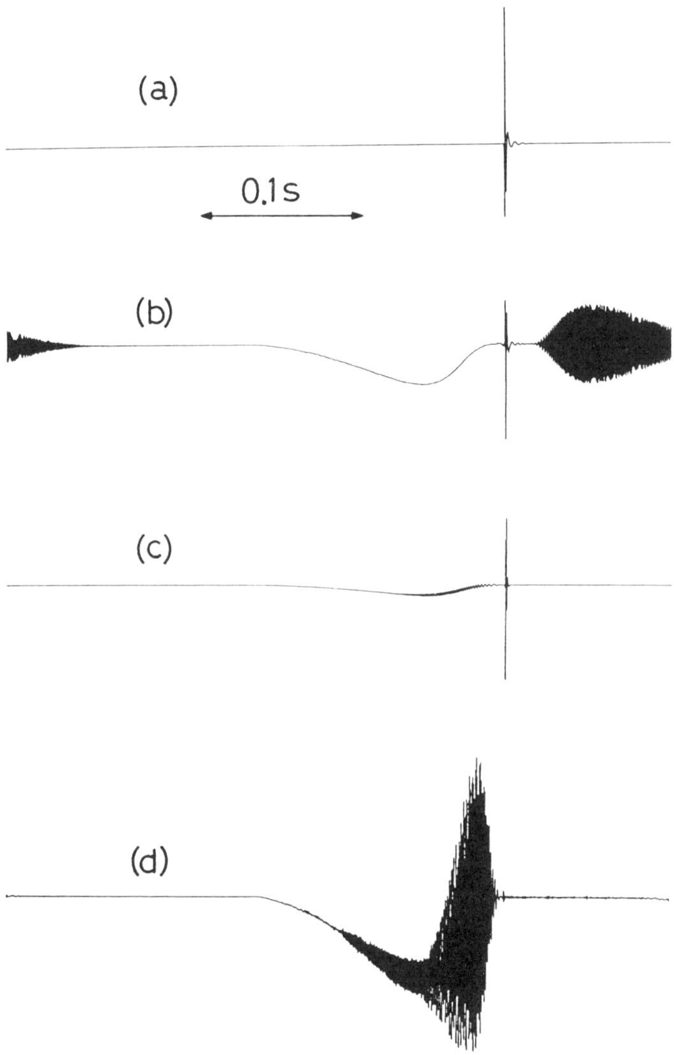

Figure 32. (a) Compression of the input signal by the 1st order filter. Compression of the output signal by the 1st (b), 2nd (c) and 3rd (d) order filter.

following mathematical operation was performed:

$$y(t) = \{x(t) + x^2(t)/64 + x^3(t)/4096\}/3 \tag{12}$$

where $x(t)$ varies in a range of -64 to +64. The signals $x(t)$ and $y(t)$ and their power spectra $X(f)$ and $Y(f)$ are shown in Figs. 33 and 34. The signal $y(t)$ is compressed by the 1st order compression filter and the result is shown in Fig. 35(a). Waveforms in Fig. 35 are cyclic signals. The linear component of $y(t)$ is compressed in the left-most part of Fig. 35(a), and then the noncompressed nonlinear components. The low frequency component appears in the right-most part, which is identical to the preceding result considering cyclic nature. Then the linear component and the low frequency component are made zero in the processor memory as Fig. 35(b) and compressed by the 2nd order compression filter. The result is shown in Fig. 35(c), in which the frequency-doubled component is compressed. Again the compressed component is eliminated as (d) and compressed by the 3rd order compression filter. The result is shown in Fig. 35(e). Figure 35(e) shows the frequency tripled component is compressed and nothing remains.

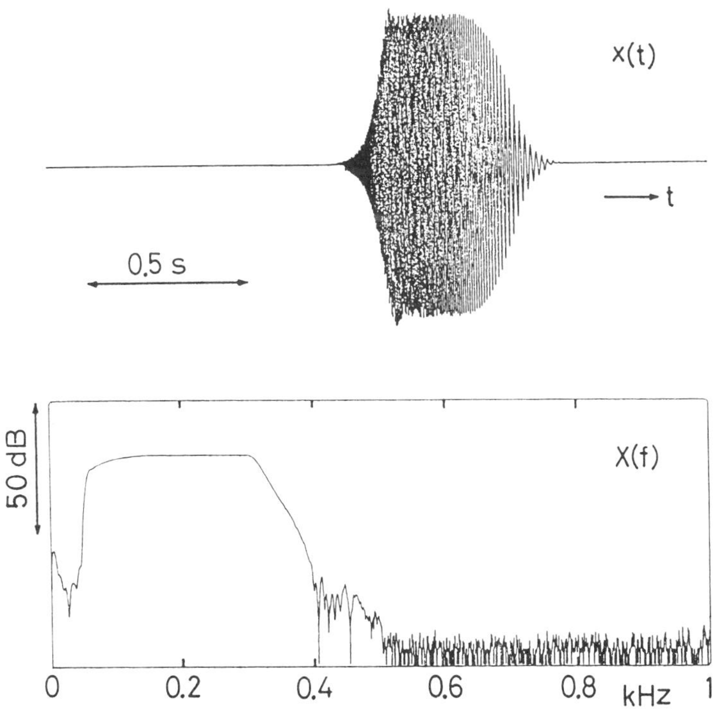

Figure 33. The input signal $x(t)$ and its power spectrum $X(f)$ of the numerical nonlinear model.

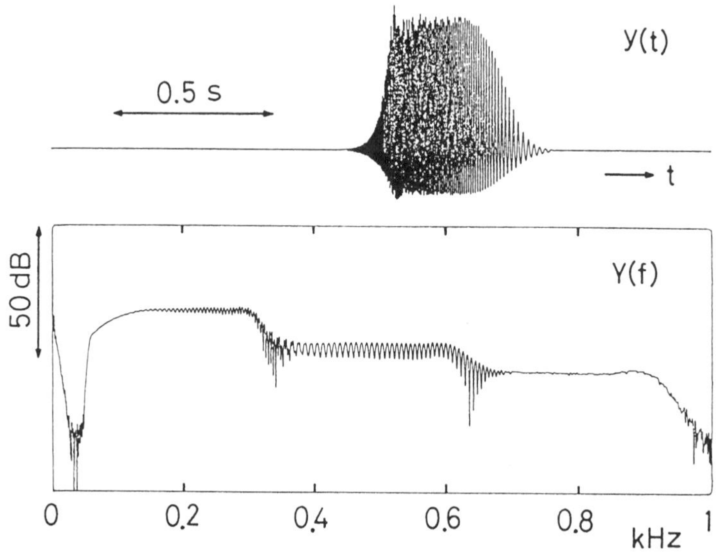

Figure 34. The output signal $y(t)$ and its power spectrum $Y(f)$ of the numerical nonlinear model.

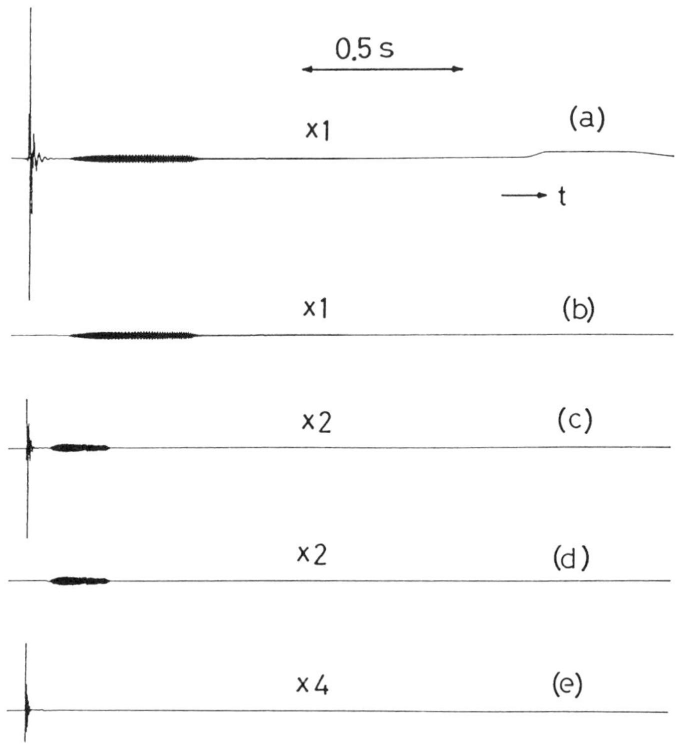

Figure 35. The compressed waveforms of $y(t)$ by 1st, 2nd and 3rd order compression filter.

References

1) Aoshima, N. :'Computer-generated pulse signal applied for sound measurement', *J. Acoust. Soc. Am.* 69, 1484-1488 (1981)
2) Aoshima, N. :'Signal compression method applied for a measurement of sound attenuation by a barrier under wind effects', *J. Acoust. Soc. Jpn.(E)* 3, 105-109 (1982)
3) Aoshima, N. :'Frequency and reverberation response of a pipe resonator', *J. Acoust. Soc. Jpn.(E)* 3, 225-230 (1982)
4) Aoshima, N. :'New method of measuring reverberation time by Fourier transforms', *J. Acoust. Soc. Am.* 67, 1816-1817 (1980)
5) Aoshima, N. :'Measurement of nonlinear vibration by signal compression method', *J. Acoust. Soc. Am.* 76, 794-801 (1984)

CHAPTER 6

DIGITAL SYSTEM CONTROLLERS FOR EXACT MODEL SIGNAL TRACKING

G. H. Hostetter
Electrical Engineering Department
University of California, Irvine
Irvine, California 92717
U.S.A.

ABSTRACT. New results are given for controller design for discrete-time tracking systems. The resulting feedback systems have transient performance selected by the designer. After the transient period, a specified transformation of the plant state exactly tracks any reference input from a class chosen by the designer. The methods are especially suitable for microprocessor-based control systems.

1. INTRODUCTION

Conventional tracking system (or servosystem) design for step-invariant discrete-time systems generally employs observer feedback with one of three methods for incorporating the reference input [1-2]:

(1) Summation with the observer feedback at the plant input, with an additional input to the observer to reflect the added plant input. In the single-input case, the reference input gain is typically chosen for unity d.c. gain to the plant output.

(2) Differencing with the plant output at the observer input. The reference input and the feedback compensation then have transmittances that differ only in algebraic sign.

(3) Coupling to the observer-controller to achieve placement of controller zeros, with zero placement determined in an ad hoc manner reminiscent of classical compensator design using root locus.

New methods are given here for the design of tracking system controllers for linear systems, especially those controlled by microprocessors. Observer feedback is used to place the composite

N. K. Sinha (ed.), Microprocessor-Based Control Systems, 105–129.

system poles as desired. The composite feedback system is then viewed as an observer, in its own right, of a class of signals specified by an imagined autonomous model system. The reference input gains to the observer-controller are selected so that the plant output observes the reference inputs. The plant output will then converge exactly to the reference input for any input signal of the model class.

The design of a linear tracking feedback controller for a known linear plant can be considered to be composed of two parts:

(1) Design for acceptable feedback system zero-input (transient) response;

(2) Design for acceptable feedback system zero-state response for a relevant or representative class of reference inputs to be tracked.

Additionally, if the plant to be controlled is continuous-time and the controller is discrete-time, it is necessary to

(3) Design for acceptable between-sample response of the plant.

Acceptable feedback system transient response is obtained by the separation property whereby design of the zero-input response of a feedback controller can be separated into two stages: State feedback is designed as if the plant state were accessible. Then, an estimate of the plant state from an observer is used for feedback in place of the actual plant state. The resulting feedback system has zero-input response that is a combination of the state feedback and the observer error responses. Particular attention is given here to an observer design method yielding minimal-order observers and requiring the solution of only linear algebraic equations for design.

When the desired zero-state performance of the feedback system can be characterized by the system's response to a family of possible input signals, the as-yet unspecified controller input gains are selected so that the plant with feedback is, itself, an observer of a suitable external input signal model.

Perhaps the most common feedback control configuration today consists of a continuous-time plant controlled through D/A and A/D converters by a discrete-time controller. When a discrete-time controller of a continuous-time plant is designed using a discrete-time model of the plant, samples of the plant response will behave as planned for the discrete-time design, but the designer must additionally insure that the between-sample response of the plant is acceptable. A number of means of improving the between-sample continuous-time plant response are discussed.

An illustrative time-invariant design example is given, then new results for the application of these methods to linear time-varying systems are presented.

2. SYSTEM MODELS

2.1 Discrete-Time State Variable Models

Linear discrete-time systems [3-4] are modeled by state equations of the form

$$\underline{x}(k+1) = A(k)\underline{x}(k) + B(k)\underline{u}(k) \tag{1}$$

where the step k is an integer, $\underline{x}$ is the n-vector system state, $\underline{u}$ is an n_u-vector of system inputs, the state coupling matrix A is n×n and the input coupling matrix B is $n \times n_u$. The order of a system is the dimension n of its state vector. The n_y-vector of measurement outputs are related to the state and inputs by a measurement equation of the form

$$\underline{y}(k) = C(k)\underline{x}(k) + D(k)\underline{u}(k) \tag{2}$$

where the state-to-measurement coupling matrix C is $n_y \times n$ and the input-to-measurement coupling matrix D is $n_y \times n_u$. The measurements $\underline{y}(k)$ are accessible for use for feedback control of the system. In addition, we introduce the n_v-vector of <u>reference outputs</u>

$$\underline{v}(k) = E(k)\underline{x}(k) \tag{3}$$

where E is $n_v \times n$. The reference outputs are not necessarily accessible to the designer (although some of these signals could be duplicated in $\underline{y}(k)$ and thereby be accessible); these are the linear combinations of state signals that the designer wishes to control.

A nonsingular change of state variables

$$\underline{x}(k) = T(k)\underline{x}'(k); \qquad \underline{x}'(k) = T^{-1}(k)\underline{x}(k) \tag{4}$$

gives state, measurement and reference equations of the same form:

$$\underline{x}'(k+1) = [T^{-1}(k)A(k)T(k)]\underline{x}'(k) + [T^{-1}(k)B(k)]\underline{u}(k)$$

$$= A'(k)\underline{x}'(k) + B'(k)\underline{u}(k)$$

$$\underline{y}(k) = [C(k)T(k)]\underline{x}'(k) + D(k)\underline{u}(k) = C'(k)\underline{x}'(k) + D(k)\underline{u}(k)$$

$$\underline{v}(k) = [E(k)T(k)]\underline{x}'(k) = E'(k)\underline{x}'(k) \tag{5}$$

In terms of the initial state $\underline{x}(0)$ and the inputs $\underline{u}(k)$, the solution for the state after step zero is the discrete convolution

$$\underline{x}(k) = \Phi(k,0)\underline{x}(0) + \sum_{i=0}^{k-1} \Phi(k,i+1)B(i)\underline{u}(i) \tag{6}$$

where the state transition matrices Φ are the n×n products of state coupling matrices

$$\Phi(k,j) = A(k-1)A(k-2)\ldots A(j-1)A(j), \qquad k > j \tag{7}$$

with

$$\Phi(k,k) = I \tag{8}$$

the n×n identity matrix. The first term in (6) is the <u>zero-input</u> component of the response, and the second term in (6) is the <u>zero-state</u> response component.

For a step-invariant linear discrete-time system, the matrices, A, B, C, D, and E do not vary with step. The state model is then

$$\underline{x}(k+1) = A\underline{x}(k) + B\underline{u}(k)$$

$$\underline{y}(k) = C\underline{x}(k) + D\underline{u}(k)$$

$$\underline{v}(k) = E\underline{x}(k) \tag{9}$$

and the solution for the state is

$$\underline{x}(k) = A^k\underline{x}(0) + \sum_{i=0}^{k-1} A^{k-i-1}B\underline{u}(i) \tag{10}$$

If the eigenvalues λ of A are distinct, there is a nonsingular constant change of state variables

$$\underline{x}(k) = T\underline{x}'(k): \qquad \underline{x}'(k) = T^{-1}\underline{x}(k) \tag{11}$$

such that the new state coupling matrix A' in

$$\underline{x}'(k+1) = \left(T^{-1}AT\right)\underline{x}'(k) + \left(T^{-1}B\right)\underline{u}(k) = A'\underline{x}'(k) + B'\underline{u}(k) \tag{12}$$

(which is a similarity transformation of A) is diagonal, with the eigenvalues of A as the diagonal elements. The new, equivalent state equations are then decoupled from one another and are of the form

$$x_j'(k+1) = \lambda_j x_j'(k) + \underline{b}_j'^T\underline{u}(k), \qquad j = 1,2,\ldots,n \tag{13}$$

where $\underline{b}_j'^T$ is the jth row of B'. The solutions for the new state variables, termed the system <u>modes</u> are

$$x_j'(k) = \lambda_j^k x_j'(0) + \sum_{i=0}^{k-1} \lambda_j^{k-i-1}\,\underline{b}_j'^T\underline{u}(i) \tag{14}$$

The original state variables $\underline{x}$, being linear combinations of the new state variables $\underline{x}'$ thus consist of linear combinations of these first order system modes. When eigenvalues of the state coupling matrix A are repeated, a change of state variables can be made that results in a new state coupling matrix in block Jordan form. For a repeated eigenvalue λ_j, the corresponding system modes involve λ_j^k, $k\lambda_j^k$, $k^2\lambda_j^k$, ...

A step-invariant discrete-time system is termed stable if and only if all of its eigenvalues are inside the unit circle on the complex plane. With zero input, the state of a stable system approaches zero with step.

A discrete-time system is completely observable if its state $\underline{x}(k)$ at every specific step k can be determined from the system model and its inputs and measurement outputs for a finite number of steps. For a step-invariant system, if it is possible to determine the state at any step, say $\underline{x}(0)$, then with a shift of step, the state at any other step can be determined in the same way. For an nth order step-invariant system (9) with n_y measurement outputs, the initial state $\underline{x}(0)$ is, in terms of the outputs and inputs, given by

$$\begin{cases} C\underline{x}(0) = \underline{y}(0) - D\underline{u}(0) \\ CA\underline{x}(0) = \underline{y}(1) - CB\underline{u}(0) - D\underline{u}(1) \\ CA^2\underline{x}(0) = \underline{y}(2) - CAB\underline{u}(0) - CB\underline{u}(1) - D\underline{u}(2) \\ \vdots \\ CA^{n-1}\underline{x}(0) = \underline{y}(n-1) - CA^{n-2}B\underline{u}(0) - \ldots - CB\underline{u}(n-2) - D\underline{u}(n-1) \end{cases} \tag{15}$$

where all but $\underline{x}(0)$ are known. This set of linear algebraic equations can be solved for $\underline{x}(0)$ if and only if the $(n_y n)\times n$ observability matrix,

$$M_0 = \text{col}\ (C,\ CA,\ CA^2,\ \ldots,\ CA^{n-1}) \tag{16}$$

is of full rank. Additional outputs do not change this result because the additional equations have coefficients involving higher powers of A which, by the Cayley-Hamilton theorem, are linear combinations of those already considered.

A discrete-time system is completely controllable if and only if, from the system model and its state $\underline{x}(k)$ at any specific step k, an input sequence $\underline{u}(k)$, $\underline{u}(k+1)$, ..., $\underline{u}(k+i)$ can be determined that will take the system to any desired later state $\underline{x}(k+i+1)$ in a finite number of steps, i. For a step-invariant system, if it is possible to move the state at any step, say $\underline{x}(0)$ to an arbitrary state at a later step, it is possible to do so starting with any initial step. For an nth

order linear step-invariant system (9) with n_u inputs and a desired state vector $\underline{x}_d$, the system state at step n is, in terms of the initial state $\underline{x}(0)$ and the inputs,

$$\underline{x}(n) = \underline{x}_d = A^n\underline{x}(0) + B\underline{u}(0) + AB\underline{u}(1) + \ldots + A^{n-1}B\underline{u}(n-1) \tag{17}$$

where all but $\underline{u}(0)$, $\underline{u}(1)$, ..., $\underline{u}(n-1)$ are known. These linear algebraic equations have a solution for the inputs if and only if the $n\times(n_u n)$ controllability matrix

$$M_c = (B, AB, A^2B, \ldots, A^{n-1}B) \tag{18}$$

is of full rank. Additional steps, giving additional equations with coefficients A^nB, $A^{n+1}B$, ... do not affect this result since these equations are linearly dependent on the others.

2.2 Continuous-Time State Variable Models

Linear continuous-time systems of order n are modeled by state equations of the form

$$d\underline{\xi}/dt = \Theta(t)\underline{\xi}(t) + \Gamma(t)\underline{\mu}(t) \tag{19}$$

where the time t is a continuous variable, $\underline{\xi}(t)$ is the n-vector state of the system, $\underline{\mu}(t)$ is an n_μ-vector of inputs, the state coupling matrix $\Theta(t)$ is $n\times n_\xi$, and the input coupling matrix $\Gamma(t)$ is $n\times n_\mu$. The n_σ-vector of measurement outputs

$$\underline{\sigma}(t) = \Omega(t)\ \underline{\xi}(t) + \Delta(t)\underline{\mu}(t) \tag{20}$$

is related to the state by a $n_\sigma\times n$ state-to-measurement coupling matrix $\Omega(t)$ and an $n_\sigma\times n_\mu$ input-to-measurement coupling matrix $\Delta(t)$. The n_β-vector of reference outputs is

$$\underline{\beta}(t) = \Pi(t)\ \underline{\xi}(t) \tag{21}$$

where Π is $n_\beta\times n$.

The solution for the state for $t \geq 0$ is given by

$$\underline{\xi}(t) = \Phi(t,0)\underline{\xi}(0) + \int_0^t \Phi(t,\tau)\Gamma(\tau)\underline{\mu}(\tau)d\tau \tag{22}$$

where the state transition matrix $\Phi(t,\tau)$ is the $n\times n$ matrix solution to

$$\partial\Phi/\partial t = \Theta(t)\Phi(t,\tau) \tag{23}$$

with

$$\Phi(\tau,\tau) = I \tag{24}$$

where I is the n×n identity matrix. The first term in (22) is the zero-input component of the state, and the second term is the zero-state component.

A time-invariant linear continuous-time system has constant matrices Θ, Γ, Ω, Δ, and Π

$$\begin{aligned} d\underline{\xi}/dt &= \Theta\underline{\xi}(t) + \Gamma\underline{\mu}(t) \\ \underline{\sigma}(t) &= \Omega\underline{\xi}(t) + \Delta\underline{\mu}(t) \\ \underline{\beta}(t) &= \Pi\,\underline{\xi}(t) \end{aligned} \tag{25}$$

The solution for the state for $t \geq 0$ is given by the convolution

$$\underline{\xi}(t) = \exp(\Theta t)\underline{\xi}(0) + \int_0^t \exp[\Theta(t-\tau)]\Gamma\underline{\mu}(\tau)d\tau \tag{26}$$

where exp denotes the matrix exponential function. If the eigenvalues λ of Θ are distinct, there is a nonsingular change of state variables

$$\underline{\xi}(t) = T\underline{\xi}'(t) \;; \qquad \underline{\xi}'(t) = T^{-1}\underline{\xi}(t) \tag{27}$$

such that the new state equations

$$\begin{aligned} d\underline{\xi}'/dt &= (T^{-1}\Theta T)\underline{\xi}'(t) + (T^{-1}\Gamma)\underline{\mu}(t) = \Theta'\underline{\xi}'(t) + \Gamma'\underline{\mu}(t) \\ \underline{\sigma}(t) &= (\Omega T)\underline{\xi}'(t) + \Delta\underline{\mu}(t) = \Omega'\underline{\xi}'(t) + \Delta\underline{\mu}(t) \\ &= (\Pi T)\underline{\xi}'(t) = \Pi'\underline{\xi}'(t) \end{aligned} \tag{28}$$

have a state coupling matrix Θ' that is diagonal, with the eigenvalues of Θ as a diagonal elements. The new, equivalent state equations are decoupled

$$d\xi_j'/dt = \lambda_j\xi_j'(t) + \underline{\gamma}_j'^T\underline{\mu}(t), \qquad j=1,2,\ldots,n \tag{29}$$

where $\underline{\gamma}_j'^T$ is the jth row of Γ' and have first order solutions (the system modes)

$$\xi_j'(t) = \exp(\lambda_j t)\xi_j'(0) + \int_0^t \exp[\lambda_j(t-\tau)]\underline{\gamma}_j'^T\underline{\mu}(\tau)d\tau \tag{30}$$

If eigenvalues λ_j of Θ are repeated, transformation of Θ to block Jordan form results in modes involving $\exp(\lambda_j t)$, $t\exp(\lambda_j t)$, $t^2\exp(\lambda_j t)$, ... The original system state variables $\underline{\xi}$, being linear combinations of the new state variables $\underline{\xi}'$, consist of linear combinations of these first order system modes.

A time-invariant continuous-time system is termed stable if and only if all of the eigenvalues of the state coupling matrix Θ are in the left half of the complex plane. With zero input, the state of a stable system approaches zero with time, since each of the system modes decays exponentially.

A continuous-time system is completely observable if its state $\underline{\xi}(t)$, at any specific time t, can be determined from the system model and its inputs and measurement outputs for a finite interval of time. It is completely controllable if, by knowing the system model and its state at any specific time, $\underline{\xi}(t_0)$, inputs $\underline{\mu}(t)$ for a finite time interval $t_0 \leq t \leq t_1$ can be determined that will take the system to any desired state at time t_1. A time-invariant continuous-time system is completely observable if and only if its observability matrix

$$M_0 = \text{col}\ (\Omega, \Omega\Theta, \ldots, \Omega\Theta^{n-1}) \tag{31}$$

is of full rank. It is completely controllable if and only if its controllability matrix

$$M_c = [\Gamma, \Theta\Gamma, \ldots, \Theta^{n-1}\Gamma] \tag{32}$$

is of full rank.

2.3 Sampled Continuous-Time Systems

Samples of the signals in a continuous-time system (19)-(21) at the ordered times t_0, t_1, t_2, ... are modeled by the appropriate discrete-time system (1)-(3), where

$$\begin{aligned} &\underline{x}(k) = \underline{\xi}(t_k); \quad \underline{y}(k) = \underline{\sigma}(t_k); \quad \underline{v}(k) = \underline{\beta}(t_k) \\ &C(k) = \Omega(t_k); \quad D(k) = \Delta(t_k); \quad E(k) = \Pi(t_k) \end{aligned} \tag{33}$$

and

$$A(k) = \Phi(t_k, t_{k-1}) \tag{34}$$

where Φ is the state transition matrix for the continuous-time system. The continuous-time and discrete-time inputs are related by

$$B(k)\underline{u}(k) = \int_{t_{k-1}}^{t_k} \Phi(t_k, \tau)\Gamma(\tau)\underline{\mu}(\tau)d\tau \tag{35}$$

In the special case of a periodically sampled time-invariant continuous-time system (25) with

$$t_k = k\Delta t \tag{36}$$

where the sampling interval Δt is constant, the corresponding discrete-time model has constant, A, C, D, and E, with

$$A = \exp(\Theta \Delta t) \tag{37}$$

and with

$$B(k)\underline{u}(k) = \exp(\Theta \Delta t) \int_{k\Delta t}^{(k+1)\Delta t} \exp(-\Theta \tau)\Gamma \underline{u}(\tau)d\tau \tag{38}$$

If the continuous-time system is driven by inputs that change only at the sample times (as is often the case when the inputs are from analog-to-digital converters), then

$$B = \exp(\Theta \Delta t)\left[\int_0^{\Delta t} \exp(-\Theta \tau)d\tau\right]\Gamma \tag{39}$$

a constant, and $\underline{u}(k)$ consists of samples of the continuous-time inputs:

$$\underline{u}(k) = \underline{\mu}(k\Delta t) \tag{40}$$

3. STEP-INVARIANT CONTROLLER DESIGN

3.1 Step-Invariant Discrete-Time Observer Design

A step-invariant observer [5-8] of a step-invariant linear discrete-time plant (9) has state equations of the form

$$\underline{z}(k+1) = F\underline{z}(k) + G\underline{y}(k) + H\underline{u}(k) \tag{41}$$

and state $\underline{z}(k)$ that estimates a linear transformation $T_z\underline{x}(k)$ of the plant. The error between the observer state and the quantity to be observed by $\underline{z}(k)$ is given by

$$T_z\underline{x}(k+1) - \underline{z}(k+1) = (MA - GC)\underline{x}(k) - F\underline{z}(k) + (MB - GD - H)\underline{u}(k) \tag{42}$$

For this observer error system to be autonomous,

$$T_zA - GC = FT_z$$
$$H = T_zB - GD \tag{43}$$

so that the error is governed by

$$T_z\underline{x}(k+1) - \underline{z}(k+1) = F[T_z\underline{x}(k) - \underline{z}(k)] \tag{44}$$

If the observer gain matrix G can be chosen so that all of the eigenvalues of F are inside the unit circle on the complex plane, the error

$$T_z\underline{x}(k) - \underline{z}(k) = F^k[T_z\underline{x}(0) - \underline{z}(0)] \tag{45}$$

will approach zero with step, so that the observer state approaches the transformation T_z of the plant state, which is denoted by

$$\underline{z}(k) \rightarrow T_z\underline{x}(k) \tag{46}$$

A <u>full order state observer</u> of the nth order plant (9) has

$$T_z = I \tag{47}$$

where I is the n×n identity matrix. Provided that the plant is completely observable, it is always possible to select the full order state observer gain matrix G such that the observer eigenvalues are all at locations arbitrarily selected by the designer. If all n eigenvalues of F are chosen to be zero, then the characteristic equation of F is

$$F^n = 0 \tag{48}$$

giving

$$\underline{z}(n) = \underline{x}(n) \tag{49}$$

That is, the observer state converges to the plant state in n steps or less. A system with all of its eigenvalues zero is termed <u>deadbeat.</u>

For observers with distinct eigenvalues, each different from the plant eigenvalues, an especially convenient method of observer design is to consider a collection of first order observers,

$$z_i(k+1) = f_i z_i(k) + \underline{g}_i^T\underline{y}(k) + \underline{h}_i^T\underline{u}(k) \tag{50}$$

each with one of the desired observer eigenvalues, f_i. The individual observed linear transformation vectors $\underline{t}_i$ are linear in the elements of the observer gains $\underline{g}_i$, which are simply chosen so that the transformation vectors are linearly independent of one another, which is always possible:

$$z_i(k) \rightarrow \underline{t}_i^T x(k) \tag{51}$$

where

$$\begin{aligned} \underline{t}_i^T &= \underline{g}_i^T C(A - f_i I)^{-1} \\ \underline{h}_i^T &= \underline{t}_i^T B - \underline{g}_i^T D \end{aligned} \tag{52}$$

A collection of n such first order observers observes

$$\underline{z}(k) \rightarrow \text{col}\ (\underline{t}_1^T, \underline{t}_2^T, \ldots, \underline{t}_n^T)\underline{x}(k) = T_z\underline{x}(k) \tag{53}$$

and, in view of the linear independence of the $\underline{t}_i$, an observer output of the form

$$\underline{w}(k) = T_z^{-1}\underline{z}(k) \tag{54}$$

observers the state $\underline{x}(k)$.

If it is desired that the observer have i repeated eigenvalues, then if that portion of the observer is taken to be in upper block Jordan form, the repeated eigenvalue subsystem observes $T_r\underline{x}(k)$ where

$$T_r = \mathrm{col}(\underline{t}_1^T, \underline{t}_2^T, \ldots, \underline{t}_i^T) \tag{55}$$

and where

$$\begin{cases} t_i^T = \underline{g}_i^T C(A - fI)^{-1} \\ \underline{t}_{i-1}^T = (\underline{g}_{i-1}^T C + \underline{t}_i^T)(A - fI)^{-1} \\ \vdots \\ \underline{t}_1^T - (\underline{g}_1^T C + \underline{t}_2^T)(A - fI)^{-1} \end{cases} \tag{56}$$

with

$$\underline{h}_j^T = \underline{t}_j^T B, \qquad j=1,2,\ldots,i \tag{57}$$

If the plant measurement outputs, which also involve linear transformations of the plant state, are used in the formulation of a state observer, the dynamic order of the observer can be reduced. For the nth order step-invariant plant (9) with the n_y rows of C linearly independent, an observer of order $n-n_y$ with n outputs

$$\underline{w}'(k) = \begin{bmatrix} 0 \\ --- \\ I \end{bmatrix} \underline{z}(k) + \begin{bmatrix} I \\ --- \\ 0 \end{bmatrix} \underline{y}(k) - \begin{bmatrix} D \\ --- \\ 0 \end{bmatrix} \underline{u}(k) \tag{58}$$

where I is the $n_z \times n_y$ identity matrix, observes

$$\underline{w}'(k) \rightarrow \begin{bmatrix} C \\ --- \\ T_z \end{bmatrix} \underline{x}(k) = N\ \underline{x}(k) \tag{59}$$

It is guaranteed that the rows of T_z and of C will be linearly independent, so that

$$\underline{w}(k) = N^{-1}\underline{w}'(k) \tag{60}$$

observes $\underline{x}(k)$.

For multiple-output plants, the required observer order for observation of a scalar transformation of the plant state (rather than the entire plant state vector) can generally be further reduced since the state transformation observed by each first order observer subsystem is a designer-determined linear combination of each of the scalar state transformations observed using a single plant output. For a plant with n_y linearly independent outputs, each first order observer subsystem observes a state transformation spanning n_y dimensions. Except in special cases, then, a single scalar linear transformation of the plant state can be observed with an observer of order n_z provided

$$(n_z + 1)n_y \geq n \tag{61}$$

3.2 Eigenvalue Placement with State and Observer Feedback

If the state $\underline{x}(k)$ of a linear, step-invariant n^{th} order plant with state equations (9) is accessible for feedback, with

$$\underline{u}(k) = K\underline{x}(k) \tag{62}$$

the feedback system is described by

$$\underline{x}(k+1) = (A + BK)\underline{x}(k) \tag{63}$$

Provided that the plant is completely controllable, the feedback gain matrix K can always be chosen so that the eigenvalues of the feedback system, those of the matrix (A + BK) are at arbitrary desired locations selected by the designer [9-11]. This is to say that the designer is completely free to choose the character of the overall system's transient performance. The design of the feedback matrix K is similar to that for an observer, where the gain G is to be chosen so that the eigenvalues of (A - GC) are at desired locations.

When an observer's estimate of the plant's state is used for feedback in place of the state itself, the feedback system has eigenvalues that are those of the observer and those that would have resulted from the state feedback, a result known as the <u>separation property</u> for observer feedback [12-13]. For a linear, step-invariant plant (9) that is both completely observable and completely controllable, an $n_z{}^{th}$ order observer, of the form

$$\underline{z}(k+1) = F\underline{z}(k) + G\underline{y}(k) + H\underline{u}(k)$$

$$\underline{w}(k) + P\underline{z}(k) + Q[y(k) - D\underline{u}(k)] \tag{64}$$

with feedback to the plant

$$\underline{u}(k) = K\underline{w}(k) \tag{65}$$

can thus always be designed such that all eigenvalues of the feedback system are at (or arbitrarily close to) locations specified by the designer. The design can proceed in two steps: For an nth order plant, state feedback is designed to place the n state feedback system eigenvalues at desired locations. Even though state feedback will not be used, these will be n of the eigenvalues of the resulting observer-feedback system. Then, the state feedback is replaced by feedback of an obsserver estimate of the same linear transformations of the state. The observer eigenvalues, which are also arbitrarily selectable by the designer, are the remaining eigenvalues of the observer-feedback system.

Expressing the feedback system (9), (64)-(65) in terms of $\underline{x}(k)$ and $\underline{z}(k)$ alone, the $(n + n_z)$ overall state equations are

$$\begin{bmatrix} \underline{x}(k+1) \\ \hline \underline{z}(k+1) \end{bmatrix} = \left[\begin{array}{c|c} A + BKQC & BKP \\ \hline GC + HKQC & F + HKP \end{array}\right] \begin{bmatrix} \underline{x}(k) \\ \hline \underline{z}(k) \end{bmatrix} \quad (66)$$

Changing variables (which does not change the eigenvalues) to a description in terms of the plant state and the observer error

$$\underline{\varepsilon}(k) = T_z\underline{x}(k) - \underline{z}(k) \quad (67)$$

gives

$$\begin{bmatrix} \underline{x}(k+1) \\ \hline \underline{\varepsilon}(k+1) \end{bmatrix} = \left[\begin{array}{c|c} A + BK & - BKP \\ \hline 0 & F \end{array}\right] \begin{bmatrix} \underline{x}(k) \\ \hline \underline{\varepsilon}(k) \end{bmatrix} \quad (68)$$

from which the separation property is obvious.

3.3 Design for Reference Input Model Tracking

Having designed a feedback structure to place all of the overall system eigenvalues to achieve the desired character of transient response, additional design freedom can then be used to obtain good tracking of relevant reference signals. Additional external input signals $\underline{r}(k)$ to the composite system may be applied at the original plant inputs and to the observer state equations so that the feedback system is, instead of (64)-(65), described by (9) and

$$\underline{z}(k+1) = F\underline{z}(k) + G\underline{y}(k) + H\underline{u}(k) + M\underline{r}(k)$$

$$\underline{w}(k) = P\underline{z}(k) + Q[\underline{y}(k) - D\underline{u}(k)] \quad (69)$$

with

$$\underline{u}(k) = K\underline{w}(k) + L\underline{r}(k) \quad (70)$$

This coupling of an external n_r-vector of inputs $\underline{r}(k)$ to the feedback system does not affect its eigenvalues. The state equations of this $(n + n_z)^{th}$ order feedback system are

$$\begin{bmatrix} \underline{x}(k+1) \\ \hdashline \underline{z}(k+1) \end{bmatrix} = \begin{bmatrix} A + BKQC & BKP \\ \hdashline GC + HKQC & F + HKP \end{bmatrix} \begin{bmatrix} \underline{x}(k) \\ \hdashline \underline{z}(k) \end{bmatrix} + \begin{bmatrix} BL \\ \hdashline HL + M \end{bmatrix} \underline{r}(k)$$

$$= \Phi \begin{bmatrix} \underline{x}(k) \\ \hdashline \underline{z}(k) \end{bmatrix} + \Lambda \underline{r}(k) \tag{71}$$

The composite system's input coupling matrix generally has B and H fixed, but L and M entirely free for the designer's choice.

A convenient class of reference signals are those that are generated by an autonomous n_η^{th} order system, of the form

$$\underline{\eta}(k+1) = \Psi\underline{\eta}(k)$$

$$\underline{r}(k) = S\underline{\eta}(k) \tag{72}$$

The output of such a <u>reference input model system</u> can consist of step, ramp, parabolic, exponential, sinusoidal and other common sequences. The general reference input model tracking problem is then for the plant and feedback controller (71) and reference input model system (72) to choose, if possible, L and M such that

$$E\underline{x}(k) \to \underline{r}(k) = S\underline{\eta}(k) \tag{73}$$

That is, we want to make the <u>feedback system</u> an observer of the fictitious reference signal model.

Considering the feedback system (71) to be an observer, through $\underline{r}(k)$, of a reference signal model (72),

$$\begin{bmatrix} \underline{x}(k) \\ \hdashline \underline{z}(k) \end{bmatrix} \to T\underline{\eta}(k) \tag{74}$$

where, using (43),

$$T\Psi - \Lambda S = \Phi T \tag{75}$$

with Φ, S, Ψ fixed, L and M in Λ free, and T only subject to (73), this constitutes a set of linear algebraic equations. If, for an initial problem formulation, there is no solution to (75), the designer may reduce the order of the reference input model and/or increase the order of the observer used for plant feedback until an acceptable solution is obtained. With this method, the viability of all design choices are tested with linear algebraic equations.

3.4 Obtaining Acceptable Between-Sample Response

The many advantages of digital over analog computation nowadays often make digital control of a continuous-time plant far more desirable than analog control. Among the potential advantages are

(1) low initial and maintenance cost

(2) low weight

(3) low power consumption

(4) high accuracy

(5) zero parameter drift with wide variations in temperature, power supply, humidity, and age

(6) ease of design and of making design changes

A basic digital feedback control arrangement has plant measurements periodically sampled by an analog-to-digital (A/D) converter and piecewise-constant control signals generated by a discrete-time observer driving a digital-to-analog (D/A) converter.

Observer feedback properly designed for the discrete-time model of a continuous-time plant and resulting in stepwise-constant continuous-time inputs to the plant will control $\underline{x}(k)$ and thus the samples of the plant state $\underline{\xi}(k)$ as desired by the designer. If the behavior of the plant between sampling times is acceptable, the design is complete: The sampled continuous-time plant is modeled as a discrete-time plant that is then feedback-controlled to achieve acceptable transient and tracking responses.

There are a number of ways of improving the between-sample response of a continuous-time plant with step-varying feedback. One of these is to increase the sample rate of the discrete-time controller. High sample rates are often expensive, however, since the amount of computation that must be performed by the controller per unit time is proportional to the sample rate. As A/D conversion typically involves several successive D/A conversions, a compromise that raises D/A but not A/D speed is to perform only some of the calculations at each step, updating the observer estimate of the plant state rapidly but incorporating corrections from output measurements at a lower rate.

Between input changes, the continuous-time plant behaves as if it has a constant input without feedback. Changing the plant's transient response, perhaps with an additional continuous-time feedback loop, can be done when it is feasible. Continuous-time filters placed between the D/A converters and the plant inputs to smooth the otherwise abrupt input transitions are helpful in many circumstances. Typically, these are low-pass filters with time constants comparable to the sampling interval. When used, they can be considered to be part of the plant, giving a higher-order augmented continuous-time plant model.

An especially attractive method of improving between-sample plant response is to reduce the extent of discontinuities in the lower-rate discrete-time control signals with higher-rate digital filters placed between the controller and the D/A conversion. For design, the discrete-time plant model is augmented with the discrete-time filter equations.

4. STEP-INVARIANT DESIGN EXAMPLE

As a simple numerical example of these results consider the single-input, single-output step-invariant plant

$$\begin{bmatrix} x_1(k+1) \\ x_2(k+1) \end{bmatrix} = \begin{bmatrix} \tfrac{1}{2} & 1 \\ \tfrac{1}{2} & 0 \end{bmatrix} \begin{bmatrix} x_1(k) \\ x_2(k) \end{bmatrix} + \begin{bmatrix} 1 \\ 1 \end{bmatrix} u(k) = A\underline{x}(k) + \underline{b}u(k)$$

$$y(k) = [1 \quad -1] \begin{bmatrix} x_1(k) \\ x_2(k) \end{bmatrix} = \underline{c}^T\underline{x}(k) \tag{76}$$

The plant characteristic equation is

$$|\lambda I - A| = \lambda^2 - 1/2\,\lambda - 1/2 = 0 \tag{77}$$

and its characteristic roots are $\lambda = 1$ and $\lambda = -1/2$. The state feedback gains

$$u(k) = [k_1 \quad k_2] \begin{bmatrix} x_1(k) \\ x_2(k) \end{bmatrix} = \underline{k}^T\underline{x}(k) \tag{78}$$

for the plant with feedback to be deadbeat are calculated as follows:

$$\underline{x}(k+1) = A\underline{x}(k) + \underline{b}u(k) = (A + \underline{b}\,\underline{k}^T)\underline{x}(k)$$

$$A + \underline{b}\,\underline{k}^T = \begin{bmatrix} (1/2 + k_1) & (1 + k_2) \\ (1/2 + 3k_1) & 3k_2 \end{bmatrix}$$

$$|\lambda I = (A + \underline{b}\,\underline{k}^T)| = \lambda^2 + (-1/2 - k_1 - k_2)\lambda + (-1/2 - k_1) \tag{79}$$

provided that $\qquad = \lambda^2$

$$\underline{k}^T = [-1/2 \quad 0] \tag{80}$$

A first order observer-controller for this plant, with eigenvalue $\lambda = 0$, has the structure

$$z(k+1) = g\,y(k) + h\,u(k) \tag{81}$$

and observes the linear transformation of the plant state given by

$$\underline{t}^T = g\underline{c}^T (A - 0)^{-1} = g[-1 \quad 3] \tag{82}$$

when

$$h = \underline{t}^T\underline{b} = 2g \tag{83}$$

Choosing g = 1, the collection of signals

$$\underline{w}'(k) = \begin{bmatrix} 0 \\ 1 \end{bmatrix} z(k) + \begin{bmatrix} 1 \\ 0 \end{bmatrix} y(k) \tag{84}$$

observes

$$\underline{w}'(k) \rightarrow \begin{bmatrix} \underline{c}^T \\ \hline \underline{t}^T \end{bmatrix} \underline{x}(k) = \begin{bmatrix} 1 & -1 \\ -1 & 3 \end{bmatrix} \underline{x}(k) = T_W \underline{x}(k) \tag{85}$$

so that

$$\underline{w}(k) = T_W^{-1}\underline{w}'(k) = \begin{bmatrix} 1/2 \\ 1/2 \end{bmatrix} z(k) + \begin{bmatrix} 3/2 \\ 1/2 \end{bmatrix} y(k) \tag{86}$$

observes $\underline{x}(k)$. When the plant is driven by the observer-controller

$$z(k+1) = y(k) + u(k) + m\, r(k)$$

$$\underline{w}(k) = \begin{bmatrix} 1/2 \\ 1/2 \end{bmatrix} z(k) + \begin{bmatrix} 3/2 \\ 1/2 \end{bmatrix} y(k) \tag{87}$$

with

$$u(k) = \underline{k}^T\underline{w}(k) + \ell\, r(k) = [-\, 1/2 \quad 0]\underline{w}(k) + \ell\, r(k) \tag{88}$$

the composite system has state equations

$$\begin{bmatrix} x_1(k+1) \\ x_2(k+1) \\ z(k+1) \end{bmatrix} = \begin{bmatrix} -(1/4) & (7/4) & -(1/4) \\ -(1/4) & (3/4) & -(1/4) \\ -(1/2) & (1/2) & -(1/2) \end{bmatrix} \begin{bmatrix} x_1(k) \\ x_2(k) \\ z(k) \end{bmatrix} + \begin{bmatrix} \ell \\ \ell \\ 2\ell + m \end{bmatrix} r(k) \tag{89}$$

which have all eigenvalues at $\lambda = 0$, as desired.

Consider now the second order reference input model system

$$\begin{matrix} \eta_1(k+1) \\ \eta_2(k+1) \end{matrix} = \begin{bmatrix} 1 & 1 \\ 1 & 0 \end{bmatrix} \begin{bmatrix} \eta_1(k) \\ \eta_2(k) \end{bmatrix} = \Psi\underline{\eta}(k)$$

$$r(k) = [1 \quad 0]\underline{\eta}(k) = \underline{\lambda}^T\underline{\eta}(k) \tag{90}$$

The reference input signal consists of an arbitrary constant plus a ramp of arbitrary slope:

$$r(k) = \eta_1(0) + k\eta_2(0) \tag{91}$$

Suppose that it is desired that the first state variable of the plant (75) track r(k), that is, that

$$[1 \quad 0]\underline{x}(k) = \underline{e}^T\underline{x}(k) \to r(k) \tag{92}$$

Then (76) becomes

$$\begin{bmatrix} 1 & 0 \\ t_{12} & t_{22} \\ t_{13} & t_{23} \end{bmatrix}\begin{bmatrix} 1 & 1 \\ 0 & 1 \end{bmatrix} - \begin{bmatrix} \ell \\ \ell \\ 2\ell + m \end{bmatrix} [1 \quad 0] = \begin{bmatrix} -1/4 & 7/4 & -1/4 \\ -1/4 & 3/4 & -1/4 \\ -1/2 & 1/2 & -1/2 \end{bmatrix}\begin{bmatrix} 1 & 0 \\ t_{12} & t_{22} \\ t_{13} & t_{23} \end{bmatrix} \tag{93}$$

which consists of six linear algebraic equations in the six variables t_{12}, t_{22}, t_{13}, t_{23}, ℓ, and m and has solution

$$\begin{aligned} t_{12} &= 1/2; \quad & t_{22} &= 1/4; \quad & \ell &= 5/4; \\ t_{13} &= 7/2; \quad & t_{23} &= -9/4; \quad & m &= 3 \end{aligned} \tag{94}$$

The completed observer-controller design then involves the external input r(k) and the observer

$$z(k+1) = y(k) + 2\,u(k) + 3\,r(k)$$

$$\underline{w}(k) = \begin{bmatrix} 1/2 \\ 1/2 \end{bmatrix} z(k) + \begin{bmatrix} 3/2 \\ 1/2 \end{bmatrix} y(k) \tag{95}$$

driving the plant (75) through

$$u(k) = [-1/2 \quad 0]\underline{w}(k) + (5/4)r(k) \tag{96}$$

If desired, these may be simplified to

$$z(k+1) = -\frac{1}{2}\,z(k) - \frac{3}{2}\,y(k) + \frac{11}{2}\,r(k)$$

$$u(k) = -\frac{1}{4}\,z(k) - \frac{3}{4}\,y(k) + \frac{5}{4}\,r(k) \tag{97}$$

The resulting feedback system is deadbeat, with a zero-input response that decays to zero in three steps or less, regardless of initial conditions. After the initial three-step transient, the first plant state variable, x_1, will exactly track any arbitrary constant-plus-ramp input r(k).

5. DEADBEAT CONTROLLERS FOR STEP-VARYING SYSTEMS

5.1 Step-Varying Deadbeat Observer Design

For an nth order step-varying plant (1) - (3), a full order state observer is another n^{th} order system, of the form

$$\underline{z}(k+1) = F(k)\underline{z}(k) + G(k)\underline{y}(k) + H(k)\underline{u}(k) \tag{98}$$

The error between the plant state and the observer state is given by

$$\begin{aligned}\underline{x}(k+1) - \underline{z}(k+1) &= [A(k) - G(k)C(k)]\underline{x}(k) - F(k)\underline{z}(k) \\ &\quad + [B(k) - G(k)D(k) - H(k)]\underline{u}(k)\end{aligned} \tag{99}$$

For the error equation to be autonomous,

$$\begin{aligned}F(k) &= A(k) - G(k)C(k) \\ H(k) &= B(k) - G(k)D(k)\end{aligned} \tag{100}$$

so that

$$\underline{x}(k+1) - \underline{z}(k+1) = F(k)[\underline{x}(k) - \underline{z}(k)] \tag{101}$$

or

$$\underline{x}(k) - \underline{z}(k) = F(k-1)F(k-2)...F(0)[\underline{x}(0) - \underline{z}(0)] \tag{102}$$

Analogous to a step-invariant deadbeat observer, it is usually desirable in the step-varying case that the observer gain sequence G(0), G(1), ..., G(n-1) be chosen so that

$$F(n-1)F(n-2)...F(0) = 0 \tag{103}$$

giving

$$\underline{z}(n) = \underline{x}(n) \tag{104}$$

An algorithm that yields (103) for plants with a single measurement output,

$$y(k) = \underline{c}^T(k)\underline{x}(k) + \underline{d}^T(k)\underline{u}(k) \tag{105}$$

whenever a solution exists at all is derived as follows [14-15]. Let $\underline{\mu}_1, \underline{\mu}_2, \ldots, \underline{\mu}_n$ be any set of basis vectors (such as the unit coordinate vectors) spanning n-dimensional space. Then (103) is satisfied if

$$\begin{cases} F(0)\underline{\mu}_1 = 0 \\ F(1)F(0)\underline{\mu}_2 = 0 \\ \vdots \\ F(n-1)F(n-2)\ldots F(0)\underline{\mu}_n = 0 \end{cases} \tag{106}$$

The first of the equations (106) is

$$[A(0) - \underline{g}(0)\underline{c}^T(0)]\underline{\mu}_1 = 0$$

$$\underline{g}(0) = \frac{A(0)\underline{\mu}_1}{\underline{c}^T(0)\underline{\mu}_1} \tag{107}$$

The second equation of (106) yields

$$\underline{g}(1) = \frac{A(1)F(0)\underline{\mu}_2}{\underline{c}^T(1)F(0)\underline{\mu}_2} \tag{108}$$

and in general,

$$\underline{g}(k) = \frac{A(k)F(k-1)\ldots F(0)\underline{\mu}_{k+1}}{\underline{c}^T(k)F(k-1)\ldots F(0)\underline{\mu}_{k+1}}, \qquad k=0,1,\ldots,n-1 \tag{109}$$

If the scalar divisor in (109) is zero at any step of the recursive gain vector calculation, it is guaranteed that reordering the <u>remaining</u> basis vectors guarantees a solution whenever the plant state can be determined from the measurements $y(k)$. This gain calculation is actually simpler to program and requires less computational effort than conventional observer gain calculation for the step-invariant case. Step-varying observers can be used to observe step-invariant plants, of course, and when this is done, the controller tends to have substantially lower-amplitude observer error as the observer is converging.

In the time-invariant case, the error in the deadbeat observer estimate of the state of the plant model is driven to zero over any interval of n steps. For time-varying observers, one strategy is to start over each time the error is ideally driven to zero, calculating a new sequence of observer gains as if the observer were being restarted, a process termed <u>cyclic observation</u>. Another method, <u>ongoing observation</u>, is to require that, at each step i after initial convergence, the preceding n gains drive the error to zero, that is

$$F(n-1+i)F(n-2+i)\ldots F(i) = 0, \qquad i=0,1,2,\ldots \tag{110}$$

In this case, the single-output observer gain algorithm becomes

$$\underline{g}(k) = \frac{A(k)F(k-1)F(k-2)\ldots F(k-n+1)\underline{\mu}_{k+1}}{\underline{c}^T(k)F(k-1)F(k-2)\ldots F(k-n+1)\underline{\mu}_{k+1}}, \quad k=n-1,n,n+1,\ldots \tag{111}$$

This choice of ongoing gains gives the observer a "moving window" property whereby at each step, the observer converges to the state of the model plant having the most recently incorporated n input and measurement samples. The computation at each step required by (111) is that of n vector-matrix products, a vector inner product, and division by a scalar. This is only slightly more than the computation associated with a single n×n matrix product. In the worst possible situation, when all but the last basis vector results in a zero scalar divisor, the equivalent n-1 matrix products and tests for zero are still only of approximately the numerical complexity of a single n×n matrix inversion.

For multiple-output systems, additional linearly independent outputs allow observer convergence in a fewer number of steps. In the case of just two plant measurement outputs, for (9) with

$$C = \text{col}\ [\underline{c}_1^T(k),\ \underline{c}_2^T(k)] \tag{112}$$

a full order state observer (98) has

$$G(k) = [\underline{g}_1(k),\ \underline{g}_2(k)] \tag{113}$$

where

$$F(k) = A(k) - \underline{g}_1(k)\underline{c}_1^T(k) - \underline{g}_2(k)\underline{c}_2^T(k) \tag{114}$$

Letting $\underline{\mu}_1, \underline{\mu}_2, \ldots, \underline{\mu}_n$ be any set of n-dimensional basis vectors, requiring that both

$$\begin{cases} F(0)\underline{\mu}_1 = [A(0) - \underline{g}_1(0)\underline{c}_1^T(0) - \underline{g}_2(0)\underline{c}_2^T(0)]\underline{\mu}_1 = 0 \\ F(0)\underline{\mu}_2 = [A(0) - \underline{g}_1(0)\underline{c}_1^T(0) - \underline{g}_2(0)\underline{c}_2^T(0)]\underline{\mu}_2 = 0 \end{cases} \tag{115}$$

gives

$$\begin{cases} [\underline{c}_1^T(0)\underline{\mu}_1]\underline{g}_1(0) + [\underline{c}_2^T(0)\underline{\mu}_1]\underline{g}_2(0) = A(0)\underline{\mu}_1 \\ [\underline{c}_1^T(0)\underline{\mu}_2]\underline{g}_1(0) + [\underline{c}_2^T(0)\underline{\mu}_2]\underline{g}_2(0) = A(0)\underline{\mu}_2 \end{cases} \tag{116}$$

where the terms in brackets are scalars. Only a 2×2 matrix must be inverted to obtain the solution of (116). Generally, it is necessary to invert a matrix of dimension equal to the number of linearly independent plant outputs used.

For the two-output example case, requiring that the observer error in two other basis vector directions be driven to zero at each successive step gives relations similar to (116), each requiring a 2×2 matrix inversion for solution for the observer gains at that step.

5.2 State and Observer Feedback

In the step-varying case, an nth order discrete-time plant (9) with state feedback

$$\underline{u}(k) = K(k)\underline{x}(k) \tag{117}$$

is governed by

$$\underline{x}(k+1) = [A(k) + B(k)K(k)]\underline{x}(k) \tag{118}$$

State transition matrices for this feedback system are

$$\Phi(i,0) = N(i-1)\, N(i-2) \ldots N(0) \tag{119}$$

where

$$N(i) = A(i) + B(i)K(i) \tag{120}$$

Choosing feedback gains K(0), K(1), ..., K(i) for deadbeat feedback, so that

$$\Phi(i,0) = 0 \tag{121}$$

For some step i, is the same mathematical problem as that of choosing observer gains for deadbeat observation and may be solved with the methods of the previous section. This is the generalization of deadbeat feedback for a step-invariant system. <u>Ongoing</u> deadbeat feedback is to require that

$$\Phi(i+k,\ k) = N(i+k-1)N(i+k-2) \ldots N(k), \qquad k=0,1,2,... \tag{122}$$

Consider an nth order step-varying plant (9), possibly having multiple outputs, and an n_z^{th} order observer

$$\begin{aligned} \underline{z}(k+1) &= F(k)\underline{z}(k) + G(k)\underline{y}(k) + H(k)\underline{u}(k) \\ \underline{w}(k) &= P(k)\underline{z}(k) + Q(k)[\underline{y}(k) - D(k)\underline{u}(k)] \end{aligned} \tag{123}$$

with observer state error

$$\underline{\varepsilon}(k) = T(k)\underline{x}(k) - \underline{z}(k) \tag{124}$$

governed by

$$\underline{\varepsilon}(k+1) = F(k)\underline{\varepsilon}(k) = F(k)F(k-1) \ldots F(0)\underline{\varepsilon}(0) \tag{125}$$

where

$$F(i) = A(i) - G(i)C(i) \tag{126}$$

Let the gains G(0), G(1), ... give acceptable, but not necessarily deadbeat, observer performance as a plant estimator:

$$\underline{w}(k) \to \underline{x}(k) \tag{127}$$

Let the state feedback gains K(0), K(1), ... in

$$\underline{u}(k) = K(k)\underline{x}(k) \tag{128}$$

give acceptable, but not necessarily deadbeat, state feedback system performance. Then the feedback system where an observer estimate of the plant state is fed back in place of the plant state itself is described by

$$\begin{bmatrix} \underline{x}(k+1) \\ \hline \underline{\varepsilon}(k+1) \end{bmatrix} = \left[\begin{array}{c|c} A(k) + B(k)K(k) & - B(k)K(k)P(k) \\ \hline 0 & F(k) \end{array}\right] \begin{bmatrix} \underline{x}(k) \\ \hline \underline{\varepsilon}(k) \end{bmatrix} \tag{129}$$

At step k, the feedback system's plant state vector is

$$\begin{aligned} \underline{x}(k) &= N(k-1)N(k-2) \ldots N(0)\underline{x}(0) \\ &\quad - [B(k-1)K(k-1)P(k-1)F(k-1)F(k-2) \ldots F(0) \\ &\quad + N(k-1)B(k-2)K(k-2)P(k-2)F(k-2) \ldots F(0) \\ &\quad + N(k-1)N(k-2)B(k-3)K(k-3)P(k-3)F(k-3) \ldots F(0) \\ &\quad + \ldots + N(k-1) \ldots N(1)B(0)K(0)P(0)]\underline{\varepsilon}(0) \end{aligned} \tag{130}$$

which is the plant's response with state feedback, plus a term of step-varying proportion to the initial observer error.

If the observer is deadbeat, converging at step i, than at step i and beyond, the plant state vector is governed by

$$\underline{x}(k+1) = N(k)\underline{x}(k) \tag{131}$$

which is then identical to the situation with state feedback. If the plant feedback is also deadbeat, with the ongoing property (122), then the observer feedback system is deadbeat, converging in $n + n_z$ steps or less.

5.3 Reference Model Signal Tracking

As in the step-invariant case, a step-varying feedback control system can be designed to, itself, observe a reference model signal model in order to achieve acceptable response to a class of reference inputs. The external inputs are coupled to the accessible plant inputs and to all state variables of the observer. Coupling gains are chosen so that a suitable transformation of the plant state observes the model signals. With step-varying external input gains, the signal model may also be step-varying. As in the step-invariant case, design choices are tested with the solution of linear algebraic equations.

6. CONCLUSION

New controller design methods for discrete-time tracking systems have been given, for which the controlled plant has prescribed transient response and will, after a transient period, exactly track any reference input signal from a class chosen by the designer. The methods are applicable to both step-invariant and step-varying systems, and require only the solution of linear algebraic equations to test design choices.

REFERENCES

1. Franklin, Gene F. and Powell, J. David, Digital Control of Dynamic Systems, Reading MA: Addison-Wesley, 1980.

2. Chen, C. T., Linear System Theory and Design, New York: Holt, Rinehart and Winston, 1984.

3. Hostetter, G. H., Savant, C. J., and Stefani, R. T., Design of Feedback Control Systems, New York: Holt, Rinehart and Winston, 1982.

4. Kailath, T., Linear Systems, Englewood Cliffs, NJ: Prentice-Hall, 1980.

5. D. G. Luenberger, 'Observing the state of a linear system," IEEE Trans. Military Electron., vol. MIL-8, pp. 74-80, Apr. 1964.

6. __________, 'Observers for multivariable systems,' IEEE Trans. Automat. Contr., vol. AC-11, pp. 190-197, Apr. 1966.

7. __________, 'An introduction to observers,' IEEE Trans. Automat. Contr., vol. AC-16, pp. 596-602, Dec. 1971.

8. Anderson, B. D. O. and Moore, J. B., Linear Optimal Control, Englewood Cliffs, NJ: Prentice-Hall, 1971.

9. Wonham, W. M., 'On pole assignment in multi-input controllable linear systems,' IEEE Trans. Automat. Contr., vol. AC-12, pp. 660-665, Dec. 1967.

10. Chen, C. T., 'A note on pole assignment,' IEEE Trans. Automat. Contr., vol. AC-13, pp. 597-598, Oct. 1968.

11. Davison, E. J., 'On pole assignment in multivariable linear systems,' IEEE Trans. Automat. Contr., vol. AC-13, pp. 747-748, Dec. 1968.

12. Luenberger, D. G., Introduction to Dynamic Systems, New York: Wiley, 1979.

13. O'Reilly, J., Observers for Linear Systems, New York: Academic Press, 1983.

14. Hostetter, G. H., 'Ongoing deadbeat observers for linear time-varying systems,' Proc. 1982 Amer. Control Conf., pp. 1099-1101, June 1982.

15. ________, 'Recursive discrete functional expansion,' IEEE Trans. Automat. Contr., vol. AC-29, pp. 654-656, July 1984.

CHAPTER 7

MICROPROCESSOR-BASED CONTROL OF DC MOTORS

Milic R. Stojic
Faculty of Electrical Engineering
University of Belgrade
Bulevar Revolucije, 73
P.O. Box 816
11001 Belgrade, Yugoslavia

ABSTRACT. In this paper, the microprocessor-based control of DC motor is considered. Structures corresponding to the digital system for DC motor speed control and digital positioning servomechanism using DC motor are proposed. For both the motor speed control system and the positioning servo possible digital control laws are discussed. The synthesis of controller parameters is carried out by the pole-placement procedure. The suggested grapho-analytical method of pole-placement, applicable to third-order digital control systems, enables the designer to obtain readily the information about effects of system constants and adjusting parameters on the accuracy, speed of response, and stability margin. The design of the real digital positioning servomechanism using DC motor is presented to illustrate the proposed technique.

1. INTRODUCTION

The state and recent developments in the control theory and applications indicate that microprocessors will revolutionize the process control. Namely, in most process applications the cost of the control system is only a relatively small fraction of the total cost, and the cost of the processor(s) is only a fraction of the control system's cost. Except in some isolated instances, there is not a large number of situations in which direct replacement of control components with a microprocessor driven version will result in either substantial cost saving or significantly important performance.

A large number of applications has been already done for a variety of unit applications, i.e., control of processes that do not interact strongly with other processes. There are typically special purpose applications for which microprocessor system has unique hardware and software. However, one area that has not received much attention, and where we have some specific interests, is in the design of the direct digital stand-alone single-loop controller. This is a general purpose application for which a degree of adjustment or programmability is required to meet the needs of specific processes and their performances.

This chapter presents a detailed procedure of microprocessor

N. K. Sinha (ed.), Microprocessor-Based Control Systems, 131–155.

application in the design and realization of digital systems for control of DC motors. In both cases the DC motor speed control and the design of DC servo control system an adopted system structure mainly depends on the employed digital control law and on the way of detecting the control variable (motor speed or angular displacement).

One way of detecting the error signal is to compare the set point to the frequency of the feedback signal from shaft encoder of tachogenerator [1]. The other is to lock the motor speed by employing a phase-locked loop [2]. For precise speed control systems both the signal from a shaft encoder or that from a tachogenerator and phase-locked loop have been used [3]. Conventional proportional-integral-derivative(PID) digital control laws, or even a modern control algorithm, have been applied in direct-digital-control (DDC) of typical processes involving processors with limited computing capabilities and memory [4]. In the paper [5], two speed control systems are considered. The first is a motor speed control system using frequency comparision and conventional digital PID-control algorithm proposed in the work [4]. The other is the application of a phase-locked loop and P-control algorithm. For both systems, the Jury stability tests [6] for second- and third-order sampled-data control systems were used to analyze the stability conditions in dependence of the sampling period and system adjustable parameters.

The system structure and the procedure of parameter optimization for the direct digital servo control system with DC motor are proposed in the work [7]. The suggested system structure is suitable for microprocessor implementation and the pole-placement procedure, proposed in the paper [7], can be advantageously applied in a broad class of third-order digital control systems. The same pole-placement procedure has been applied in the work [8] to design a digital motor speed control system using a counting-type angle encoder and an integral control algorithm eliminating the steady-state error.

In this paper, the design of digital motor speed control systems and servo control systems using DC motors is outlined. The structures of the microprocessor-based motor control systems are suggested in Section 2. The Section also includes ways of converting a motor speed and an angular position into corresponding digital representations, discussion about possible digital control laws, and some additional comments on the sampling rate. The suitable grapho-analytical procedure of synthesizing system constants and control parameters is presented in Section 3. The design of a real digital servo control system is given in Section 4. An assembly program for the servo system under consideration, and the implementation using the microprocessor INTEL 8080 are given in Section 5.

2. SYSTEM DESCRIPTION

In both cases of the digital positioning servo and the motor speed control, the typical structure of a microprocessor-based DC motor control system is shown in Fig. 1. The system comprises the continuous part and the digital portion including digital control law, parameter setting, timing, etc.

2.1. System Continuous Part

As it is shown in Fig. 1, the continuous portion of the system usually consists of the digital-to-analog converter (DAC), power amplifier (PA) having output saturation at ±U, and the DC motor.

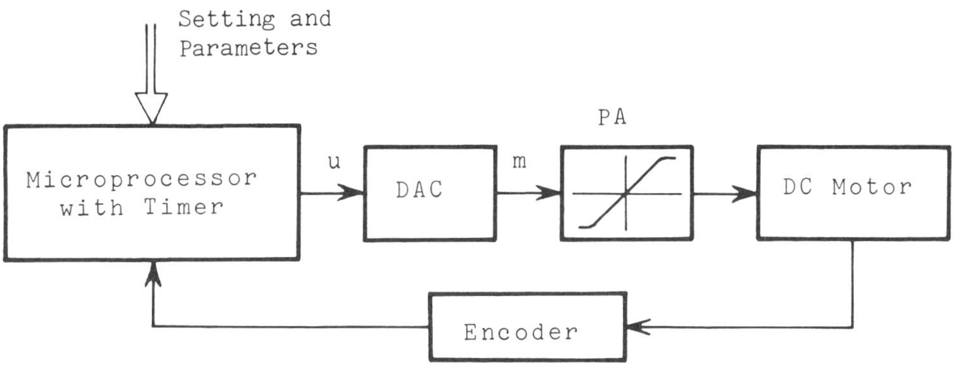

Fig. 1. Structure of a microprocessor-based system for dc motor control

The transfer function of a separately excited DC motor may be approximated by

$$G_m(s) = \frac{K_m}{T_m s + 1} \text{ (rad/s·V)} \tag{1}$$

where the transfer gain K_m and the mechanical time constant T_m are usually given in advance; otherwise, they can be measured by a simple experiment.

Compared with the mechanical part of the system, the power amplifier or the driving circuit is assumed to be of sufficiently large bandwidth so that the amplifier, in a linear regime of operation, may be described by a pure adjustable gain K_A.

The analog-to-digital converter may be treated as a zero-order hold having the transfer function

$$G_h(s) = K_h \frac{1 - e^{-Ts}}{s} \text{ (V/count)} \tag{2}$$

The motor speed and angular position may be detected in different ways. Incremental shaft encoders may be used as a pulse source for the generation of a signal proportional to angular velocity [9]. A typical circuit for such an application is shown in Fig. 2. The circuit will provide a continuous indication of the pulse rate averaged over a selected time interval. The circuit utilizes a shift register containing n bits and a bidirectional counter. The contents of the shift register are shifted at a constant frequency f producing a delay between the information input and the output of the shift register given by

$$\tau = \frac{n}{f} \tag{3}$$

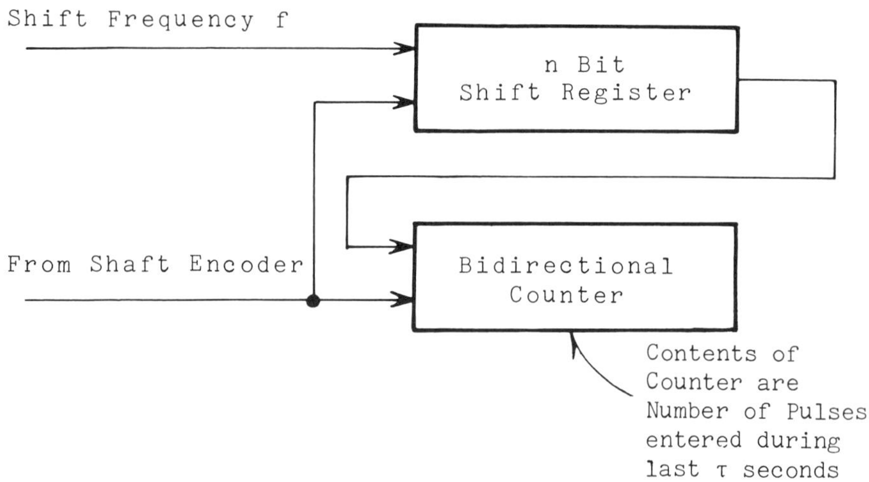

Fig. 2. Pulse rate measuring technique

The input pulse train coming from the incremental shaft encoder at a frequency f_i also enters the shift register. If $f > f_i$, the number of pulses in the shift register will be the number of pulses that entered during the preceding time interval τ. The average input pulse rate is, therefore,

$$R_{av} = \frac{P}{\tau} \tag{4}$$

where P is the number of pulses counted in the interval τ.

Since each input pulse enters the shift register immediately, and since its contents are shifted at a constant frequency, the average pulse rate will be continuously monitored, and any change in the input pulse rate will have an immediate effect on the value of R_{av}.

The bidirectional counter is employed to indicate the number of pulses that are in the shift register. In order to accomplish this task, the input pulses that enter the shift register are also used to increase the value of the counter, and the pulses that exit from the shift register are used to decrease the value of the counter. At a given time $t = T$ the value of the counter is

$$R_T = \sum_{i=0}^{T} P_i - \sum_{i=0}^{T-\tau} P_i = \sum_{t=T-\tau}^{T} P_i = P \tag{5}$$

where P_i is the instantaneous pulse rate, or, if (4) is used,

$$R_T = \tau \cdot R_{av} \tag{6}$$

Thus we see that the value of the counter gives a reading that is proportional to the average input pulse rate.

An another simpler way of detecting simultaneously the motor speed and angle increments consists of using an ordinary binary counter instead of a bidirectional one. This is one of the simplest methods of converting a shaft displacement into a digital representation. This so-called counting-type angle encoder consists of two basic components, a plastic disc and a binary counter. On the disc periphery, quantum marks are engraved with uniform spacing. When the disc rotates continuously, these quantum marks produce clock pulses in the reading head, and these clock pulses load the binary counter. At each sample instant the microprocessor disables counting, reads the counter contents, resets the counter, and enables counting again. In this manner, adding counter contents one by one, the total of output angle increments is detected, and by reading counter contents successively, the output angle rate is generated. The microprocessor can pick counter contents without stopping counting process. In this way, applied in the paper [7], at each sample instant the microprocessor reads the counter twice, except if the counter contents are equal; otherwise, readings are repeated.

Hence, in the case of a counting-type angle encoder, the feedback variable $b(t)$ may be expressed as

$$b(t) = \frac{K_n}{2\pi}\left[\theta(t) - \theta(t-T)\right] \quad \text{(counts) for time interval,} \tag{6}$$

where $\theta(t)$ denotes the output angle in radians, and K_n is the total number of quantum marks on the disc.

The Laplace transform of (6) is

$$B(s) = \frac{K_n}{2\pi} \frac{1 - e^{-Ts}}{s} \Omega(s) \quad \text{(counts/sample period)} \tag{7}$$

and thus the transfer function of the feedback element becomes

$$H(s) = \frac{B(s)}{\Omega(s)} = K_n^* \frac{1 - e^{-Ts}}{s} \quad \text{(counts/sample period)/(rad/s)} \tag{8}$$

where $K_n^* = K_n/2\pi$ denotes the number of pulses generated per radian and $\Omega(s)$ is the Laplace transform of controlled variable (motor speed) in rad/s.

2.2. Digital Control Laws

The comparision of different digital PID control laws with their analog versions were considered in papers published in the early stage of DDC. The P, I, and D actions can be made independent of each other, but they are subject to the length of sampling period [10]. Unlike continuous time control in which high frequency noise is normally filtered out, it may cause law frequency fluctuations in digital systems [11]. Digital control laws basically similar to the analog PID versions with some modifications and improvements were proposed [12] as well as tuning rules for digital algorithms [13-16].

Since the portion of information is lost due to the sampling process, it is inevitable that the digital version of the PID algorithm will perform as well as analog PID control law unless the sampling period is very small, which under some circumstances is not economical for sampled--data control systems. Digital systems, on the other hand, allow considerable flexibility in the types of algorithms that can be used to apply newer control theories for direct digital process control systems, and to exploit the software flexibility while keeping any disadvantages of digital control to a minimum [4].

2.2.1. Motor speed control system. The block-diagram of a digital system for DC motor speed control is shown in Fig. 3. The transfer functions $G_m(s)$, $G_h(s)$, and $H(s)$ appearing in the continuous portion of the system are given by (1), (2), and (8), respectively.

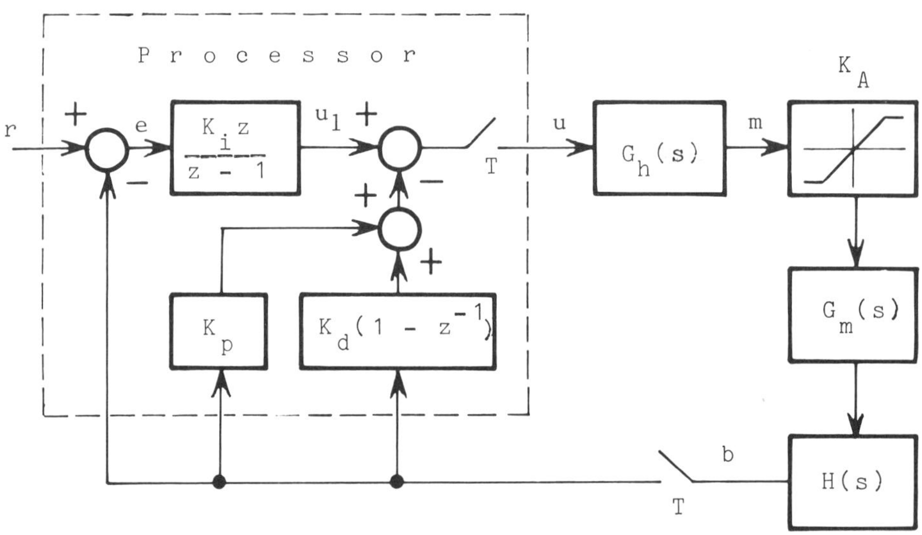

Fig. 3. Block diagram of a microprocessor-based motor speed control system

The suggested PID control algorithm can be written in the velocity form

$$u(kT) = K_p\{b[(k-1)T] - b(kT)\} + K_i[r(kT) - b(kT)] + \\ + K_d\{2b[(k-1)T] - b[(k-2)T] - b(kT)\} \tag{9}$$

or in the position form

$$u(kT) = -K_p b(kT) + K_i \sum_{i=0}^{k} [r(iT)-b(iT)] - K_d\{b(kT)-b[(k-1)T]\} \tag{10}$$

where $b(kT)$ and $r(kT)$ are sampled-data of the feedback variable (encoder output) and the set point, respectively, while $u(kT) = u(kT) - u[(k-1)T]$ represents the incremental change of the control variable $u(kT)$. The coefficients K_p, K_i, and K_d appearing in (9) or (10), are respectively, the proportional, integral, and derivative gains. To avoid sudden changes in the control variable when the set point is changed, notice that the set point signal $r(kT)$ is included only in the I--action in (9) and (10). Consequently, the P-action and D-action form local minor loops in the block-diagram of the system in Fig. 3, where only the I-action eliminating the steady state error is left in the main loop.

The proposed PID control algorithm can be rewritten by three equations more suitable for microprocessor implementation:

$$e(kT) = r(kT) - b(kT) \tag{11a}$$

$$u_1(kT) = K_i e(kT) + u_1[(k-1)T] \tag{11b}$$

$$u(kT) = -K_p b(kT) - K_d\{b(kT) - b[(k-1)T]\} + u_1(kT) \tag{11c}$$

where $e(kT)$ and $u_1(kT)$ represent sample-data of the error signal and output of cascade integral compensator, respectively.

When the counting-type angle encoder is used as a feedback element, the desired motor speed in the stationary state (set point) is assigned by the system reference input $r = N$ representing a necessary number of counts during the sampling interval. For example, if a desired motor speed Ω_{ss} in the steady-state is given in rpm then

$$N = \frac{K_n \Omega_{ss} T}{60} \quad \text{(counts)} \tag{12}$$

Recall that K_n represents the total number of quantum marks on the disc.

When the driver works in a linear regime, the characteristic equation of the system can be easily derived according to the block-diagram in Fig. 3 as

$$z(z-1) + [K_p z(z-1) + K_i z^2 + K_d(z-1)^2]G_p(z) = 0 \tag{13}$$

where the transfer function $G_p(z)$ is calculated using the Table of z-transform [17]:

$$G_p(z) = Z\left[G_h(s)K_A G_m(s)H(s)\right]$$

$$= K(1 - z^{-1})^2 Z\left[\frac{1}{s^2(T_m s + 1)}\right]$$

$$= K\left[\frac{T}{z} - \frac{T_m(1 - A)(z - 1)}{z(z - A)}\right] \tag{14}$$

where $K = K_h K_A K_m K_n^*$ and $A = e^{-T/T_m}$.

Substituting $G_p(z)$ from (14) into (13) and then reducing (13) into the polynomial form, it is obtained, after simple manipulations,

$$z^4 + a_3 z^3 + a_2 z^2 + a_1 z + a_0 = 0 \tag{15}$$

where

$$\begin{aligned} a_3 &= -(1+A) + K(T-T_m+AT_m)(K_p+K_i+K_d) \\ a_2 &= A - KTA(K_p+K_i+K_d) - KT(K_p+2K_d) + KT_m(1-A)(2K_p+K_i+3K_d) \\ a_1 &= KTA(K_p+2K_d) - KT_m(1-A)(K_p+3K_d) + KK_dT \\ a_0 &= KK_d(T_m-AT_m-AT) \end{aligned} \tag{16}$$

Notice, when $K_d = 0$, then the coefficient a_0 is zero and the system characteristic equation becomes

$$z^3 + A_2 z^2 + A_1 z + A_0 = 0 \tag{17}$$

with

$$\begin{aligned} A_2 &= -(1+A) + K(T-T_m+AT_m)(K_p+K_i) \\ A_1 &= A - KTA(K_p+K_i) - KTK_p + KT_m(1-A)(2K_p+K_i) \\ A_0 &= KTAK_p - KT_mK_p(1-A) \end{aligned} \tag{18}$$

In both cases (15) and (17) coefficients are functions of the system constants and the controller parameters that are to be adjusted with respect to the desired dynamic behaviour of the system.

2.2.2. Positioning servo system. The structure of the digital servo control system, proposed in the paper [7], is shown in Fig. 4. In the system, the counting-type angle encoder having the transfer function (8) is very suitable for simultaneous generation of digital signals proportional to the motor speed and the angular position.

The suggested digital control law includes the cascade lead compensation with two adjustable parameters D and H, and the local velocity

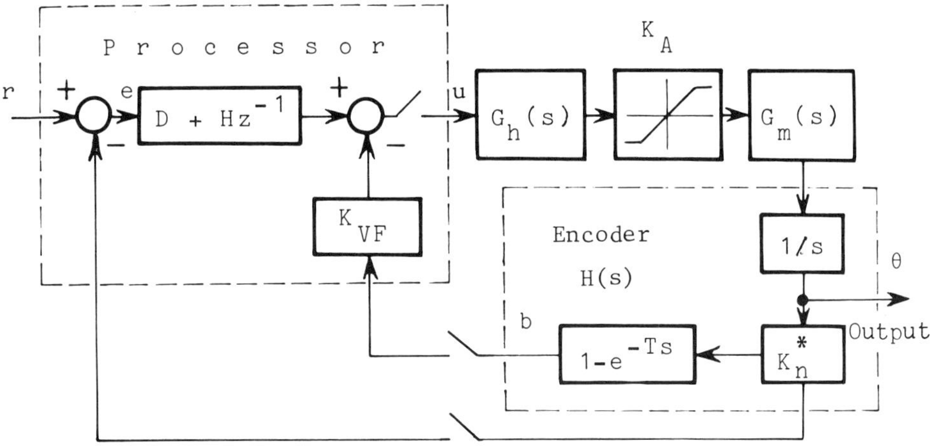

Fig. 4. Block diagram of a microprocessor-based servo control system

feedback with the adjustable gain K_{VF}. Thus the system manipulative (control) variable $u(kT)$ can be written as

$$U(kT) = De(kT) + He[(k-1)T] - K_{VF}b(kT) \tag{19}$$

where $e(kT) = r - K_n^*\theta(kT)$.

The desired output angle θ_r (set point) is assigned by the number of pulses $r = N$ corresponding to the number of quantum marks on the disk within the angle θ_r, i.e.,

$$N = K_n^*\theta_r \tag{20}$$

where θ_r is given in radians.

Introducing $G_s(s) = G_h(s)K_AG_m(s)/s$, and using the z-transform, we obtain, in a linear regime,

$$G_s(z) = K(1 - z^{-1})Z\left[\frac{1}{s^2(T_ms + 1)}\right]$$

$$= K\left[\frac{1}{z - 1} - \frac{T_m(1 - A)}{z - A}\right] \tag{21}$$

where, as in equation (14), $K = K_hK_AK_mK_n^*$ and $A = e^{-T/T_m}$.

According to the block-diagram in Fig. 4, the system characteristic equation is derived as

$$1 + \frac{Dz + H}{z} G_s(z) + K_{VF} \frac{z - 1}{z} G_s(z) = 0 \tag{22}$$

Substituting $G_s(z)$ from (21) into (22), the equation can be reduced into the polynomial form (17), where

$$\begin{aligned} A_2 &= K(D+K_{VF})\left[T-T_m(1-A)\right] - (1+A) \\ A_1 &= K(D+K_{VF})\left[T_m(1-A)-TA\right] + K(K_{VF}-H)\left[T_m(1-A)-T\right] + A \\ A_0 &= K(H-K_{VF})\left[T_m(1-A)-TA\right] \end{aligned} \tag{23}$$

In a particular system, the digital-to-analog converter, power amplifier, DC motor, and sampling period are usually selected in advance so that the parameters T, K_A, K_m, and T_m appearing in (23) may be considered as given system parameters, while the controller parameters D, H, and K_{VF} are to be synthesized in order to meet the desired system performance with respect to the accuracy, speed of continuous-time response, and stability margin.

3. SELECTION OF THE DESIRED CLOSED-LOOP POLE POSITIONS

If the control parameter adjustments are to be carried out using the pole-placement procedure, the designer is required to specify desired target locations for the closed-loop poles. This is achieved by specifying the coefficients of the system characteristic equation which are necessary to ensure that the corresponding identity has a solution. In practice, however, the desired closed-loop pole-set is best specified in terms of lower order characteristic equation in z. This, in turn, can be reflected back through the inverse z-transform to relate to the desired bandwidth and transient response of the closed-loop continuous--time response. It is therefore important to realize that the discrete--time desired pole-set must be specified in conjunction with the sampling period T. In some cases, as for example in the self-tuning approach, the desired pole-set and sampling period can be set-up experimentally and various schemas quickly tried. However, for preliminary guidance the following points have been found useful:

(i) Decide upon the achievable closed-loop bandwidth of the continuous system f_c, bearing in mind the various engineering constraints upon the speed of transient response.

(ii) Pick the sampling rate three or four times greater than f_c; that is

$$T \simeq \frac{1}{3f_c} \tag{24}$$

(iii) Specify the desired closed-loop pole-set and determine the coefficients of the discrete-time characteristic equation using the z-transform version of the desired continuous-time response.

For third-order digital control systems the characteristic equation has the form (17), and the proper root-set of this equation can be achieved by employing the charts in Fig. 5 or 6. The construction of the charts and their application in solving the pole placement problem for the third-order digital control systems are explaned in more details in the following developments.

First, we introduce the new complex variable $\bar{z}$ by putting $z = A_2\bar{z}$ into (17) to obtain

$$\bar{z}^3 + \bar{z}^2 + \beta\bar{z} + \alpha = 0 \tag{25}$$

where only two variable coefficients

$$\alpha = A_0/A_2^3 \quad \text{and} \quad \beta = A_1/A_2^2 \tag{26}$$

appear.

Setting $\bar{z} = (\bar{\omega}_n)_z e^{j\bar{\psi}} = (\bar{\omega}_n)_z\bar{\zeta}_z + j(\bar{\omega}_n)_z\sqrt{1 - \bar{\zeta}_z^2}$, with $\bar{\psi} = \arccos\bar{\zeta}_z$ and $-1 \leq \bar{\zeta}_z \leq 1$, into equation (25), then separating the real and imaginary parts of this equation, and equating these parts to zero, equation (25) may be rewritten as two simultaneous equations in two unknowns α and β:

$$(\bar{\omega}_n)_n^3 T_3(\bar{\zeta}_z) + (\bar{\omega}_n)_z^2 T_2(\bar{\zeta}_z) + \beta(\bar{\omega}_n)_z T_1(\bar{\zeta}_z) + \alpha T_0(\bar{\zeta}_z) = 0 \tag{27}$$

$$(\bar{\omega}_n)_n^3 U_3(\bar{\zeta}_z) + (\bar{\omega}_n)_z^2 U_2(\bar{\zeta}_z) + \beta(\bar{\omega}_n)_z U_1(\bar{\zeta}_z) + \alpha U_0(\bar{\zeta}_z) = 0 \tag{28}$$

where $T_k(\bar{\zeta}_z)$ and $U_k(\bar{\zeta}_z)$ are the Chebyshev functions of the first and second kind, respectively [18].

By using the well-known additional theorem

$$T_k(\bar{\zeta}_z) = \bar{\zeta}_z U_k(\bar{\zeta}_z) - U_{k-1}(\bar{\zeta}_z) \tag{29}$$

equation (27) can be replaced by

$$(\bar{\omega}_n)_z^3 U_2(\bar{\zeta}_z) + (\bar{\omega}_n)_n^2 U_1(\bar{\zeta}_z) + \beta(\bar{\omega}_n)_z U_0(\bar{\zeta}_z) + \alpha U_{-1}(\bar{\zeta}_z) = 0 \tag{30}$$

Equations (28) and (30) can be easily solved for variable coefficients α and β to get

$$\begin{aligned} \alpha &= (\bar{\omega}_n)_z^2\left[(\bar{\omega}_n)_z U_2^*(\bar{\zeta}_z) + U_1^*(\bar{\zeta}_z)\right] \\ \beta &= -(\bar{\omega}_n)_z\left[(\bar{\omega}_n)_z U_3^*(\bar{\zeta}_z) + U_2^*(\bar{\zeta}_z)\right] \end{aligned} \tag{31}$$

where $|\bar{\zeta}_z| < 1$ and $U_1^*(\bar{\zeta}_z)$, $U_2^*(\bar{\zeta}_z)$, and $U_3^*(\bar{\zeta}_z)$ are the Chebyshev polynomials of the second kind [18] defined by

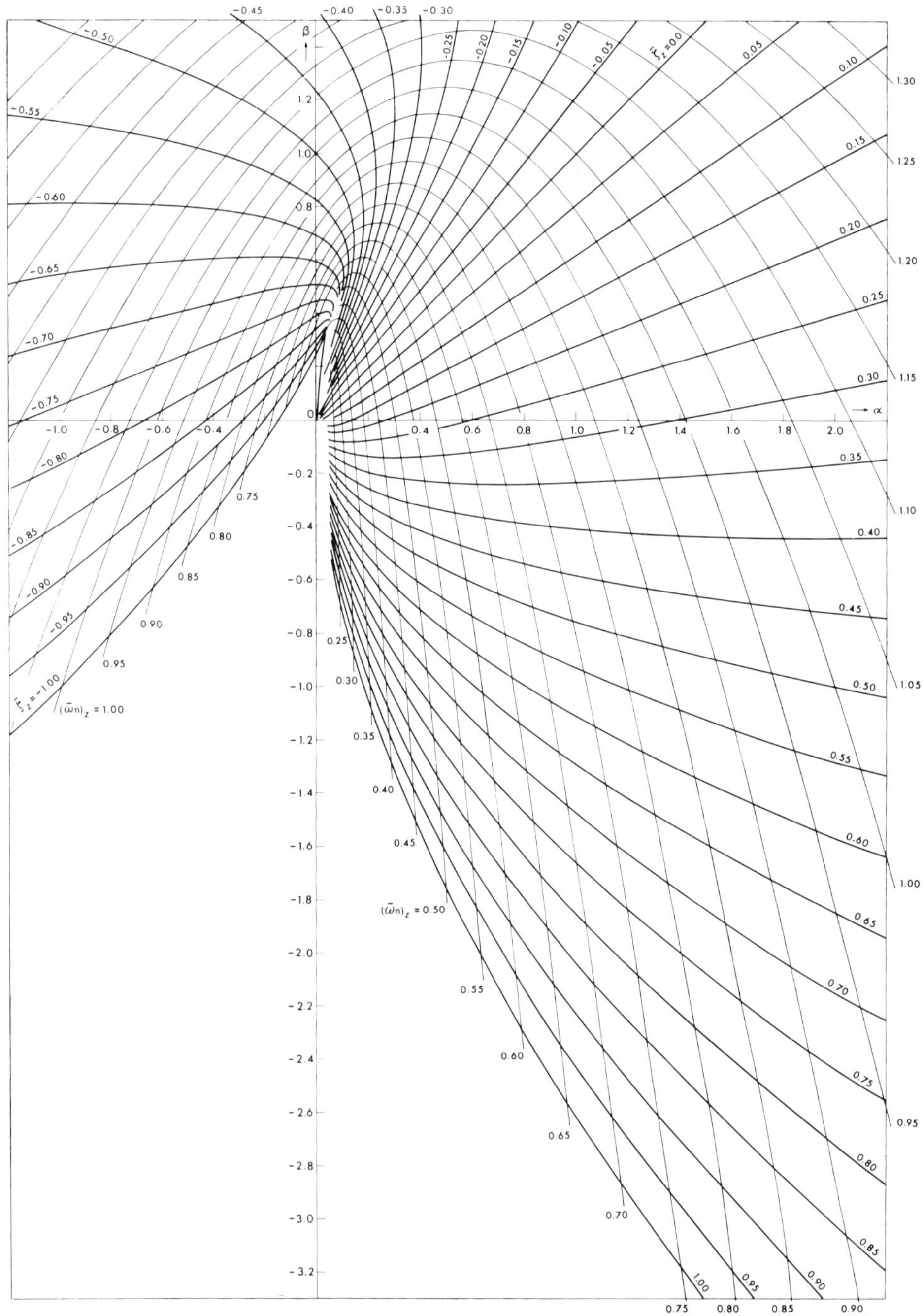

Fig. 5. Root loci of complex-poles in the parameter plane

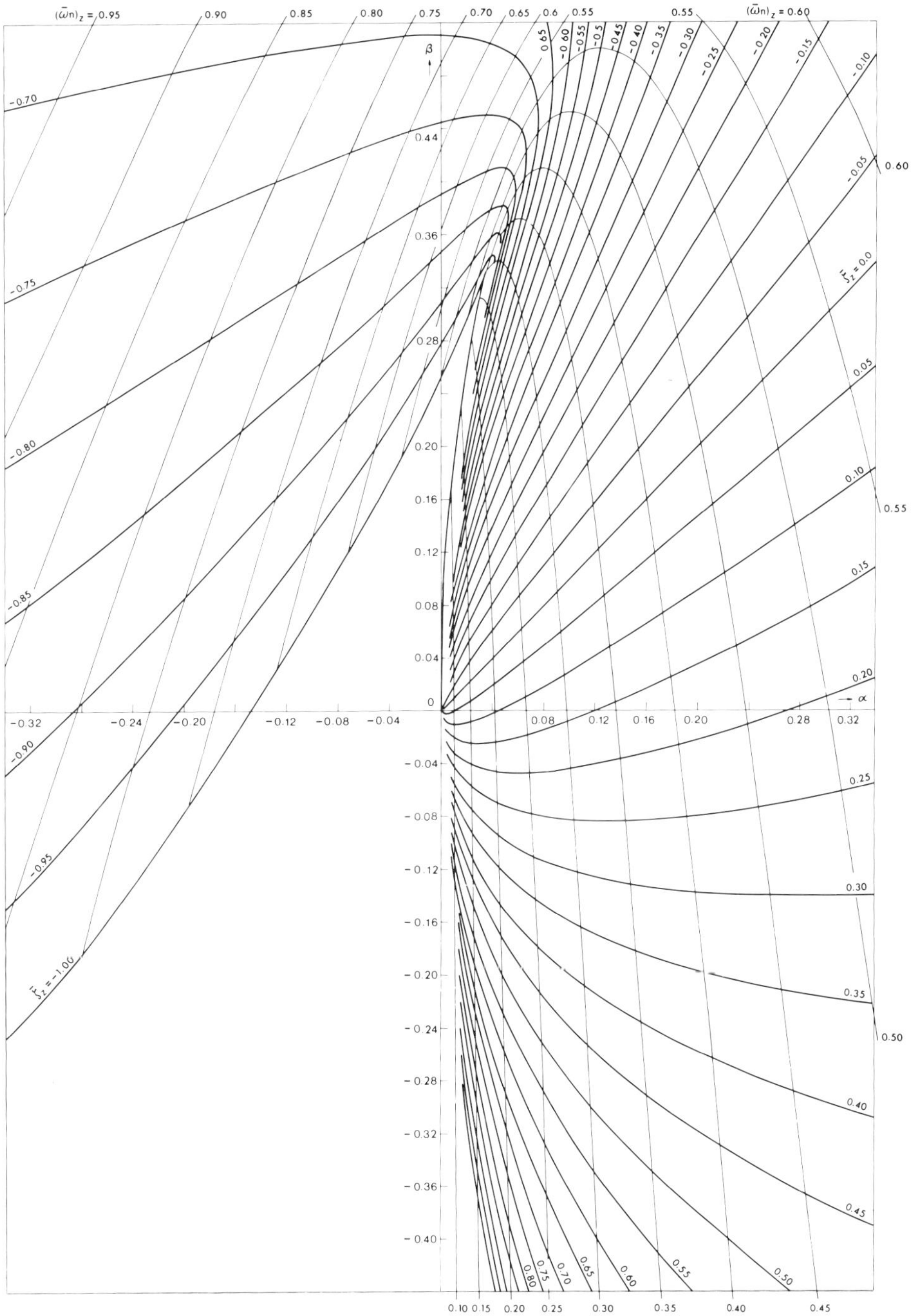

Fig. 6. Enlarged region of Fig. 5 around the origin

$$U_k^*(\bar{\zeta}_z) = \frac{\sin(k\cdot\arccos\bar{\zeta}_z)}{\sqrt{1-\bar{\zeta}_z^2}}$$

Recall that these polynomials are easily calculated using the well-known reccurence formula $U_{k+1}^*(\bar{\zeta}_z) - 2\bar{\zeta}_z U_k(\bar{\zeta}_z) + U_{k-1}^*(\bar{\zeta}_z) = 0$ with $U_0^*(\bar{\zeta}_z) \equiv 0$ and $U_1^*(\bar{\zeta}_z) \equiv 1$.

To relate z- and $\bar{z}$-plane, we first substitute $z = (\omega_n)_z\zeta_z + j(\omega_n)_z\sqrt{1-\zeta_z^2}$ and then, according to relation $z = A_2\bar{z}$, we conclude that

$$(\omega_n)_z = |A_2|(\bar{\omega}_n)_z \tag{32a}$$

$$\zeta_z = (\mathrm{sgn}A_2)\bar{\zeta}_z \tag{32b}$$

$$\sigma_z = A_2\bar{\sigma}_z \tag{32c}$$

where (32a) and (32b) relate parameters defining complex-pole locations in the planes of the complex variables z and $\bar{z}$, while σ_z and $\bar{\sigma}_z$ are real roots of equations (17) and (25), respectively.

Since $z = e^{sT}$, the relations between the undamped natural frequency ω_n and the relative damping coefficient ζ from the continuous-time response and the parameters $(\omega_n)_z$ and ζ_z determining corresponding complex-pole locations in the z-plane are readily obtained as

$$(\omega_n)_z = e^{-\omega_n\zeta T} \quad \text{and} \quad \zeta_z = \cos(\omega_n T\sqrt{1-\zeta^2}) \tag{33}$$

where, in the principal strip or in the Nyquist frequency band,

$$0 \leqq \omega_n \leqq \frac{\pi}{T\sqrt{1-\zeta^2}} \tag{34}$$

Using equations (31), two families of curves corresponding to different values of $\bar{\zeta}_z = \text{const}$ and $(\bar{\omega}_n)_z = \text{const}$ are plotted in Fig. 5. Fig. 6 shows the enlarged region of Fig. 5 around the origin. Notice that the curves in the (α,β)-plane in Figs. 5 and 6 represent the loci of complex poles of equation (25) as the parameters α and β vary. It is also to be noted here that the charts in Figs. 5 and 6 are general and hence these charts can be advantageously used in factoring third-order polynomials, thus enabling the pole placement for any third-order digital feedback control system.

Note that the outlined procedure of deriving (31) is only a special case in the more general method of analysis and synthesis of sampled-data feedback control systems in the parameter plane. The method has been primarily presented in the papers [19], [20], and [21].

In the proposed procedure of synthesizing controller parameters it is necessary first to specify the relative damping coefficient ζ and undamped natural frequency ω_n to match desired system characteristics with respect to the stability margin and the speed of continuous-time response.

In virtue of (18) or (23), the sign of the coefficient A_2 is to be adopted and then $\bar{\zeta}_z$ is determined by $\bar{\zeta}_z = (\mathrm{sgn}A_2)\zeta_z$. Once $\bar{\zeta}_z$ is determined, along the curve $\bar{\zeta}_z$ = const of Figs. 5 or 6 the point $M(\alpha_m,\beta_m)$ corresponding to the desired $(\bar{\omega}_n)_z$ may be placed. According to the Viète rule, the third real root of (25) is calculated from

$$\bar{\sigma}_z = -1 - 2\bar{\zeta}_z(\bar{\omega}_n)_z \tag{35}$$

The stability condition requires that all roots of (17) are to be inside the unit circle; hence, the absolute value of A_2 must satisfy the constraints $(\omega_n)_z < 1$ and $\sigma_z < 1$, or

$$|A_2| < \frac{1}{|\bar{\sigma}_z|} \qquad \text{and} \qquad |A_2| < \frac{1}{(\bar{\omega}_n)_z} \tag{36}$$

The other two coefficients A_1 and A_0 in (17) are calculated using (26), so that

$$A_1 = \beta_m A_2^2 \qquad \text{and} \qquad A_0 = \alpha_m A_2^3 \tag{37}$$

Putting numerical values of A_2, A_1, and A_0, as well as the values of system constants K, T_m, and T into equations (23), we obtain three equations which can be solved for adjustable or controller parameters D, H, and K_{VF}.

Note that the proposed procedure may be also applied for the synthesis of controller parameters K_p, K_i, and K_d of the motor speed control system in Fig. 3. In order to employ the charts in Figs. 5 or 6, it is first necessary to reduce the system order by putting $K_d = 0$. Then the system characteristic equation is given by (17) with the coefficients (18). In such a way, and using the outlined procedure of pole--placement it is possible to fit the appropriate values of proportional and integral gains. To improve the speed of system continuous-time response, the derivative gain K_d can be adjusted later on.

As it has been mentioned, the grapho-analytical method presented in this section may be applied in the parameter synthesis of any third--order control system. Thus the method has been advantageously employed in the design of microprocessor-based servomechanisms [7] and digital systems for DC motor speed control [8], [22].

4. APPLICATION

Now, a real digital servo control system having the structure shown in Fig. 4 is considered. The printed armature DC servo motor G16M4 characterized by a very small mechanical time constant T_m = 0.02 s was applied. The other relevant motor characteristics are: nominal voltage 60 V, power output W_0 = 244 W, n = 2350 rpm, EMF constant K_E = 23.0 mV/rpm. Hence, the gain of the motor transfer function is $K_m = 1/K_E$ = 4.55 rad/V.s.

In the system, the 8-bit DAC and feedback counter were used. The employed driving circuit with adjustable gain has the output saturation on ±25 V. For the detection of motor speed and angular position, the plastic disc giving $K_n = 800$ pulses per revolution is used.

The I/O ports arrangements, realization of the digital control law (19), and interactive software are suited by the INTEL 8080 microprocessor.

In the servo system under consideration, the motor corner frequency $f_c = 1/T_m = 50$ Hz, so that an appropriate sample period $T = 10$ ms may be adopted.

Setting the numerical values $T = 0.01$, $K_m = 4.55$, $K_n^* = K_n/2\pi = 127$, $K_h K_A = 1$ (or $K = K_h K_A K_m K_n^* = 577.85$) into (23), we arrive at

$$A_2 = 1.2308(D + K_{VF}) - 1.6065 \tag{38a}$$

$$A_1 = 1.0430(D + K_{VF}) - 1.2308(K_{VF} - H) + 0.6065 \tag{38b}$$

$$A_0 = 1.0430(H - K_{VF}) \tag{38c}$$

After specifying the stability margin and the speed of system continuous-time response to correspond to the pair of closed-loop complex poles inside the principal strip of the s-plane with $\zeta = 0.7$ and $\omega_n = 85$ rad/s, we calculate $\zeta_z = \cos(\omega_n T\sqrt{1 - \zeta^2}) = 0.82$. Then, in virtue of equation (38a), the negative sign for A_2 is adopted and thus $\bar{\zeta}_z = (\mathrm{sgn} A_2)\zeta_z = - 0.82$. Finally, interpolating between curves on the

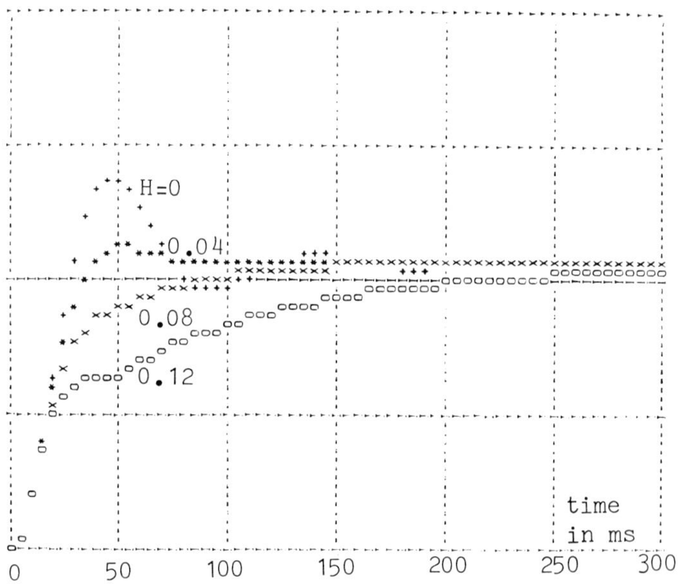

Fig. 7. Simulation results corresponding to changes of the controller parameter H

chart in Fig. 6, the point M(0.04,0.4) corresponding to $\bar{\zeta}_z = -0.82$ and $(\bar{\omega}_n)_z = 0.53$ is placed. The third real root of equation (25) is calculated from $\bar{\sigma}_z = -1 - 2\bar{\zeta}_z(\bar{\omega}_n)_z = -0.131$. Bearing in mind the stability condition (36), an appropriate magnitude $|A_2| = 1.376$ (or $A_2 = -1.376$) is adopted, and the coefficients $A_1 = 0.749$ and $A_0 = -0.108$ are calculated from (37). The complex-root pair and the real root of equation (17) are determined by $\zeta_z = (\mathrm{sgn}A_2)\bar{\zeta}_z = 0.82$, $(\omega_n)_z = |A_2|(\bar{\omega}_n)_z = 0.73$, and $\sigma_z = A_2\bar{\sigma}_z = 0.18$. Setting the numerical values of A_2, A_1, and A_0 into (38) and solving equations (38) for controller parameters, we obtain $D = 0.16$, $H = -0.04$, and $K_{VF} = 0.1$.

In order to verify the results of the foregoing grapho-analytical procedure and to investigate the influence of the power amplifier saturation, the servo control system has been simulated by using the block-oriented simulation language CSMP. Results of the simulation run, showing efects of variations of the controller parameter H, is presented in Fig. 7. It is seen that the system performance is highly sensitive with respect to changes of the controller parameters. On the other hand, the presence of the driver output saturation has a small influence on the servo system continuous-time response; it only increases the system settling time.

5. PROGRAMMING

In this section an assembly program for the microprocessor INTEL 8080 is given in details. In connection with the program, some additional comments are necessary.

In digital control systems the microprocessor computational time as well as the analog-to-digital conversion time reflects as a transport lag within the system closed loop. Recall that the presence of transport lag in a continuous-time control system generally complicates the system analysis, increases the phase lag through the system, and hence tends to decrease the stability margin. On the other hand, the presence of transport lag in a sampled-data system does not create any new problem [21]; nevertheless in the presence of transport lag some interesting phenomena generally occur. It may be shown that in sample-data control systems a considerably small transport lag has practically the same influence upon the system stability as a phase-lag network. However, whenever the transport lag exceeds a certain amount of the sample interval, it produces the same effect as in a continuous-data feedback control system, i.e., it decreases the stability margin. Therefore, in most cases the computational time of the microprocessor is to be made small as much as possible. It is to be pointed out that the computational time of a microprocessor, which has no hardware facilities for multiplication and division, depends, to a large extent, upon a number of necessary multiplications and consequently it is desirable to develope the multiplication subroutine, which requires a small computational time. In the system under consideration, such subroutines for both the single and double precision are developed, and the time saving is achieved by multiplying the multiplier with 256 and taking the result from the higher-significant-bit register of the accumulator. In 8-bit microprocessors, it is equivalent to the division by 256. In such a way, we save the time necessary for successive roundings that are to be performed within the accumulator registers, when the multiplication is accomplished by the standard multiplication algorithm [23]. It is the reason why we introduce in the program the new effective parameters DPAR = D*256 = 41, HPAR = H*256 = 10, and KVFPAR = KVF*256 = 26.

The source program and the associate subroutines with necessary explanations are given below.

```
S O U R C E   S T A T E M E N T

        OPTION 8080

**********************************************************************
*       PROGRAM EQUATES
**********************************************************************
ROM     EQU     #400          *FIRST ADDRESS OF ROM
RAM     EQU     #1000         *FIRST ADDRESS OF RAM
STACK   EQU     #1400         *STACK ADDRESS
DPAR    EQU     41            *DPAR=D*256, D=0.16, DPAR=41
HPAR    EQU     10            *HPAR=H*256, H=0.04, HPAR=10
KVFPAR  EQU     26            *KVFPAR=KVF*256, KVF=0.1, KVFPAR=26
**********************************************************************
*       MICROPROCESSOR READ ADDRESSES
**********************************************************************
FEEDB   EQU     #2000         *FEEDBACK COUNTER
CLOCK   EQU     #2800         *SAMPLING CLOCK
SYNCH   EQU     #3000         *DISPLAY SYNCHRONISATION
**********************************************************************
*       MICROPROCESSOR WRITE ADDRESSES
**********************************************************************
DAC     EQU     #2000         *OUTPUT DAC
COUNT   EQU     #2800         *FEEDBACK COUNTER ENABLE/CLEAR CONTROL
TEST    EQU     #3000         *DISPLAY DAC
**********************************************************************
*       P O S I T I O N I N G   S E R V O   P R O G R A M
**********************************************************************
        ORG     ROM
*
START   LXI     SP,STACK
        MVI     A,DPAR        *CALCULATE KDPAR
        LHLD    KPAR
        CALL    MULTD
        MOV     A,L
        STA     KDPAR
        MVI     A,HPAR        *CALCULATE KHPAR
        LHLD    KPAR
        CALL    MULTD
        MOV     A,L
        STA     KHPAR
```

```
        MVI     A,KVFPAR       *CALCULATE KKVFPAR
        LHLD    KPAR
        CALL    MULTD
        MOV     A,L
        STA     KKVFPAR
        MVI     A,#80
        STA     DAC
        MVI     A,2
        STA     COUNT          *ENABLE FEEDBACK COUNTER
        LXI     H,0
        SHLD    INPUT          *(INPUT)=0
        SHLD    POSN           *(POSN)=0
        SHLD    HERRORP        *(HERRORP)=0
        CALL    RDFDBCK        *UPDATE CNTRP
LOOP    CALL    SMPCLCK        *SYNCHRONISATION WITH THE SAMPLING CLOCK
        XCHG
        LHLD    POSN
        DAD     D
        SHLD    POSN           *(POSN)=(POSN)+(ANINCR)
        XCHG
        LHLD    INPUT
        XCHG
        CALL    SUBD           *ERROR=(INPUT)-(POSN)
        SHLD    ERRORP
        LDA     KDPAR
        CALL    MULTD          *DERROR=ERROR*KDPAR/256
        XCHG
        LHLD    HERRORP
        CALL    SUBD           *OUTCAS=DERROR-(HERRORP)
        SHLD    OUTCAS
        LHLD    ANINCR
        LDA     KKVFPAR
        CALL    MULTD          *(H,L)=(ANINCR)*KKVFPAR/256
        XCHG
        LHLD    OUTCAS
        XCHG
        CALL    SUBD           *COMMAND=(OUTCAS)-(ANINCR)*KKVFPAR
        DAD     H
        DAD     H              *COMMAND=(COMMAND)*4
        CALL    LIMIT
        STA     DAC
        STA     COMMAND
        LHLD    ERRORP
        LDA     KHPAR
        CALL    MULTD
        SHLD    HERRORP        *(HERRORP)=(ERRORP)*KHPAR/256
```

```
**********************************************************************
*       S U B R O U T I N E S
**********************************************************************
*
**********************************************************************
*       MULTIPLICATION  -  DOUBLE PRECISION
*
*       (H,L)=(H,L)*(A)/256
*       (H,L) : SIGNED NUMBER IN 2' COMPLEMENT FORM
*         (A) : POSITIVE INTEGER ONLY
*       DIVISION INCLUDES ROUNDING
**********************************************************************
MULTD   MOV     E,A           *SAVE MULTIPLIER
        MOV     A,H
        ORA     A
        PUSH    PSW           *SAVE MULTIPLICAND SIGN
        CM      TWOCMP        *IF MINUS, NEGATE
        MOV     C,H
        PUSH    H             *SAVE MULTIPLICAND
        MOV     A,E
        CALL    MULTS         *MULTIPLY (H)*(A)
        XTHL                  *SAVE (H)*(A), LOAD MULTIPLICAND
        MOV     C,L
        MOV     A,E
        CALL    MULTS         *MULTIPLY (L)*(A)
        MOV     A,L
        ORA     A             *ROUNDING
        JP      MD1
        INR     H
MD1     MOV     A,H
        POP     H             *RECOVER MULTIPLICAND
        ADC     L
        MOV     L,A
        MVI     A,0
        ADC     H
        MOV     H,A
        POP     B             *RECOVER MULTIPLICAND SIGN
        MOV     A,B
        ORA     A
        CM      TWOCMP        *IF MINUS, NEGATE
        RET
**********************************************************************
*       MULTIPLICATION  -  SINGLE PRECISION
*
*       (H,L) = (C)*(A)
*       (C),(A) : POSITIVE INTEGERS ONLY
**********************************************************************
MULTS   MVI     B,0
        LXI     H,0
        MVI     D,8
```

```
MS1     DAD     H
        RLC
        JNC     MS2
        DAD     B
MS2     DCR     D
        JNZ     MS1
        RET
**********************************************************************
*       NEGATION  -  DOUBLE PRECISION
*
*       (H,L) = 2'COMPLEMENT OF (H,L)
**********************************************************************
TWCOMP  MOV     A,L
        CMA
        INR     A
        MOV     L,A
        MOV     A,H
        CMA
        JNZ     TC1
        INR     A
TC1     MOV     H,A
        RET
**********************************************************************
*       LIMIT
*
*       (A) = + 127 IF (H,L) > + 127
*       (A) = (H,L) IF + 127 ≥ (H,L) ≥ - 128
*       (A) = - 128 IF (H,L) < - 128
*       (A) IS IN THE OFFSET BINARY CODE
**********************************************************************
LIMIT   MOV     A,H
        ORA     A
        JM      LM1
        JNZ     LM2             *JUMP IF (H,L) > + 255
        MOV     A,L
        ORA     A
        JP      LM3             *JUMP IF (H,L) ≦ + 127
LM2     MVI     A,#FF           *(A) = + 127 (OFFSET BINARY CODE)
        RET
LM1     CPI     #FF
        JNZ     LM4             *JUMP IF (H,L) ≦ - 256
        MOV     A,L
        ORA     A
        JM      LM3             *JUMP IF (H,L) ≧ - 128
LM4     XRA     A               *(A) = - 128 (OFFSET BINARY CODE)
        RET
LM3     RAL                     *CONVERSION TO THE OFFSET BINARY CODE
        CMC
        RAR
        RET
```

```
***********************************************************************
*       SUBSTRACTION  -  DOUBLE PRECISION
*
*       (H,L) = (D,E) - (H,L)
*       (H,L) = (D,E) - (H,L)
***********************************************************************
SUBD    MOV     A,E
        SUB     L
        MOV     L,A
        MOV     A,D
        SBB     H
        MOV     H,A
        RET
***********************************************************************
*       READ  FEEDBACK
*
*       READ FEEDBACK COUNTER TWICE, ACCEPT IF THE READINGS ARE
*       EQUAL, OTHERWISE TRY AGAIN
*       (H,L) = ANGULAR INCREMENT IN RESPECT TO THE PREVIOUS READING
*       UPDATE CNTRP
***********************************************************************
RDFDBCK LXI     H,FEEDB
        MOV     A,M
        MOV     B,M
        CMP     B
        JNZ     RDFDBCK
        LXI     H,CNTRP
        MOV     B,M
        MOV     M,A
        SUB     B
        MOV     L,A
        MVI     H,0
        ORA     A
        RP
        MVI     H,#FF
        RET
***********************************************************************
*       SYNCHRONIZATION WITH THE SAMPLING CLOCK
*
*       EXIT FROM THE LOOP ON 0 TO 1 TRANSITION OF THE CLOCK
*       (H,L), (ANINCR) = ANGULAR INCREMENT DURING THE SAMPLING INTERVAL
***********************************************************************
SMPCLCK LDA     CLOCK
        ORA     A
        JM      SMPCLCK
        CALL    RDFDBCK
        SHLD    ANINCR
SC1     LDA     CLOCK
        ORA     A
        JP      SC1
        CALL    RDFDBCK
```

```
        XCHG
        LHLD    ANINCR
        DAD     D
        SHLD    ANINCR
        RET
***********************************************************************
*       DATA STATEMENTS
***********************************************************************
*       ORG     RAM
INPUT   DS      2           *ANGULAR POSITION REQUEST
POSN    DS      2           *CURRENT POSITION IN RESPECT TO THE
*                            INITIAL POSITION
HERRORP DS      2           *(HERRORP) = (ERRORP)*KHPAR/256
CNTRP   DS      1           *FEEDBACK COUNTER IN THE PREVIOUS INTERVAL
ANINCR  DS      2           *ANGULAR INCREMENT
ERRORP  DS      2           *ERROR IN THE PREVIOUS INTERVAL
OUTCAS  DS      2           *OUTPUT OF THE CASCADE COMPENSATOR
COMMAND DS      1           *INPUT TO DAC
KDPAR   DS      1           *KDPAR = DPAR*KPAR/256
KHPAR   DS      1           *HPAR*KPAR/256
KKVFPAR DS      1           *KKVFPAR = KVFPAR*KPAR/256
KPAR    DW      366         *KPR = 0.7*KPAR/256; 1≤ KPAR ≤ 768 :
*                            KPR = 0 - 2.15; KPR = 1 : KPAR = 366
***********************************************************************
        END
```

To visualize the motor control and to enable easy communication with the digital servo control system, several interactive programs have been developed and associated. For the sake of brevity, these programs are omitted here.

The hardware configuration enclosing the source and destination of the information and the control information for identification and timing have been designed usign standard digital components and devices. Note that the servomechanism in testing and in normal operation reveals the dynamic performance, which is in agreement with the results of system simulation.

6. CONCLUSION

Simple structures of a microprocessor-based motor speed control system and a positioning servo system using DC motor have been suggested. In both systems using DC motors a common counting-type incremental encoder is employed for the measuring of the motor speed and angular position.

Two suitable digital control laws requiring relatively small microprocessor computational time are suggested: PID-DDC algorithm for the motor speed control and combined (cascade and parallel) compensation for the positioning servo. Proposed control laws eliminate the steady--state error and enable easy adjustments of controller parameters with respect to the desired accuracy, the speed of continuous-time response, and the stability margin.

The grapho-analytical procedure of pole-placement and the charts presented in this chapter are general. Namely, the charts can be advantageously used in factoring third-order polynomials, thus enabling the adjustment of controller parameters by the pole-placement technique in any third-order digital feedback control system.

The presented assembly program may be also of a general purpose since portions of the program may be used in any DDC stand-alone single--loop controller.

The presented theoretical results have been employed in the design of several digital motor speed control systems and positioning servos and they all run very well.

ACKNOWLEDGEMENT

The author wishes to acknowledges the support of his colleague and friend Dr. Predrag Jovanović of the Physics Department, University of Birmingham. The author also wishes to thank the University of Birmingham for providing the computing resources.

REFERENCES

[1] A.C. Lin and W.W. Koepsel, 'A microprocessor speed control system,' *IEEE Trans. Ind. Electron. Contr. Instum.*, vol. IECI-**24**, no. 3, August 1977.

[2] D.H. Smithgall, 'A phase-locked loop motor control system,' *IEEE Trans. Ind. Electron. Contr. Instrum.*, vol. IECI-**22**, no. 4, November 1975.

[3] H.C. Wang and J.N. Chen, 'A microprocessor-based digital system for DC motor speed control,' in *Proc. 1979 Symp. on Automatic Control*, Taiwan, December 1979.

[4] D.M. Auslander, Y. Takahashi, and Tomizuka, 'Direct digital process control: Practice and algorithms for microprocessor application,' *Proc. IEEE*, vol. **66**, no. 2, February 1978.

[5] H.C. Wang, 'Sampling period and stability analysis for the micro-computer-based motor control systems,' *IEEE Trans. Ind. Electron. Contr. Instrum.*, vol. IECI-**28**, no. 2, May 1981.

[6] B.C. Kuo, *Digital Control Systems*, New York: SRL Publ., 1977.

[7] M.R. Stojić and P. Jovanović, 'Direct digital servo control system: Design, practice, and algorithms for microprocessor application,' *Automatika*, Yugoslav Journal of ETAN, no. 5-6, December 1981.

[8] M.R. Stojić, 'Design of the microprocessor-based digital system for DC motor speed control,' *IEEE Trans. on Industr. Electron.*, vol. IE-**31**, no. 3, August 1984.

[9] J.T. Beckett, 'Analysis of incremental digital positioning system with digital rate feedback,' *ASME* paper 64-WA/AUT-3.

[10] R.R. DeBolt and R.E. Powell 'A natural 3-mode controller algorithm for DDC,' *ISA-J*, pp. 43-53, September 1966.

[11] K.W. Goff, 'Dynamics in direct digital control, I and II,' *ISA-J*, pp. 44-54, November 1966.

[12] E.H. Bristol, 'Designing and programming control algorithms for DDC systems,' *Control Engineering*, vol. **24**, no. 1, pp. 24-26

[13] E.B.Dahlin, 'Designing and tuning digital controllers, Part I and II,' *Instrum. Contr. Syst.*, vol. 41, pp. 77-83, 87-91, June and July 1968.

[14] Y. Takahashi, C.S. Chan, and D.M. Auslander, 'Parametereinstelung bei linearen DDC-Algorithmen,' *Regelungstechnik und Process-Datenverarbeitung*, vol. **19**, no. 6, pp. 237-284, 1971.

[15] K.C. Chiu and C.L. Smith, 'Digital control algorithms, Part 1, Dahlin algorithm, Part 2, Kalman algorithms, Part 3, Tuning PI and PID controllers,' *Instrum. Contr.*, pp. 57-59, 55-58, 41-43, Oct.,Nov., and Dec. 1973.

[16] P.E. Wellstead, J.M. Edmunds, D.L. Prager, and P. Zanker, 'Self tuning pole-zero assignment regulators,' *Int. J. Control*, vol. **30**, pp. 1-26, 1979.

[17] J.T. Tou, *Digital and Sampled-data Control Systems*, New York: McGraw-Hill, 1959.

[18] Tables of Chebyshev's Polynomials $S_n(x)$ and $C_n(x)$, National Bureau of Standards, Applied Mathematics Series, no. 9, U.S. Gavernment Printing Office, Washington, DC, 1952.

[19] D. Mitrović, 'Graphical analysis and synthesis of feedback control systems, Pt. III - Sampled-data feedback control systems,' *AIEE Trans. (Application and Industry)*, no 77, January 1959.

[20] D.D. Siljak, 'Analysis and synthesis of feedback control systems in the parameter plane, Pt. II - Sampled-data systems,' *IEEE Trans. (Application and Industry)*, no. 83, November 1964.

[21] M.R. Stojić, 'Sampled-data control systems with transport lag by the parameter plane method,' *Presses Academique Europeenes, A - - Revue*, Bruxelles 6, pp. 105-112, July 1966.

[22] M.R. Stojić, 'Direct digital speed control system: Design and microprocessor realization,' *Publ. of Electrical Engineering Faculty, Series: ETA*, University of Belgrade, no. 142-149, pp. 57-69, 1982.

[23] J.W. Waler, V.A. Shatzel, and H.Y. Nice, *Practical Microcomputer Programming: The INTEL 8080, A User's Guide to Application Programming for the 8080*, Evanston: Norther Technology Books, ILL 60204.

CHAPTER 8

MICROPROCESSOR-BASED AC MOTOR CONTROL

Min H. Park
Department of Electrical Engineering
Seoul National University
Shinrim-Dong, Kwanak-Ku
Seoul 151, Korea

ABSTRACT. The developments in Power Electronics and LSI technology have made the microprocessor-based adjustable speed ac motor drives become promising. This chapter is concerned with the principle, hardware and software implementations, and some processing techniques in inverter-driven ac motor systems which are controlled by microprocessors. Discussed are the slip frequency control, PWM inverter control, field oriented control and finally, energy optimization control strategy.

1. INTRODUCTION

The advent of the silicon-controlled rectifier, or thyristor, that can handle large power has made the revolutionary developments in the area of control, equipments, and applications of electric machines. In particular, ac motors which have been regarded as constant speed motors now become competitive with dc motors in variable speed drive systems. Since ac motors have several advantages over dc motors in the respects of cost, robustness, weight and maintenance, it seems that ac motors will replace dc motors in many industrial applications.

On the other hand, recent achievements in LSI technology led to practical utilization of the microprocessor. The microprocessor makes it possible to realize very complicated control strategy that it is difficult or even impossible to realize with conventional analog circuitry. When the microprocessor is used in variable speed motor drives, it gives high accuracy surmounting the nonlinearities involved in motor characteristics, power conversion equipments and sensing devices, and providing against the drifts of temperature and source voltage. Moreover, the protection capability of the microprocessor, those are, on and off-line testing, data recording at faults and parallel operation of multiple processors, etc., result in the more reliable control/supervisoring systems.

This chapter deals with the microprocessor-based ac motor drive systems. There are too many materials related to this topic to include all of them in this short chapter. Thus only a few representative control schemes are discussed, laying main emphasis upon the control of

N. K. Sinha (ed.), Microprocessor-Based Control Systems, 157–181.

inverter-fed induction motor drives.

2. ADJUSTABLE SPEED AC MOTOR DRIVES

The adjustable speed drive systems of ac motors, fed from inverters or cycloconverters, have been commercially utilized in industrial applications. Recent developments in the field of power electronics and micro-electronics make it possible to replace dc motors by ac ones in variable speed motor applications. In this section, a brief review of some concepts related to adjustable speed ac motor drives is given.

2.1. Power Semiconductor Circuits

2.1.1. Inverter circuits. The core of variable speed ac motor drive is the frequency control of power source. Among the many candidates of frequency changers, inverters are most widely used to drive ac motors. Forced-commutated inverters are largely categorized in two types of power circuit configurations, those are, the voltage source inverter (VSI) and the current source inverter(CSI). Schematic diagrams of the inverters are shown in Figure 1. The output voltage(current)of VSI (CSI) will have six-step waveform. The VSI circuit in Figure 1(a) with some modifications, replacing controlled rectifier by diode bridge, may

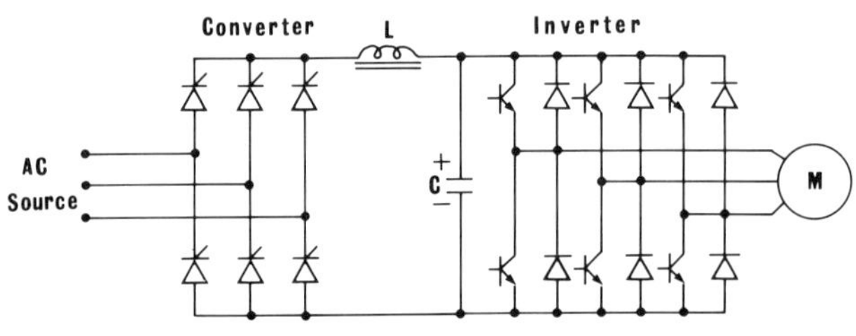

(a)

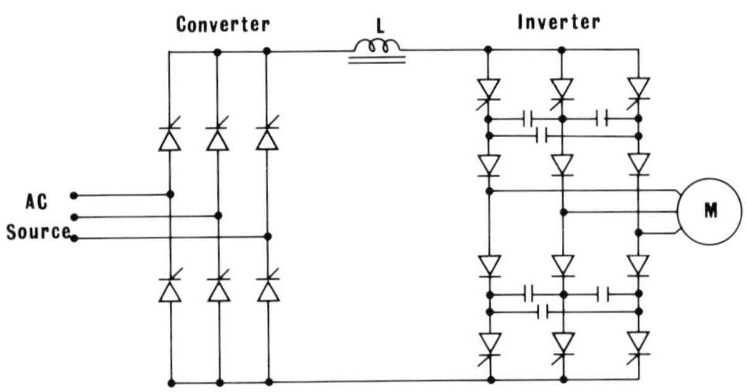

(b)

Figure 1. Schematic diagrams of (a) voltage source inverter and (b) current source inverter.

also serve as a Pulse Width Modulated(PWM) inverter.

VSI offers stable operation over a wide speed range and is capable of driving multiple motors simultaneously. But in case of regenerative braking, it needs one more converter to permit the current to flow in reverse direction. A large filtering capacitance in the dc link normally causes the time delay and thus the VSI may be often unsuitable to the applications where high speed response of the system is required.

In a PWM inverter, the output voltage is composed of multiple pulses and the voltage can be controlled by adjusting the width of the pulses. Since both the frequency and voltage are simultaneously controlled in the inverting part of the system, the time delay caused in filtering circuits has negligible effects on the system response. This improves the relative stability in a closed control loop. The harmonic effects on the motor performance can be minimized by optimal modulation techniques while the switching losses increase with the number of pulses.

The CSI is much simpler in power circuit than the VSI. It offers the less derating factor of circuit components and the ease of torque control. Moreover, the CSI system has the capability of four quadrant operation without any further modifications in power circuit[1].

2.1.2. Power semiconductors. The inverter systems in Figure 1 are constructed of switching devices some of which are described in the following. What device is to be chosen is dependent on the voltage and current ratings, operating frequency, and many other circumstantial conditions of a particuar application.

The thyristor is a good switching device which can be turned on by small amounts of gating energy. But it needs auxilliary commutation circuit to be turned-off because the gate of the thyristor loses control once the thyristor begins conducting. The devices that can be commutated by themselves such as power transistors, GTOs, and power MOS-FETs save considerable amounts of power circuit complexity and overall system costs. Although these devices currently available have relatively lower power gains and smaller ratings than thyristors, it is expected their performance will exceed those of thyristors in the near future[2].

2.2. Speed Control of Induction Motors

2.2.1. Constant flux operation. If the air-gap flux of an induction motor is maintained constant, the torque of the motor will be approximately proportional to the slip frequency. While the synchronous motor rotates exactly the same speed dictated by the inverter frequency, the speed of an induction motor decreases by the amounts of slip to meet the load torque. Normally the speed is controlled in open loop as depicted in Figure 2 by solid lines.

To maintain the air-gap flux constant throughout the speed range, the output voltage of the inverter should be varied linearly with frequency. Since the air-gap flux depends on the air-gap voltage, rather than the terminal volgage, the terminal voltage is slightly increased in low frequency region to compensate the stator voltage drop. This is achieved by incorporating the nonlinear relationship between the inverter frequency f and the reference voltage V_{ref} that controls the converter output voltage, as shown in Figure 2. When the more precise voltage regulation is required, the feedback loop represented in dashed lines may also be included[3].

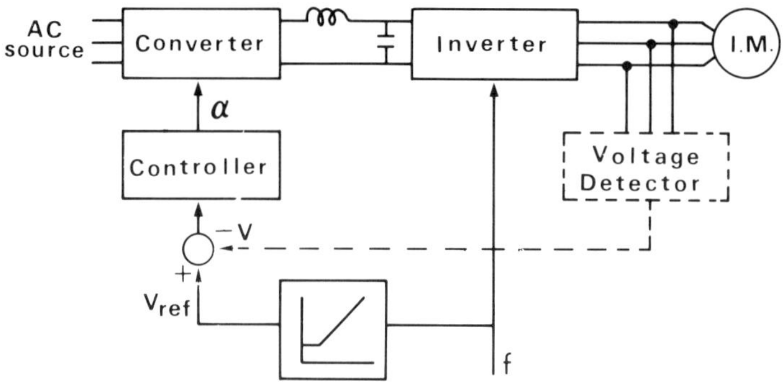

Figure 2. Block diagram of constant flux drive with VSI.

2.2.2. Slip frequency operation. An advanced method for an inverter-fed induction motor control, named slip frequency control, is shown in Figure 3. The inverter frequency is determined by the sum of the motor speed, f_M, and slip frequency, f_{sl}. The gating angle of the converter is controlled by the current demand that reflects the load torque. If the reference and/or the actual motor speed deviate from a steady state operating point, the error between them activates the controller to control the slip and current, and the motor speed is adjusted to its reference value[4].

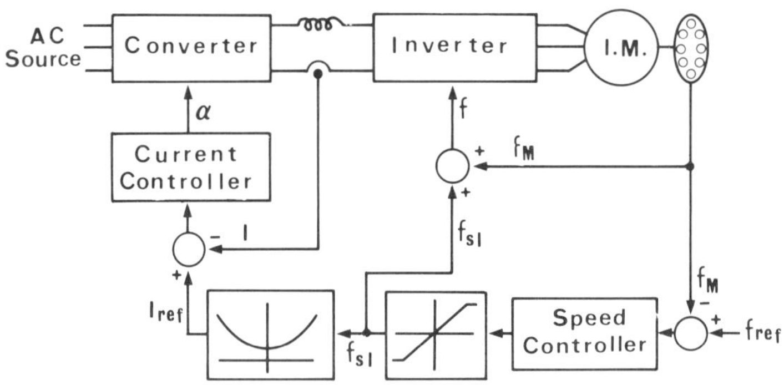

Figure 3. Block diagram of slip frequency drive with CSI.

2.2.3. Field oriented control. An elegant control method that has received wide attentions in recent days is the field oriented control. Generally, the stator current of an induction motor, i_s, is divided into two components - the field component and the torque component. If they are decoupled, the field and the torque of the motor can be controlled independently by adjusting the associated components of the current. Figure 4 shows the vector diagram of an induction motor with the field-oriented coordinate (T-M axis) and the stator reference coordinate(α-β axis). From the figure, the relationships between the components of the reference axes can be derived and are used to establish the control algorithm.

In the steady state, two components i_M and i_T are perpendicular to each other and are equivalent to the field and armature current in a dc motor respectively. Thus, in the field oriented control, an ac motor is controlled in a similar manner of dc motor control. Though the decoupling procedure requires complicated control algorithm the field oriented control has shown itself the powerful method to achieve excellent performance[5]. A typical control block diagram will be given in Section 4(see Figure 18).

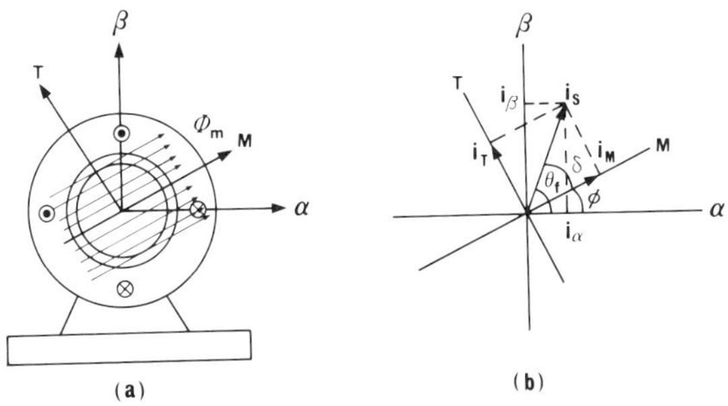

(a) Physical interpretation. (b) Vector diagram.
i_M : Field current component.
i_T : Torque current component.
i_α, i_β : α-β axis components of stator current.

Figure 4. Current relationship of two phase induction motor.

2.3. Synchronous Motor Drive

Microprocessor-based inverter control systems discussed so far are constrained to the drive of induction motors. There is no reason why they cannot be used to drive synchronous motors. But there may exist the control scheme that is peculiar to the synchronous motor drive as the slip frequency control is to the induction motor drive.

It is well known that the inverter can be line-commutated if the leading reactive power for commutation is supplied from ac lines to the inverter. This is logical deduction from the inverting operation of the phase controlled rectifier. The inverter of this type is named "line-commutated inverter", and is extensibly used in HVDC transmission/distribution systems and static Scherbius drive systems. On the other hand, a synchronous motor can be driven in unity or leading power factor by adjusting the supply voltage and field current. Thus it is quite natural to drive the synchronous motor by the line-commutated inverter maintaining the leading power factor of the motor. This is referred to as "brushless motor". In this case, however, the terminology "Load-Commutated Inverter (LCI)" is preferred since the reactive power for commutation is supplied by the load, not by the ac lines.

In the LCI drives, the output voltage of the inverter leads the output current by the angle β, the sum of available turn-off angle for the thyristor, μ, and the commutation overlap angle, δ. Figure 5 shows the block diagram of the LCI-driven synchronous motor system where the turn-off angle is maintained constant to ensure stable commutation throughout the whole speed range. This is achieved by measuring the shaft positon of the motor with which the gating signals to the inverter are synchronized. The overlap angle is measured from the line voltage and the leading angle is determined as the sum of the overlap angle and predetermined turn-off angle. If the air-gap flux is constant, the developed torque of the motor is approximately proportional to the input current. Thus the torque of the motor is controlled by regulating the dc link current. In order to maintain leading power factor regardless of the load, the field current I_f is made to be increased with the armature current I to compensate the armature reaction, by adding $K_f I$ to I_{fo} (the no-load field current)[6].

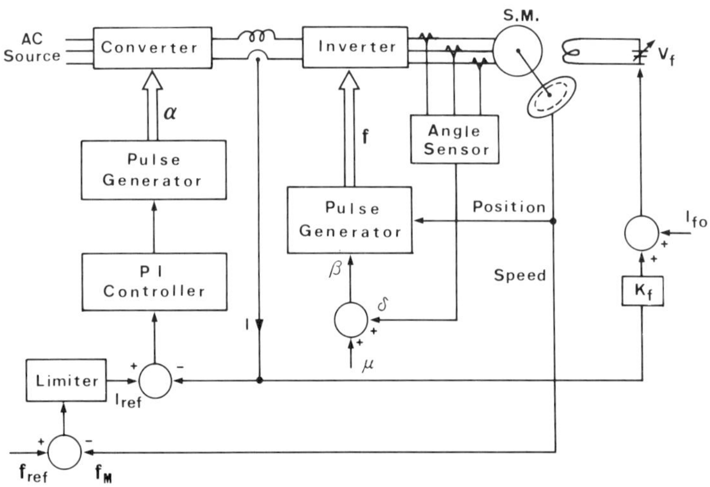

Figure 5. Speed control system of a synchronous motor fed from load commutated inverter.

3. SIGNAL PROCESSING USING A MICROPROCESSOR AND PERIPHERAL ICs

Most signal transducers provide the measured values of light, pressure, or temperature, for example, in the form of voltage or current level-the analog signals. But since a microprocessor works only with digital data, such analog signals must be converted into a digital form before they are used in a microprocessor. The reversed process, the conversion of digital signals into analog ones, is also needed to use the results of the microprocessor in real world. Therefore, in a microprocessor-based control systems, the conversion devices for analog-to-digital(A/D) or digital-to-analog(D/A) conversion are indispensible[7]. This section discusses some conversion techniques valuable in adjustable speed ac motor drive systems.

3.1. Position and Speed Measurements

It is required to know the shaft position in some synchronous motor drive systems, and the rotational speed of the shaft in most adjustable speed ac motor drive systems. The shaft position(speed) of a motor can be measured in digital form by combining a potentiometer(tachogenerator) with a A/D conversion unit. A pulse generator, which is coupled to the shaft of a motor and generates pulse train as the motor rotates, is also in use to measure the position and speed. It obviates the need of A/D conversion since the speed or position is converted directly into a digital form. Several methods to calculate the speed from the pulse train are discussed in the following. To obtain a position, only some modifications of software will be needed.

3.1.1 Counting pulses during sampling time. The simplest way to measure the speed is to count the pulses during a specified time interval, the sampling time. This method is simple, robust to noise and, in addition, ensures the uniform sampling time. But it requires relatively long sampling time or the large number of pulses per revolution, i.e., high output frequency of the pulse generator, to obtain a good speed resolution.

3.1.2. Measuring time interval between pulses. The interval between two adjacent pulses is inversely proportional to the motor speed. Thus the speed can be evaluated by measuring the interval, counting a high-frequency clock pulses from one encoder pulse to the next one. This method is relatively simple and offers high accuracy. However, when the speed is low, the counter is required to have a large number of bits. Since the interval between pulses depends on the speed of a motor it is impossible to guarantee an uniform sampling time. In addition, a small noise between pulses may cause the failure in measuring the speed.

3.1.3. Combination of two methods. The difficulties involved in above two methods can be overcome in cost of some complexities. Assume that the pulse generator generates P-pulses per revolution and let the frequency of the reference clock pulse be f_c, which is sufficiently high compared to that of pulse generator output. The time chart of these two pulse trains is shown in Figure 6. Two counters are employed and over the sampling interval T_s, they respectively count the clock pulses and the pulse generator output. In general, the sampling instant is not coincident to the arrival of an encoder pulse and the counters continue to count until the next pulse arrives. Thus the "effective" sampling time becomes T_d, as indicated in Figure 6.

If the encoder and clock pulses are counted as M_1 and M_2, repectively, the rotational speed of the motor, N, is given by

$$N = \frac{60 f_c M_1}{P\ M_2} \quad [\text{rpm}] \tag{1}$$

For N and P of the order of one hundred, ΔT, the fluctuation of sampling time is about a few millisecond. As P increases, ΔT is reduced further and becomes negligible compared to sampling time. This method requires rather complicated hardware and software but offers accurate speed measurement over a wide speed range[8].

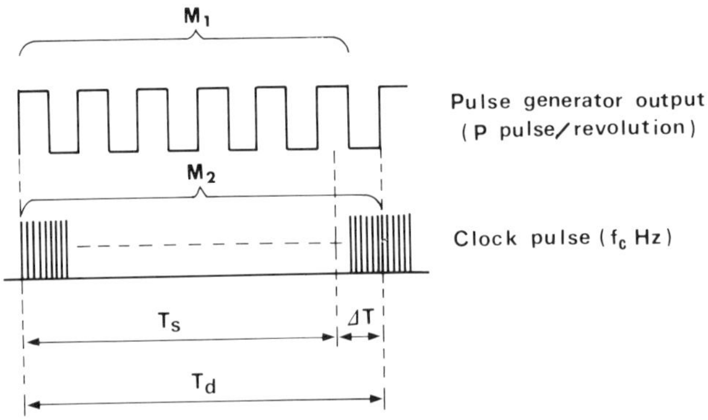

Figure 6. Time chart of pulse trains.

3.2. Analog Singnal Control Using a Microprocessor

In CSI and six-step VSI systems, dc link current or voltage fed to the inverter is regulated by adjusting the gating angle of the controlled rectifier. And the inverter frequency is controlled by adjusting the conduction time of the switching devices of the inverter. In microprocessor-based systems, the phase angle of the controlled rectifier and the output frequency of the inverter is determined in the microprocessor and should undergo the digital-to-analog conversion process. But in the final stage of control, the signals are in the form of "pulse" which turn on or off the switching devices. This gives the possibility of dispensing with D/A conversion in some cases.

3.2.1. Phase angle control of controlled rectifier. Six thyristors in the rectifier are triggered sequentially, being synchronized with the source voltage, so as their phase relationships result in the rectified dc voltage which is proportional to the cosine value of the phase delay angle, α. Conventional methods use D/A converter to convert the output voltage command from the microprocessor into a proper voltage level which gives the phase angle at which the associated thyristor begins to conduct. This is shown in Figure 7(a). As poined out earlier, gating pulses can also be obtained directly from the digital circuit as shown in Figure 7(b). In Figure 7(b), the crossover points of line voltage should be informed to the microprocessor in order to synchronize the gating signals with the line voltage. The microprocessor then generates gating pulses delayed by α from the crossover points.

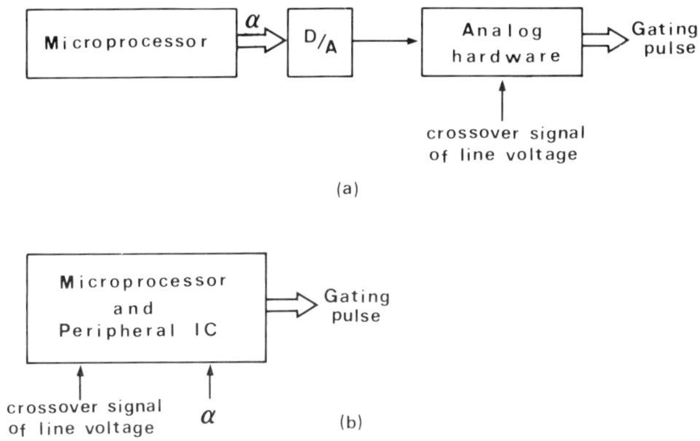

Figure 7. Block diagram of the converter gating angle control.

3.2.2. Inverter frequency control. Conventionally, the inverter frequency command signal is given by a voltage level which is converted into a pulse train through a voltage to frequency converter or VCO. This pulse train is then fed to a ring counter which triggers six thyristors sequentially. The process is shown in Figure 8 where 120° conduction

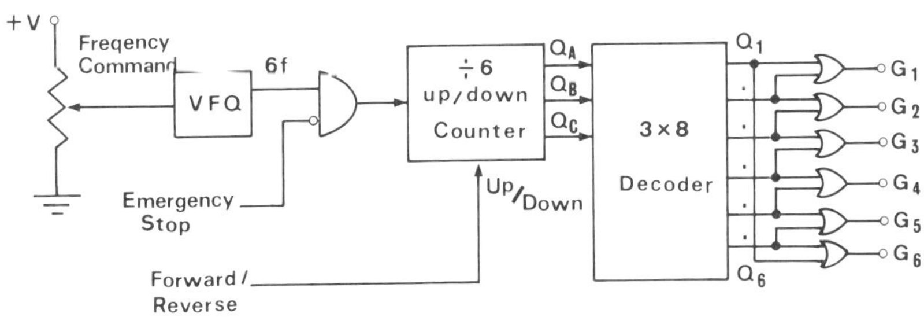

Figure 8. Conventional inverter frequency control scheme.

interval of the thyristors are assumed. 180° conduction is achieved simply by connecting one more decoder output to each OR gate, for example, Q_3 in addition to Q_1 and Q_2 to the first OR gate for the gating signal G_1. A reverse direction operation is accomplished by changing the up/down mode of the ring counter and so changing the phase sequence of the inverter gating signals. When the emergency stop line is set the pulse train from VCO is blocked. Then the motor is dc-excited and fastly slowed down to stop[9].

In a microprocessor-based version, the VCO in Figure 8 can be replaced by the latch and divide-by-N counter as shown in Figure 9. The microprocessor outputs the reciprocal of a required frequency, that is,

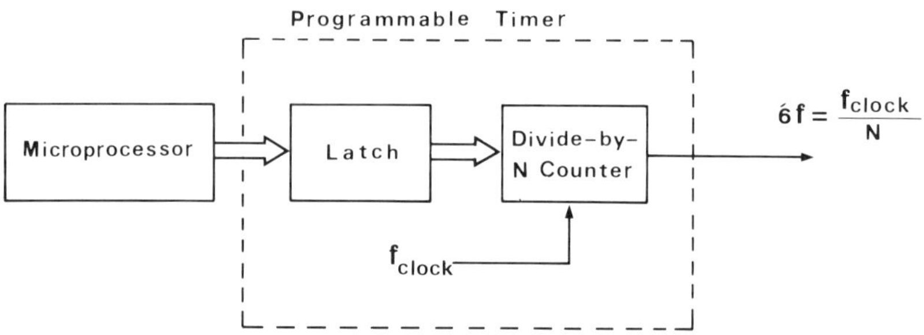

Figure 9. Microprocessor-based inverter frequency control.

the period of one cycle of the inverter output. The same function is provided by a Programmable Interval Timer(PIT) which is programmed to output gating signals in the frequency given by the microprocessor.

A more advanced scheme, shown in Figure 10, uses PIT as an interrupt source and makes the microprocessor directly control the gating signals.

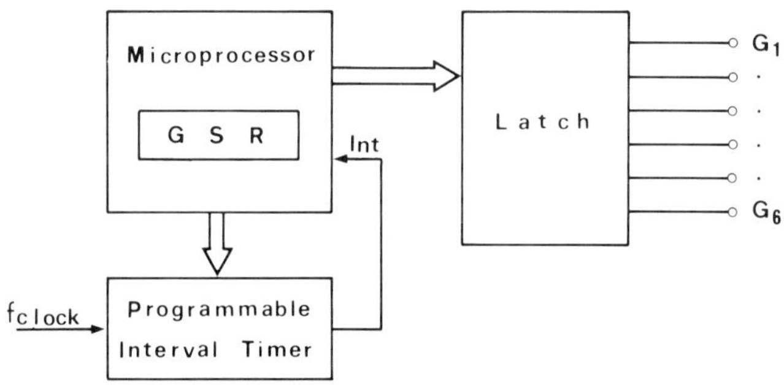

Figure 10. Inverter frequency control using PIT.

rupt source and makes the microprocessor directly control the gating signals. The software for this scheme is given in Figure 11. The content of GSR(Gating Sequence Register) represent the current state of gating signals with the first bit corresponding to G_1, second bit to G_2 , and so on. Last two bits are not used. The main program in Figure 11 assigns the GSR with the initial value, 03H for 120°conduction or 07H for 180° conduction(Suffix H denotes the Hexadecimal code). It reads required frequency f, given by some other control source, calculates N, the period of one cycle, and then set PIT by N. N corresponds the clock frequency divided by 6T, i.e., $f_c/6T$. PIT interrupts the microprocessor by every N-clock pulses thus resulting in the output frequency of 6f.

In the interrupt service routine, when the emergency stop line is activated, the current gating signal is maintained so that the inverter delivers dc output and the motor goes down to stop. In normal operations the contents of GSR is shifted left or right according to the forward or reverse operation, respectively.

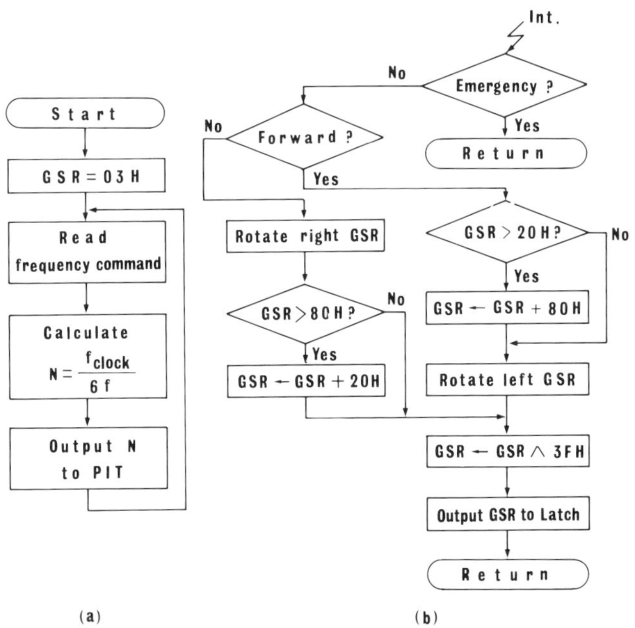

Figure 11. The flowchart for inverter frequency control using PIT.

4. AC MOTOR DRIVES WITH MICROPROCESSORS

In this section, techniques for hardware and software implementations of the ac motor drive systems described in Section 2 are discussed. Involved are feedback loop algorithms, hardware configurations, and some practical aspects of digital control systems.

4.1. The Current Source Inverter System

A current source inverter, as its name implies, supplies a motor with current regulated in phase-controlled converter. When it serves to drive an induction motor, an equivalent circuit study reveals that the motor operates on the unstable portion of the torque-speed relationship to ensure non-saturated condition. Thus, in order to stabilize the system, the feedback loop for speed regulation should be included in addition to the current regulation loop[10].

In Figure 3, was shown a control scheme in which the slip frequency of the motor, f_{sl}, controls dc link current, I. The control inputs to the microprocessor are the reference speed, actual speed and dc link current, with the outputs, the converter gating angle and inverter frequency.

4.1.1. System hardware. Figure 12 shows the hardware configuration of the CSI control system. The Z-80 microprocessor is assumed in consideration of its popularity in current process control applications. Two CTC (Counter and Timer Circuit)s are used to read dc link current and speed, under the assumption of the voltage-to-frquency conversion in the earlier stage of measurements. CTC1 is used as a "Counter" to count the speed and current pulses while CTC2 as a "Timer", to interrupt CPU at presumed instants on which CPU reads the current and speed from CTC1.

Outputs of the microprocessor are available through two ports of PIO. Converter gating angle, in 8-bit word, is converted into analog voltage level to feed the gating circuit of the converter. Inverter frequency, on the other hand, is 16-bit word and converted into the gating pulses through a divide-by-N counter. The sampling time of the system is given by the interrupt cycle of CTC2. Through a CRT terminal users input the reference speed and may display the measured current and speed.

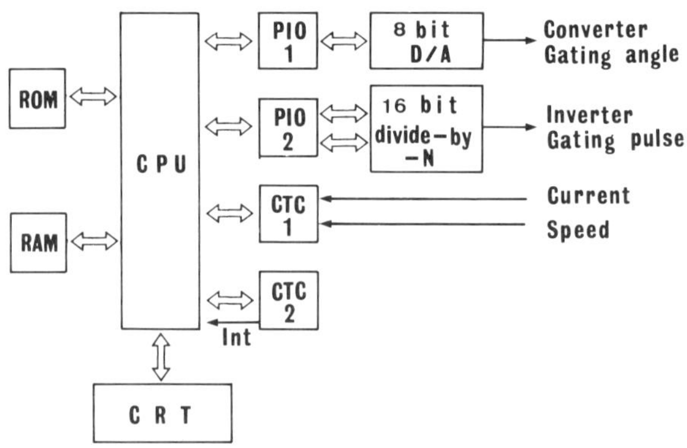

Figure 12. System hardware for CSI driven induction motor.

4.1.2. System software. Flowcharts for the feedback control are shown in Figure 13. The main program at first initializes memory, CTC, and PIO. At the sampling instant it reads reference speed from the CRT terminal and the speed and current from the input buffer. With these values, it calculates the converter gating angle and inverter frequency throughout the feedback algorithm. It then outputs the results and waits for next sampling time. The interrupt service routine reads the speed and current from CTC1 and save them in input buffer.

The feedback algorithm does realize the components of the feedback loop in Figure 3. The nonlinear relationship between the dc link current and the slip frequency can be implemented either by the look-up table or by the on-line computation of the nonlinear function. Two controllers are used, the speed controller and the current controller. P(Proportional), PI(Proportional-plus-Integral)and PID(Proportonal-plus-Integral-plus-Differential) controllers which have been employed in conventional analog feedback control systems may well be used in digital control strategies. There are two ways in implementing them by a microprocessor, the position form and the speed form. The latter is preferred in most cases since the former occupies excessive memories. The general PID

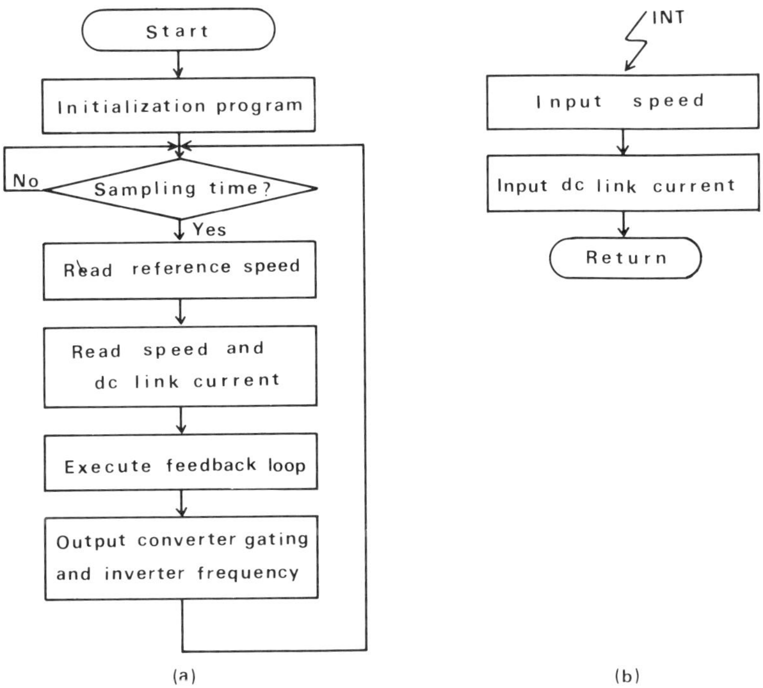

Figure 13. Software flowchart for current source inverter control.

controller, in speed form, is expressed as

$$e_2(n) = e_1(n-1)+(K_p+K_iT+K_d/T)e_1(n)-(K_p+2K_d/T)e_1(n-1)+ \\ +(K_d/T)e_1(n-2) \quad (2)$$

(K_p ; Proportional gain, K_i ; Integral gain, K_d ; Differentiall gain) where e_1 and e_2 represent the error and controller output at the instant indicated by corresponding index and T is the sampling time. In above expression, K_d=0 and K_i=K_d=0 will give the PI and P controller, respectively. In most cases, tuning the controller gains largely depends on designer's experience. The systematic way to determine the gains that result in best dynamic performance is often highly involved, and is beyond the scope of this chapter[11].

4.2. PWM Inverter System

The conduction time of the switching elements in a PWM inverter should be controlled in such a manner that resulting ac voltage resembles a sinusoidal waveform. And, when the PWM inverter feeds an ac motor, a linear relationship between the voltage and frequency is required. There are many ways proposed in matching these requirements. The most popular one is the sinusoidal pulse-width modulation, which is shown in Figure 14.

In Figure 14(a), the three-phase reference waves are shown with the high-frequency carrier wave. The intersections between them give the switching points of the inverter, resulting in output voltage and current as shown in Figure 14(b). The fundamental components of the output voltage waveforms follow the reference waves, and one can control the

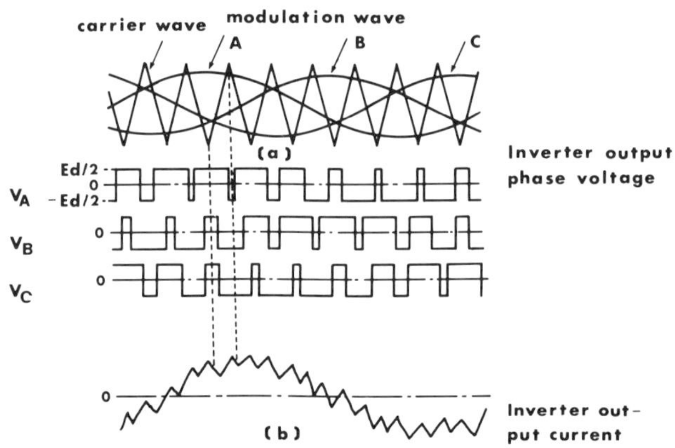

Figure 14. Sinusoidal pulse width modulation.

frequency and rms value of the output voltage by adjusting the reference wave appropriately.

The realization of PWM strategy in Figure 14 with analog circuitry involves the reference wave generator, the triangular carrier wave generator and the electronic circuits to compare them and output gating signals. Some or all of these components can be eliminated when a microprocessor is used. Due to its flexibility and adaptability, the microprocessor gives versatile possibilities in realizing PWM control strategy. Two typical applications of the microprocessor to PWM inverter control are discussed in the following.

4.2.1. Pattern retrieval method. Consider an arbitrary PWM waveform shown in Figure 15 where, for simplicity, only three switchings per one quater cycle is assumed. If the memory of the microprocessor contains the value of the switching angles, or the switching "pattern", only the retrieval of this pattern from memory is needed to generate proper gating signals. The choice of switching angles is purely optional; one may want output voltage waveform with minimum total harmonic contents and the other may prefer minimum harmonic power loss in his specific load conditions. In any case, it is necessary to derive the mathematical expressions with which the gating angles are evaluated. Generally these expressions result in a set of nonlinear equations and numerical methods with optimization techniques should be employed to solve them.

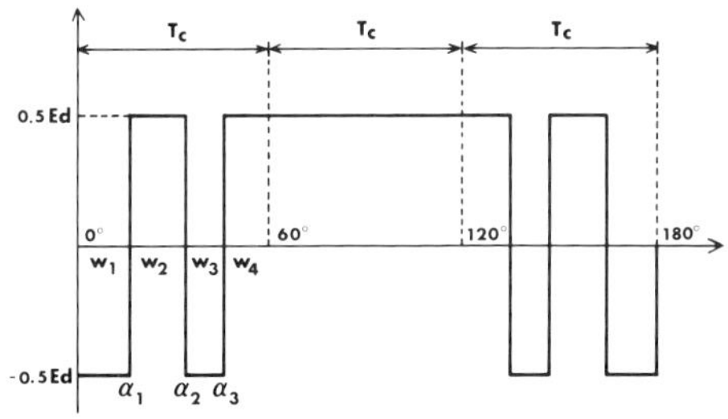

Figure 15. PWM waveform with arbitrary switching instants.

A more practical approach is to determine switching instants so as the output waveform of the inverter contains no harmonics of specified orders. This is termed harmonic elimination method. A general Fourier series represenation of an arbitrary function f(wt) is,

$$f(wt) = \sum_{n=1}^{\infty} (A_n \mathrm{Sin}(nwt) + B_n \cos(nwt)) \tag{3}$$

If f(wt) is quater-wave symmetric and an odd function, the Fourier coefficients of the series is given by

$$A_n = \frac{4}{n\pi}\left[1 + 2\sum_{k=1}^{M}(-1)^{\alpha k}\cos(n\alpha_k)\right] \tag{4}$$

$$B_n = 0$$

where α_k represents the k-th switching angle.

Clearly, the PWM waveform in Figure 15 is a quater-wave symmetric odd function and thus above equations hold in this case. If we wish to have a fundamental component of the magnitude V and to eliminate fifth- and seventh-order harmonics, the switching angles must satisfy the following equations.

$$A_1 = \frac{4}{\pi}(1 - 2\cos\alpha_1 + 2\cos\alpha_2 - 2\cos\alpha_3) = V$$

$$A_5 = \frac{4}{5\pi}(1 - 2\cos 5\alpha_1 + 2\cos 5\alpha_2 - 2\cos 5\alpha_3) = 0 \tag{5}$$

$$A_7 = \frac{4}{7\pi}(1-2\cos 7\alpha_1+2\cos 7\alpha_2-2\cos 7\alpha_3) = 0$$

The aboves constitute a set of three nonlinear equations with three unknown variables α_1, α_2 and α_3. It should be noted that switching N-times per one quater cycle results in N consistent equations. Thus one can control N harmonics independently, including the fundamental component, by adjusting N switching angles appropriately[12].

Once the switching angles are evaluated they are transformed into corresponding pulse widths from w_1 to w_4 as shown in Figure 15, and then stored in ROM table. The hardware configuration of the system is shown in Figure 16. Through the I/O port, CPU reads the frequency with which it calculates the output voltage to maintain constant air-gap flux density. The voltage specifies the adress of ROM table. CPU then reads the pulse widths and stores them in I/O buffer. The system employs two

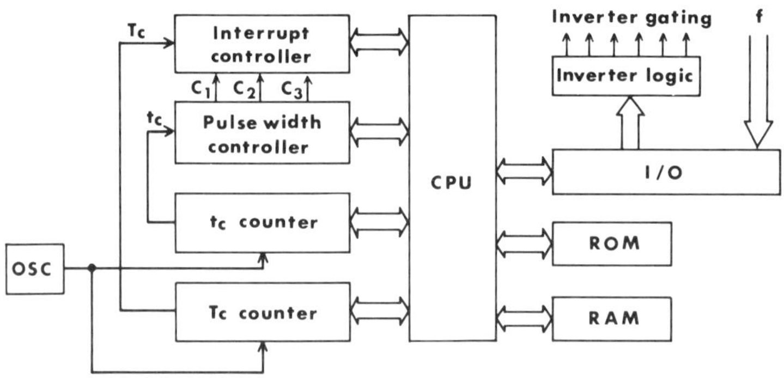

Figure 16. System hardware for pattern retrieval method.

counters, one for timing pulses, t_c, and the other for phase shifting, or updating pulses, T_c, in three phase system. In this case T_c corresponds to 60 degrees as shown in Figure 15. When the contents of the t_c counter reaches the value of the pulse width, CPU is interrupted by the programmed interrupt controller and output the gating signal through the I/O port. Above procedure is repeated by every T_c time, and t_c and T_c are updated in every loop.

The pattern retrieval method considerably alleviates the computational burden of the microprocessor. But the output voltage is necessarily quantized, thus the relatively large ROM is needed to improve the resolution of the voltage in precise control[13].

4.2.2. <u>On-Line computation method</u>. The on-line computation method can be regarded as a "simulation" of the sinusoidal pulse-width modulation in Figure 14, in a microprocessor. The microprocessor generates the reference wave and carrier wave in itself, and calculates the matching points between them. It would be desirable that the reference wave be

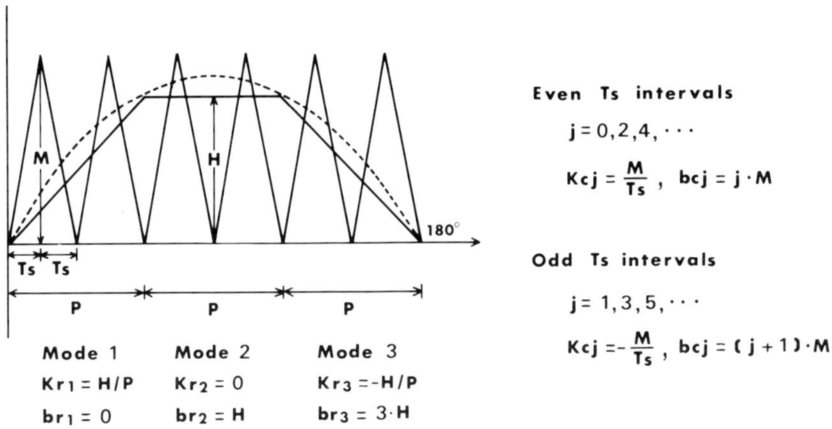

Figure 17. Quasi-sine wave modulation in a microprocessor

a pure sinusoid but in a practical way, the reference wave is approximated a trapezoidal wave as shown in Figure 17. Of course, the ouput voltage will contain more harmonics than the case of the pure sinusoidal reference wave.

The reference wave in Figure 17 is assumed to have two breakpoints in positive half cycle, at 60° and 120°. The frequency of the carrier wave, f_c is given by

$$f_c = 2nf \tag{6}$$

where f is the frequency of the reference wave and n is the number of carrier pulses during one half cycle which is equal to or greater than unity. To evaluate the intersection points, following two equations can be given;

$$\begin{aligned} y &= K_{ri}x + b_{ri} && \text{for reference wave} \\ y &= K_{cj}x + b_{cj} && \text{for carrier wave} \end{aligned} \tag{7}$$

The values of K_{ri}, K_{cj}, b_{ri}, and b_{cj} in each mode are given in Figure 17. Above equations give the abscissa of the intersection as

$$x = \frac{b_{cj} - b_{ri}}{K_{ri} - K_{cj}} \tag{8}$$

which is to be evaluated in a microprocessor.

In a three phase system, the mode of the reference wave of each phase is uniquely determined during the period of 60°, and the intersections for each phase are calculated by every T_s time. To output gating pulses it requires four counters as shown in Figure 18. First three counters are used to generate gating signals at the points given by the CPU while the last one, Counter 4, is used to generate a timing signal which informs the CPU on the time T_s, and resets the other counters by every T_s.

Figure 19(a) shows the main program. After initialization, CPU reads the input frequency and calculates the constants K_{ri}'s and b_{ri}'s which characterize the three phase reference waves for a given 60°

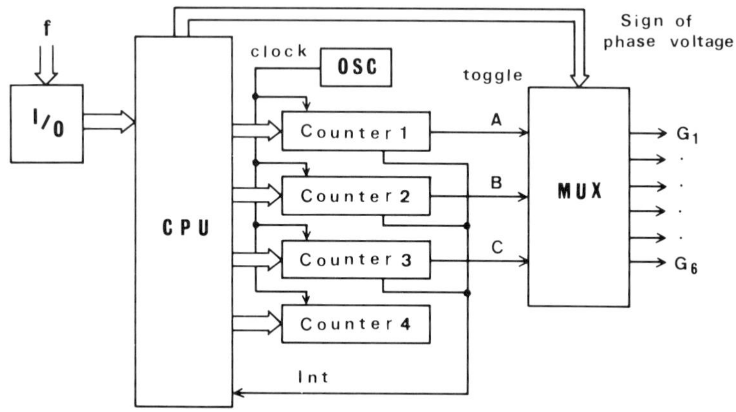

Figure 18. System hardware of on-line computation method.

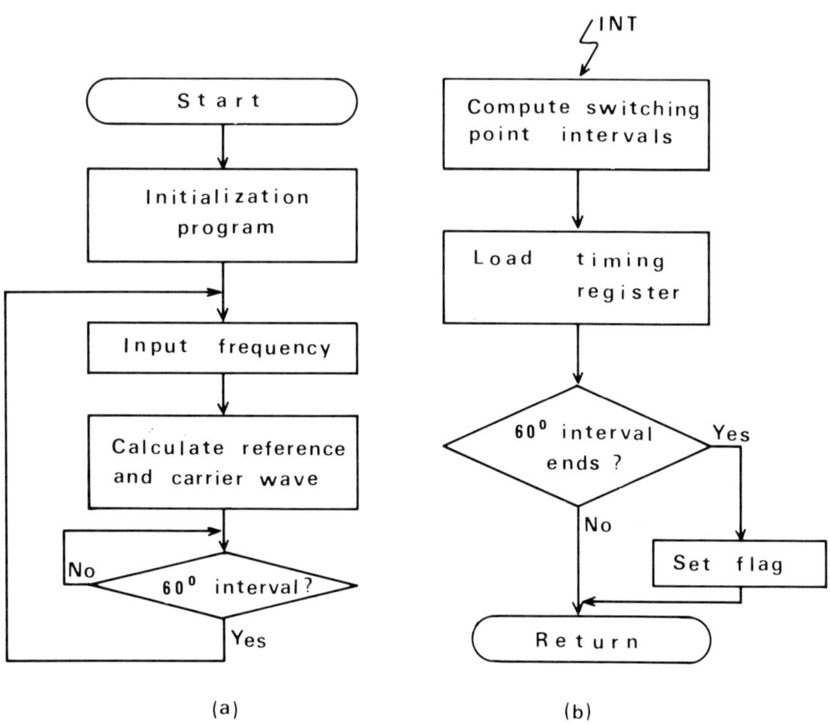

Figure 19. System software of on-line computation method.

interval of i-th mode. It also determines the carrier wave frequency, or T_s in turn, with which the constants K_{cj} and b_{cj} for every j=0,1,.... are calculated. The main routine repeats itself by every interval of the reference mode. When CPU is interrupted by counter 4, it executes the interrupt pulse generation routine in Figure 19(b). The switching points of three phases are computed and loaded in timing registers which are connected to first three counters in Figure 18. Counters from 1 to 3 generate the gating signals at the instants specified in associated timing registers in CPU. The multiplexer in Figure 18 determines whether the signal for each phase is in positive or negative half cycle, and delivers six gating signals along to the sequence that is already tabulated in the memory of the microprocessor[14].

4.3. Field Oriented Control System

In Figure 4, was shown the vector relationship between the stator current components. There are many possibilities in implementing the field oriented control strategy. A relatively simple way is to evaluate and control the torque angle θ_f and the main current i_s as shown in Figure 20. The magnitude of the input current is controlled in the converter circuit while the torque angle is controlled by the inverter frequency. This scheme is based on the relationships

$$
\begin{aligned}
i_s &= (i_M{}^2 + i_T{}^2)^{\frac{1}{2}} \; e^{j(\phi+\beta)} \\
&= (i_M{}^2 + i_T{}^2)^{\frac{1}{2}} \; e^{j\theta_f} \qquad (9)
\end{aligned}
$$

which can be easily derived referring to Figure 4.

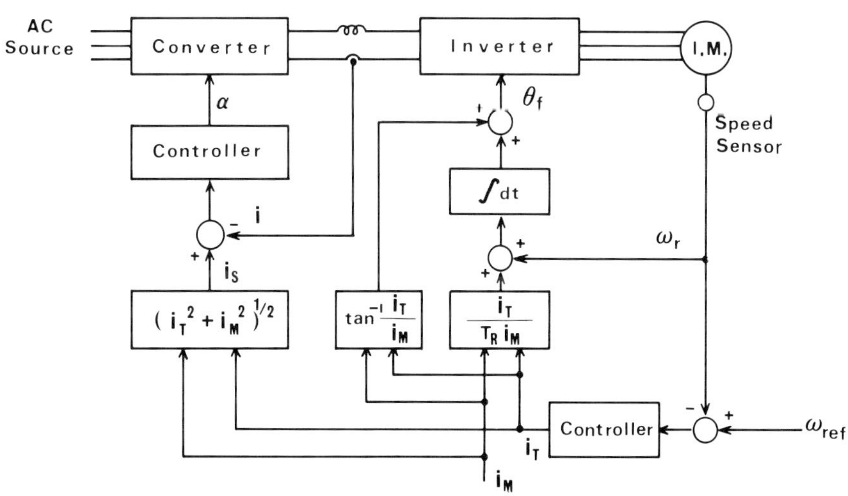

Figure 20. Block diagram of field oriented control.

Since the field oriented control involves complicated arithmetic operations such as square root and arctangent, it is very difficult to implement them with analog circuitry and thus the microprocessor is indispensible. Taylor series expansions are useful to evaluate such nonlinear functions in a microprocessor. But it is time-consuming task to compute them with Taylor series, and in most cases of real time control the sampling time is too short to meet the reasonable tolerance error bounds even for 16-bit microprocessors. Therefore, the look-up table is usually employed or the coprocessor, only for arithmetic operations, is used in parallel to the main processor. Both are described in the following.

4.3.1. Look-up table method using ROM. Figure 21 shows the hardware block diagram for the field oriented control using a 8086 16-bit microprocessor. The speed and the current of the motor are already converted into pulse trains and they are counted by two 8253 counters. Internal timing counter interrupts CPU periodically, then CPU reads the speed and current from the 8253 counters. During the execution of the feedback control algorithm nonlinear functions are evaluated by referring the ROM tables. The ROM size is determined according to the number of bits of the input and output of the table. If a full 16-bit input/output format is used, the ROM size becomes excessively large. Thus the data should be truncated appropriately to trade-off with the ROM size.

The system software for the field oriented control is shown in Figure 22. At the sampling time, the reference speed is read and i_M, which gives the air-gap flux level, is determined. Then i_T is evaluated from the reference and actual speed of the motor. $i_M^2 + i_T^2$ and i_T/i_M are calculated, with which the values of $\sqrt{i_M^2 + i_T^2}$ and $\tan^{-1}(i_T/i_M)$ are obtained from respective ROM tables. The results are fed to the controllers to give the converter gating angle and the inverter frequency. After the execution of one cycle, CPU waits for next interrupt[15].

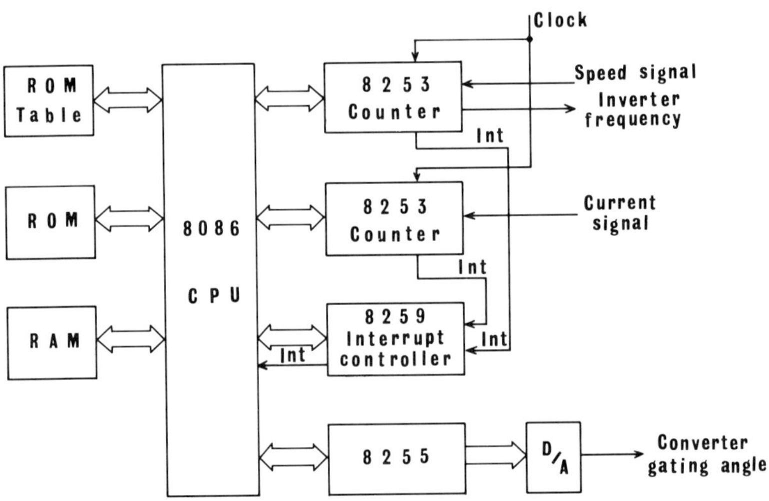

Figure 21. Hardware block diagram of field oriented control.

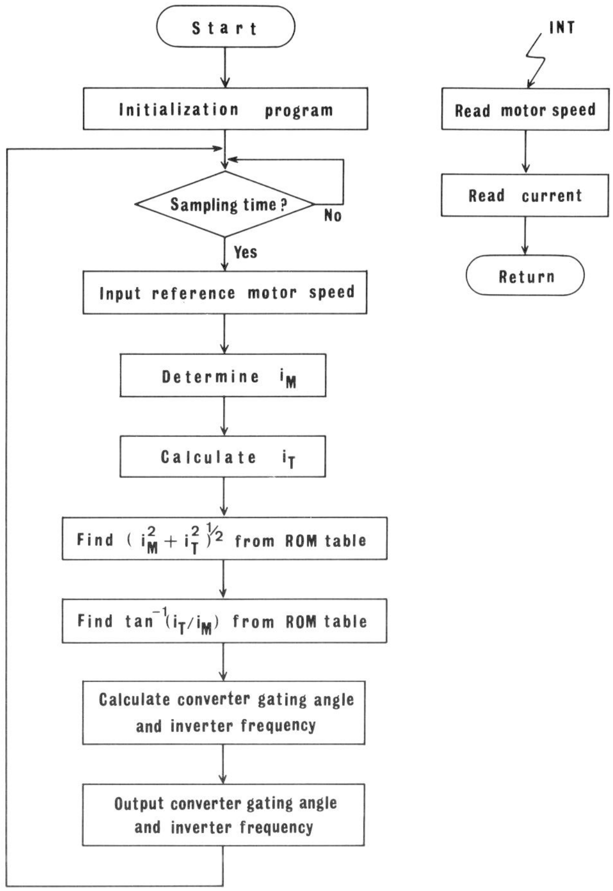

Figure 22. Software flowchart of field oriented control.

4.3.2. Field oriented control using a coprocessor. As pointed out earlier, limitations on the ROM size inevitably result in truncation and this cause adverse effects on the system performance. Thus sometimes is used the coprocessor which takes full charge of complicated arithmetic operations. Most manufacturers provide users with the coprocessor for each microprocessor system, for example, 8087NDP(Numeric Data Processor) for the 8086 microprocessor. 8087NDP performs the arithmetic process 100 times as fast as 8086. 8086 refers all complicated arithmetic calculatons to 8087NDP and goes along with the results of it.

When a coprocessor is used the Bus Controller(8288 for 8086 and 8087NDP) should be used in addition. The Bus Controller checks the status of the main processor and generates control signals that link the main processor and coprocessor and, enable them to communicate with each other. If the coprocessor of the other family than that of the main processor is employed, proper interfacing techniques, of course, will be necessary.

4.4. Energy Optimization Control

For an inverter-induction motor drive system controlled by a microprocessor, the microprocessor gives feasibility of not only adjusting the motor speed but also improving the efficiency of the motor. This section gives an example of an application of the microprocessor to the control of an induction motor in conjunction with energy savings.

4.4.1. Control algorithm. In conventional drive systems induction motors have been driven under the constant voltage to frequency ratio, which implies the constant air-gap flux. Under this condition, there can be only one torque-speed curve that meets the load torque at a given reference speed. If the motor is allowed to run under variable V/f ratio, there exist infinitely many torque-speed curves that satisfy the same load requirements. Each voltage-frequency pair will give the same torque and speed but resulting in different slip.

The efficiency of an induction motor is the function of slip, and there exists a slip that gives the maximum efficiency. We refer to it as the maximum-efficiency slip, or the optimal slip. The optimal slip can be calculated from the equivalent circuit of an induction motor. But it is usually very difficult to calculate the optimal slip exactly due to the harmonics of supply voeltage and the nonlinear characteristics of the motor. A simple and exact alternative is the direct measurement of the efficiency. Fortunately, the optimal slip is nearly constant at a given torque and speed condition. Once the optimal slip is obtained at various speeds, it can be easily implemented by look-up table.

Figure 23 shows the control block diagram of an optimal efficiency drive system. Main control loops are based on slip control strategy, but with some differences. The look-up table tracks the optimal slip for a

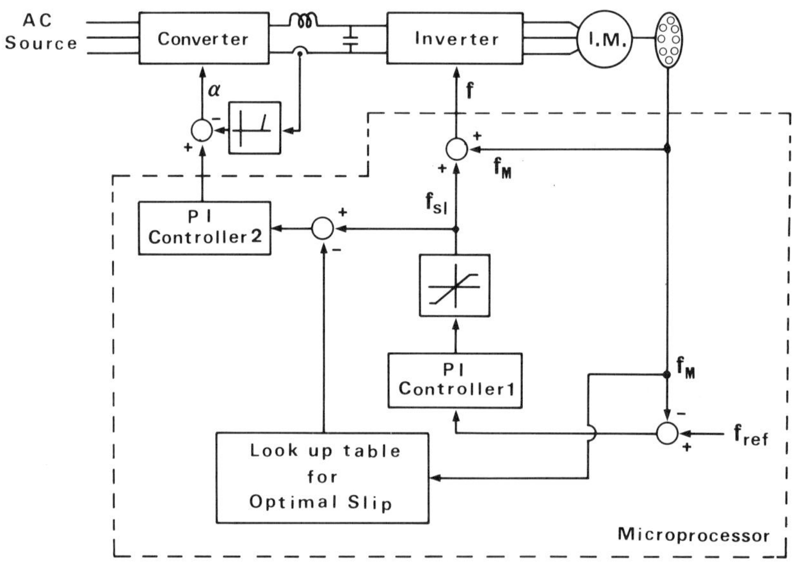

Figure 23. Control block diagram of energy optimization control.

given motor speed and, the error between the optimal slip and the output of the controller 1, f_{sl}, actuates the controller 2. Thus the controller 2 adjust the output voltage of the converter so as to reduce the error to zero while controller 1 acts as a speed controller as in normal slip control schemes.

4.4.2. Implementation. To realize the algorithm, a Z-80 microprocessor is used. The portion enclosed in dashed lines in Figure 23 corresponds to the microprocessor software. The hardware configuration is similar to that shown in Figure 12. The flowchart of the main program is shown in Figure 24(a). The purpose of the main program is to initialize the system giving the initial values of I/O ports and program constants. When the CPU is interrupted, it executes the control programs, shown in Figure 24(b) and (c), to give the inverter frequency and the converter gating angle. The speed sampling time is about 20 msec and the sampling time of the slip tracking loop is 100 msec.

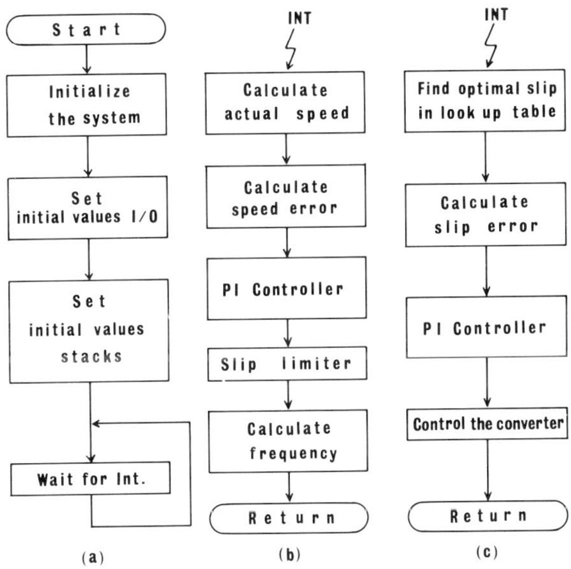

Figure 24. Software flowchart for energy optimization control.

When the energy optimization control described above is applied, saving of input powers about ten per cent or more can be achieved compared with the conventional constant flux control, at a quarter of full load. It will be more effective if the PWM inverter is used in conjunction with the energy optimization control to drive the induction motor, especially under low load factor. Provided the inverter-fed induction motor speed control system is already in use, the algorithm can be applied without any further hardware cost. Only a little modifications in software will be needed[16].

6. CONCLUSIONS

This chapter presented basic concepts related to the microprocessor-based ac motor control in conjunction with some practical application examples. Introducing the microprocessor into ac motor drive systems improves the reliability as well as the performance of the systems. As a result, highly advanced control schemes such as field oriented control with others discussed throughout this chapter are extensibly utilized from robots or NC machines which are required to be extremely accurate, to large scale energy-effective pump or fan drives.

Judging from current tendency, the state-of-the art will be significantly changed in the near future. And ac motors, driven by power semiconductor circuits, seem to be promising as long as the microprocessor and control technology continue to advance.

ACKNOWLEDGEMENT

The author appreciates the assistance and useful comments given by S. K. Sul, T. W. Chun and S. G. Jeong, candidates of Ph.D. in Department of Electrical Engineering, Seoul National University, during the preparation and the revision of the manuscript.

REFERENCES

[1] B. K. Bose, Adjustable Speed AC Drive Systems, IEEE Press, 1981.
[2] R. Sittig and P. Roggwiller, Semiconductor Deveices for Power Conditioning, Plenum Press, 1981.
[3] J. M. D. Murphy, Thyristor Control of AC Motors, Pergamon Press, 1973.
[4] K. P. Phillips, "Current Source Converter for Induction Motor Drives", IEEE Trans. Ind. Appl., Vol. IA-8, pp.679-683, Nov./Dec. 1972 .
[5] F. Blaschke, "The Principle of Field Orientation as Applied to the New TRANSVEKTOR Closed-Loop Control System for Rotating Field Machines", Siemens Review, Vol. 34, pp.217-220, May 1972.
[6] S. Nishikata and T. Kataoka, "Dynamic Control of a Self-Controlled Synchronous Motor Drive System", IEEE Trans. Ind. Appl., Vol. IA-20, pp. 598-604, May/June 1984.
[7] Eugene L. Zuch, Data Acquisition and Conversion Book, Datel-Intersil, 1980.
[8] T. Ohmae, et al., "A Microprocessor-Controlled High Accuracy Wide Range Speed Regulator for Motor Drives", IEEE Trans. Ind. Elec., Vol. IE-29, pp.267-271, Aug. 1982.
[9] P. C. Sen, Thyristor DC Drives, John Wiley & Sons Inc., New York, 1981.
[10] T. A. Lipo and E. P. Cornell, "State-Variable Steady-State Analysis of a Controlled Current Induction Motor Drive", IEEE Trans. Ind. Appl., Vol. IA-11, pp.704-712, Nov./Dec. 1975.
[11] R. Isermann, Digital Control Systems, Springer-Verlag, 1981.

[12] H. S. Patel and R. G. Hoft, "Generalized Techniques of Harmonic Elimination and Voltage Control in Thyristor Inverters; Part I-Harmonic Elimination", IEEE Trans. Ind. Appl., Vol. IA-9, pp.310-317, May-June 1973.

[13] B. K. Bose and H. A. Sutherland, "A High-Performance Pulsewidth Modulator for an Inverter-Fed Drive System Using a Microcomputer", IEEE Trans. Ind. Appl., Vol. IA-19, pp.235-243, March/April 1983.

[14] G. S. Buja and P. Fiorini, "A Microcomputer-Based Quasi-Continuous Output Controller for PWM Inverters", IEEE IECI Proc., Ind. Appl. of Microprocessors, pp. 107-111, 1980.

[15] M. Akamatsu, K. Ikeda and H. Tomei, "High Performance IM Drive by Coordinate Control Using a Controlled Current Inverter", IEEE Trans. Ind. Appl., Vol. IA-18, pp.382-392, July/Aug. 1982.

[16] M. H. Park and S. K. Sul, "Microprocessor-Based Optimal-Efficiency Drive of an Induction Motor", IEEE Trans. Ind. Elec., Vol. IE-30, pp.69-73, Feb. 1984.

CHAPTER 9

MICROPROCESSOR-BASED NONLINEAR CONTROLLER FOR THE ROLL MOTION OF A MODEL AIRCRAFT

Nicholas J. Krikelis
Control Systems Division
National Technical University
10682 Athens, Greece

ABSTRACT. The use of a microprocessor to control the roll motion of a model aircraft for wind tunnel experimentation is presented. A nonlinear model for the plant dynamics is employed. The controller is implemented as a non-linear control algorithm resulting in a stepwise control signal. The proposed controller is capable of handling efficiently rather fast varying disturbances which would be detrimental in the preservation of stability. To reduce the computation time in the microprocessor unit, a 3-byte fixed point arithmetic has been employed.

1. INTRODUCTION

In recent years, with the advancement of less expensive and more versatile microprocessor systems, many industrial control applications involving microprocessors have evolved.

Microprocessor control applications range from step motors [1], PWM controllers [2], wattmeters [3], thyristor converters [4], and process controllers [5], all the way to mechanical linkage controllers [6]-[8] and automotive engine control [9],[10]. Conventional proportional-integral-derivative (PID) control laws, or even modern control algorithms, have been implemented on general purpose microprocessors with limited computing capabilities and memory [11].

This paper considers the roll motion dynamics of an aircraft and presents the microprocessor implementation of a nonlinear controller, which attains excellent stability, accuracy and fast speed of response.

The aircraft, as a rigid body in space, has 6 degrees of freedom (Fig.1):

- ability to move along the three axes x, y, z
- ability to rotate about the same axes.

Assuming that the x axis coincides with the longitudinal axis of the aircraft, the purpose of the controller studied here-

N. K. Sinha (ed.), Microprocessor-Based Control Systems, 183–204.

in is primarily to enhance the stability about this axis, namely the roll motion of the aircraft.

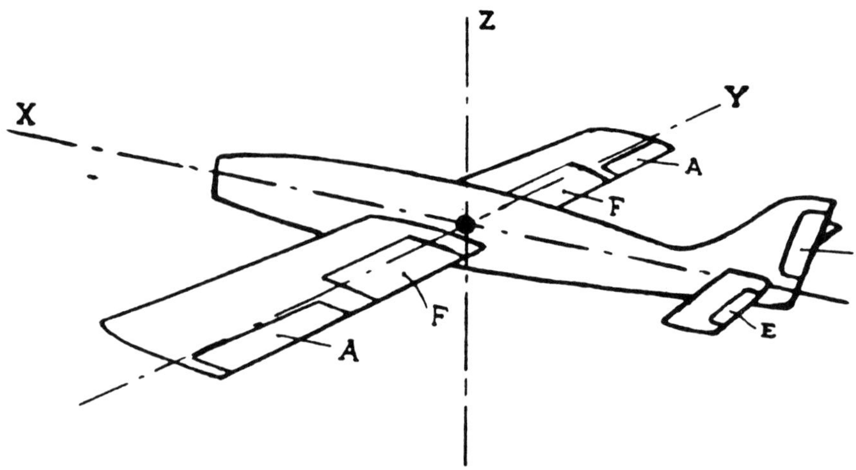

A: Ailerons
E: Elevators
R: Rudder
F: Flaps

Fig.1: Schematic of the aircraft wings.

Generally speaking, the roll motion affects and in return is affected by the other motions of the aircraft [12], but we are going to neglect any such interactions (except for one case, which is discussed later in the text). Thus, one is led to the consideration of a one-degree-of-freedom model, representing the roll motion of the aircraft.

The aircraft to be controlled has a mass of 5 kg and is flying horizontally at a constant speed of 10 m/sec. It is subjected to atmospheric disturbances and the task of the controller is to keep the aircraft always horizontal, by moving its ailerons. The dynamic system was simulated on a Cyber-171 computer at the National Technical University of Athens.

2. ROLL MOTION EQUATION

The roll motion equation results from the equilibrium of the torques acting on the aircraft with respect to the axis x and is given as:

$$M_\alpha + M_\tau + M_\vartheta = M_\delta + M_k \tag{1}$$

Using the characteristic quantities of the aircraft, each of the terms of the above equation is expressed below:

$$M_{\alpha} = I_{xx} \ddot{\vartheta} \tag{2}$$

$$M_{\tau} = \frac{1}{4} C C_{D} \rho \ell^{4} \dot{\vartheta}^{2} \tag{3}$$

$$M_{\vartheta} = \frac{1}{4} m g \ell \sin\varphi \sin(2\vartheta) \tag{4}$$

$$M_{\delta} = \frac{1}{4} C_{D} \rho C \ell v_{d}^{2} \cos^{2}(\varphi+\vartheta) \tag{5}$$

$$M_{k} = 2\pi R_{f} \ell_{f} C \rho V_{\infty}^{2} \left(1 - \frac{\varphi_{1}}{\pi} + \frac{\sin\varphi_{1}}{\pi}\right) \delta \tag{6}$$

The derivation of Eqs. (3), (4), (5) and (6) appears in the Appendix A.

When each torque term contributing to the rolling is expressed as a function of characteristic quantities of the aircraft, the roll motion equation can be written as follows:

$$A\ddot{\vartheta} + B\dot{\vartheta}^{2} + C \sin(2\vartheta) = Dv_{d}^{2} \cos^{2}(\varphi+\vartheta) + E\delta \tag{7}$$

Where:

$$A = I_{xx} \tag{8}$$

$$B = \frac{1}{4} C C_{D} \rho \ell^{4} \tag{9}$$

$$C = \frac{1}{4} m g \ell \sin\varphi \tag{10}$$

$$D = \frac{1}{4} C C_{D} \rho \ell^{2} \tag{11}$$

$$E = 2\pi R_{f} \ell_{f} C \rho V_{\infty}^{2} \left(1 - \frac{\varphi_{1}}{\pi} + \frac{\sin\varphi_{1}}{\pi}\right) \tag{12}$$

The inertial moment I_{xx} of the aircraft can be calculated by considering it as a cylinder of radius r = 60 mm having a mass of m_1 = 2 kg uniformly distributed on its circumference, and an attached rectangular wing with a span of 1.50 m and a chord of 0.22 m, having a mass of m_2 = 3 kg uniformly distributed. Solving the equation (7) for $\ddot{\vartheta}$, one obtains:

$$\ddot{\vartheta} = \frac{D}{A} v_{d}^{2} \cos^{2}(\varphi+\vartheta) + \frac{E}{A} \delta - \frac{B}{A} \dot{\vartheta}^{2} - \frac{C}{A} \sin(2\vartheta) \tag{13}$$

Replacing the constants A, B, C, D and E with their arithmetic values, the following equation is obtained:

$$\ddot{\vartheta} = 0.0669 v_d^2 \cos^2(0.0698+\vartheta) + 10.4261\,\delta - 0.0377\,\dot{\vartheta}^2 - 0.5775 \sin(2\vartheta) \quad (14)$$

It should be noted that this is the equation to be used for simulating the aircrafts'roll motion; it is an approximation to the actual equation of motion.

3. PROPOSED NON-LINEAR CONTROL SCHEME

The roll motion equation of the previous section is clearly non-linear. As such, use of classical controllers, such as the P-I-D, would be inadequate. On the contrary, use of a non-linear controller would permit a more efficient control of the aircraft's roll motion. The control law adopted is the following:

$$S = a_1\vartheta + a_2\dot{\vartheta} + a_3\vartheta\dot{\vartheta} + a_4\vartheta^2 + a_5\dot{\vartheta}^2 \quad (15)$$

where S is the control signal, expressing the position (in positive or negative steps from the idle position) that the ailerons should take.

In the above equation, the first two terms represent the "linear part" of the controller, while the last three reflect the "non-linear" part.

The coefficients a_1 through a_5 are determined by the following approach: The output of the plant is the angular position, ϑ. Regarding the output as a function of the parameters a_1 through a_5, the controller's objective is to keep its absolute value as close to zero as possible, i.e.,

$$|\vartheta_T(a_1, a_2, a_3, a_4, a_5)| \leq \varepsilon, \quad t \geq T, \quad \varepsilon > 0 \quad (16)$$

In the above relation, ε is a positive number having a small value. The subscript T denotes the values of ϑ after some time T. The value of T could be chosen small or large, depending on the speed of response requirements. The values adopted in the present study are T=2 sec and ε=1.5 degrees. Therefore, the problem is to keep the absolute value of the angular position ϑ_T of the aircraft below 1.5 degrees, for any time t greater than or equal to T=2 sec.

In the following, the symbol ϑ instead of ϑ_T will be used for simplicity. The values of the parameters $a_1,\ldots,a_5$, which satisfy the stated condition for ϑ, will be calculated by an iterative procedure. The superscript n is used to denote the respective quantities after the n<u>th</u> iteration. The gradient of the angular position ϑ is given by

$$\text{grad } \vartheta = \frac{\partial \vartheta}{\partial a_1} da_1 + \frac{\partial \vartheta}{\partial a_2} da_2 + \frac{\partial \vartheta}{\partial a_3} da_3 + \frac{\partial \vartheta}{\partial a_4} da_4 + \frac{\partial \vartheta}{\partial a_5} da_5 \quad (17)$$

Substitution of the differentials by small variations (for computational reasons) yields the following expression for the nth iteration:

$$(\text{grad } \vartheta)^n = \sum_{j=1}^{5} \left(\frac{\partial}{\partial a_j}\right)^n \Delta a_j^n \quad (18)$$

But

$$\left(\frac{\partial \vartheta}{\partial a_j}\right)^n = \frac{\vartheta_j^n - \vartheta^n}{d^* a_J^n}, \quad j = 1, \ldots, 5 \quad (19)$$

Where:

$$\vartheta^n = \vartheta^n(a_1^n, a_2^n, a_3^n, a_4^n, a_5^n)$$

$$\vartheta_j^n = \vartheta_j^n(a_1^n, a_2^n, \ldots, a_j^n + d^* a_j^n, \ldots, a_5^n)$$

$d^* a_j^n$ is a very small quantity given by $d^* a_j^n = a_j^n da$, where da is also a very small, but constant quantity.
Also,

$$\Delta a_j^n = a_j^{n+1} - a_j^n \quad (20)$$

and

$$a_j^{n+1} = a_j^n + \Delta a_j^n \quad (21)$$

Let Δa_j^n be such that

$$\Delta a_j^n = -\left(\frac{\partial \vartheta}{\partial a_j}\right)^n . K \quad (K > 0) \quad (22)$$

The "minus" sign signifies that in every new iteration the variation proceeds towards a smaller value of ϑ.

The steps of the iterative procedure to satisfy the requirements of Eq.(16) are summarized below:

1) Choose the values of da and K
2) Choose the constants a_j^o $(j=1,\ldots,5)$
3) Set n=0
4) Calculate: $\vartheta^n = \vartheta^n(a_1^n, a_2^n, a_3^n, a_4^n, a_5^n)$
5) Calculate: $da_j^{*n} = a_j^n \, da$ $(j=1,\ldots,5)$
6) Calculate: $a_j^{*n} = a_j^n + da_j^{*n}$ $(j=1,\ldots 5)$
7) $\vartheta_j^n = \vartheta_j^n(a_1^n, a_2^n, \ldots, a_j^{*n}, \ldots, a_5^n)$ $(j=1,\ldots,5)$
8) $\left(\frac{\partial \vartheta}{\partial a_j}\right)^n = \frac{\vartheta_j^n - \vartheta^n}{da_j^{*n}}$ $(j=1,\ldots,5)$

9) $\Delta a_j^n = -K(\frac{\partial \vartheta}{\partial a_j})^n \qquad (j=1,\ldots,5)$

10) Calculate the new a_j's: $a_j^{n+1} = a_j^n + \Delta a_j^n \qquad (j=1,\ldots,5)$

11) $n = n+1$

12) Repeat the iterations starting from step 4.

The iterations stop when the desired result is obtained. It should be pointed out that the method described above is rather slow in the search of the absolute minimum of the function ϑ. However, the rate of convergence was noticeably improved by changing either the rate of descend to the minimum or the sensitivity of the method by means of K and da respectively.

4. MICROPROCESSOR IMPLEMENTATION OF THE CONTROLLER

The flow-chart of the controller (Fig.2) is readily implemented, since the only input to the microprocessor is the angular position of the aircraft. Its angular velocity is then obtained from the first difference between two consecutive "reads" of the angular position.

Details for the flow chart are now presented. The variable OLDPOS contains the angular position of the aircraft at the instant t-dt, while the variable NEWPOS contains the same quantity at time t. The quantity DT symbolizes the time interval dt between two consecutive samplings of the angular position of the aircraft. In the actual program, this constant is included in the coefficients a_2, a_3 and a_5 in an attempt to minimize the computational task of the microcomputer. THDOT represents the angular velocity.

When the microcomputer initiates its control function, the variables OLDPOS, NEWPOS and THDOT are set equal to zero. Thus, during the first run of the program, the controller "thinks" that the aircraft was initially horizontal (OLDPOS =0), and attains at time dt the angular position that is "read" through its position encoder, which, in general, leads to a wrong determination of the aircraft angular velocity. In turn, this also gives a wrong control signal to the step motor, which drives the ailerons. This anomaly lasts only for a time dt, until the next data input. Then, the controller works normally. The ailerons can move within an interval of ±10 steps from their idle position (each of these steps corresponds directly to the resolution of the step motor), thus the controller must check whether they have reached, or not, the limit of their movement and, consequently, whether they should stop waiting for a signal to move to the opposite direction or not. This stepwise motion is represented

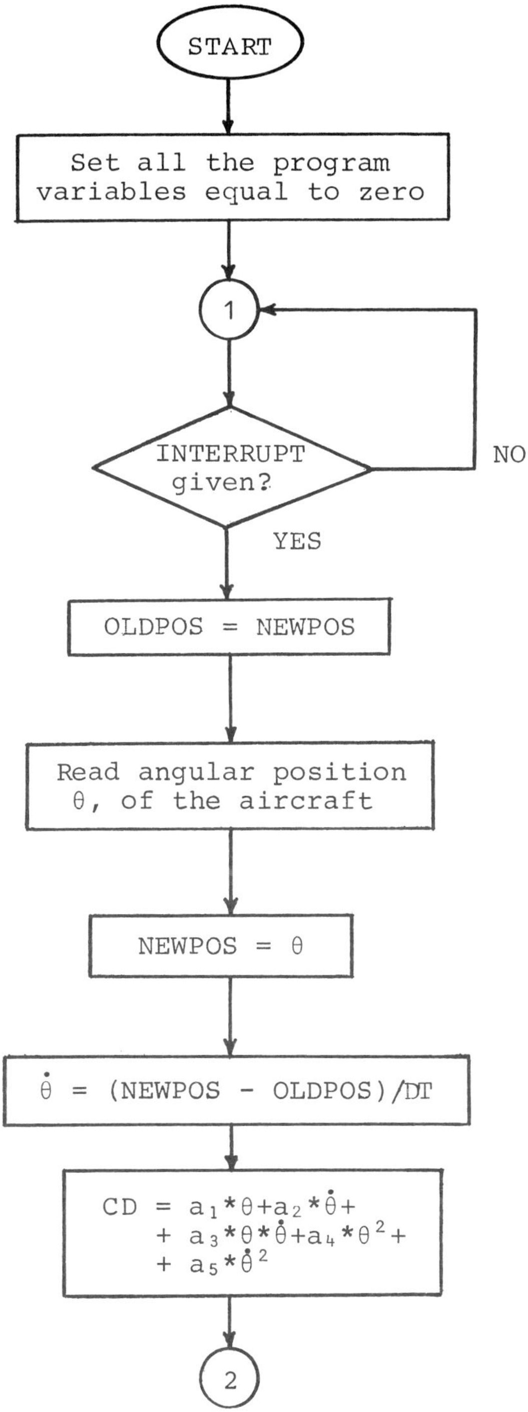

Fig.2. Flow-chart of the controller (continued in next page).

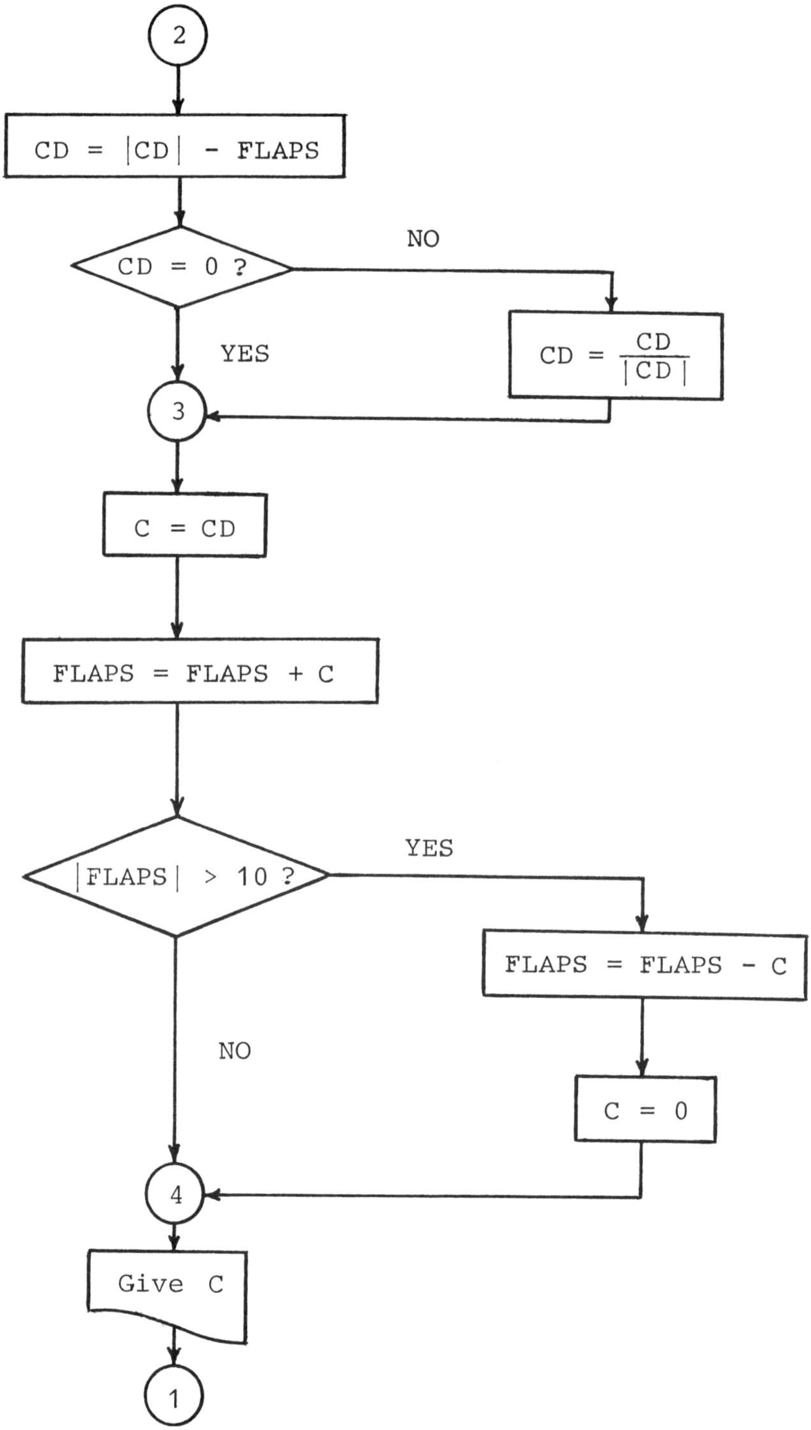

Fig.2. Flow-chart of the controller.

by a variable called "FLAPS". Since the variable FLAPS is zero at the first run of the program, it is required that the ailerons are at their idle position at the initial time that the microprocesser takes control over the aircraft.

Clearly, the time dt allowed for the step motor movement is not long enough to drive it to the calculated position. For this reason, each run of the program simply determines the direction the step motor should be moved and drives it to the right position one step at a time. It is assumed, that the noise affecting the commands of the microprocessor is negligible, thus, no provision is taken in the block diagram to adjust for an incorrect positioning of the step motor.

The capabilities of the microcomputer used [13] are as follows:

ROM capacity	:	2048 bytes
RAM capacity	:	256 bytes
I/O capabilities	:	2 eight-bit individually programmable input or output ports 1 six-bit programmable input or output port
Internal clock frequency	:	2.5 MHz

The I/O ports in this microcomputer can be either memory or I/O mapped. We are going to memory map these ports, which means that we can input from or output data to other devices using the normal instructions provided for reading from or writing into memory respectively.

For the complete working system, the following components are also needed : a high resolution angular position meter, which measures the angular position of the aircraft (radians) in binary form, a programmable timer which produces a pulse at certain intervals (every 10 msec) to signal the microcomputer to sample data and process it, and a driver that may move the step motor, by one step, according to the microcomputer commands.

The period of the internal clock frequency of the microcomputer (state), is 400 ns. Just because the time dt is chosen equal to 10 msec, the program's execution time should not exceed 25000 states.

To reduce the computation time, the fixed point arithmetic was adopted and implemented. Three-byte binary numbers are being used. It is assumed that their 10 most significant bits are for the integer part, while the remaining 14 less significant bits represent the fractional part of the numbers. This scheme helps to represent decimal numbers in the range of-512 to +511.9999, with an accuracy of 4 decimal digits. This is considered acceptable for our purpose, since it is easy to verify that we never deal with numbers outside this range, while the accuracy of the 4 decimal digits is more than satisfactory.

The program presented in this section makes use of the two 8-bit ports, as the input to the microcomputer and the 6-bit port as the output to the driver of the step motor.

The addresses of the 2 input ports are INPO and INP1 employing the two least significant bytes. The numbers represented by these two bytes have two integer and 14 fractional bits, thus being in the range of -2 to +1.9999, which is acceptable, since the aircraft should not be deflected by more than ±1.57 rad. Since the microcomputer makes use of 3-byte numbers, one extra byte is attached to the number representing the input, so that it does not affect its value. The attached byte is treated as the most significant byte of the number.

The address of the output port is OUTP. If in the two least significant bits the binary number 00 is present, the driver does not rotate the step motor. When the binary 01 is present, the driver turns the step motor one step towards the positive direction. Finally, when the number is the binary 10, the driver moves the step motor one step towards the negative direction.

The memory addresses used are the following:

NEWPOS 0, NEWPOS 1, NEWPOS 2	:	The three bytes always contain the angular position measured at time t.
OLDPOS 0, OLDPOS 1, OLDPOS 2	:	These addresses contain the angular position measured at time t-dt.
THDOT 0, THDOT 1, THDOT 2	:	Here the angular velocity of the aircraft, computed for time t, is stored.
IMT 0, IMT 1, IMT 2	:	Intermediate results during the computation are placed here.
A1, A2, A3	:	Here the addable, the subtrahend or the multiplicand is stored.
B1, B2, B3	:	The second addable, the minuend or the multiplier is placed in these three bytes.
C1, C2, C3	:	The result of a 3-byte arithmetic operation is placed here.

The main program "CONTROL" initializes the registers, the memory addresses, the stack and the output port. Then, it gets into a waiting loop, waiting for the pulse from the timer, which will act on an interrupt pin of the microprocessor (RST 7.5), and originate the execution of the control subroutine "NLCTL". After the execution of this subroutine, the program "CONTROL" returns to the waiting loop for the next interrupt.

The subroutine "NLCTL" computes the angular velocity, using the formula:
THDOT = (NEWPOS - OLDPOS)/DT
Since DT is a constant, equal to 0.01 sec, the formula becomes:
THDOT = (NEWPOS - OLDPOS)100.

In the last expression, instead of multiplying the result of the subtraction by 100, thus wasting computation time, it is preferable to shift it seven bits to the left. This way, it is multiplied by 128. Naturally, this should be divided afterwards by 1.28, to obtain the correct angular velocity. However, this division is avoided by incorporating the factor 1.28 in the coefficients a_2, a_3 and a_5, which act on the variable THDOT. Namely, in the program, the coefficients a_2, a_3 and a_5 used are modified, the first two having been resulted from a division of the initial coefficient by 1.28 and the third one from a division by $(1.28)^2$, consistent with the equation of the controller.

In addition, for the multiplication by 128, instead of performing seven consecutive shifts of the number to the left, it is much simpler to make a shift by one byte to the left and then make another shift by one bit to the right. This is what actually happens in the subroutine NLCTL.

The coefficients that have resulted from the simulation of the aircraft are (see section 5):

$$a_1 = -200$$
$$a_2 = -40$$
$$a_3 = -2$$
$$a_4 = -4$$
$$a_5 = -2$$

The modified coefficients used in the microcomputer program are:

Decimal	Binary	Hexadecimal
a_1 = -200	= 1100111000.00000000000000	= CE 00 00
a_2 = -31.25	= 1111100000.11000000000000	= F8 30 00
a_3 = -1.5625	= 1111111110.01110000000000	= FF 9C 00
a_4 = -4	= 1111111100.00000000000000	= FF 00 00
a_5 = -1.2207	= 1111111110.11000111100000	= FF B1 E0

These coefficients are symbolized in the program by Ajk, where j is an integer from 1 to 5 used to indicate the coefficient, while k is an integer from 0 to 2 used to identify the order of the byte of a certain coefficient. Thus, byte 1 of a_3 will be represented as A31.

The program in assembly language [14] uses the subroutines ADDY, SUBBY and MULTY to perform 3-byte addition, subtraction and multiplication respectively.

The maximum execution time of this program is 26660 states, equal to 10.664 msec, for a clock period of 400 nsec, and could not be acceptable, since it should be less than 10 msec. However, the actual execution time complies with the above requirement, since the execution time of the multiplication subroutine MULTY, which is called five times per run of the program, depends upon the zero bits of the absolute value of the multiplier, being reduced by 42 states for every fractional zero bit and by 87 states for every integer zero bit.

The reduction of the execution time obtained for the multiplications a_3.THDOT, a_4.NEWPOS and a_5.THDOT is easily found to be equivalent to 3861 states (the absolute values of a_3, a_4 and a_5 contain a total of 36 fractional and 27 integer zero bits). Thus, the actual execution time of the program is 22799 states (9.1196 msec with a clock period of 400 nsec). This is further decreased when the remaining two multiplications are performed in the program. Unfortunately, this decrease cannot be calculated, since the multipliers NEWPOS and THDOT are variables.

5. ROLL MOTION SIMULATION RESULTS

The flow chart for the roll motion simulation is shown in Fig.3.

The plant to be controlled is a model aircraft with the following characteristics:

Mass	(m)	: 5	(kg)	
Overall Length	(L)	: 1.10	(m)	
Wing Span	(2ℓ)	: 1.50	(m)	
Dihedral Angle	(φ)	: 4	(^{0})	
Wing Shape		: Rectangle		
Wing Chord	(C)	: 0.22	(m)	
Aileron Chord	(C_f)	: 0.044	(m)	(=0.20 C)
Aileron Length	(ℓ_f)	: 0.30	(m)	
Distance between the middle of the aileron and the longitudinal axis x of the aircraft	(R_f)	: 0.60	(m)	
Wing Section Angle of Attack	(α)	: 12	(^{0})	
Velocity of the Aircraft at infinity	(V_∞)	: 10	(m/sec)	

In addition, the following characteristics of the environment have been assumed:

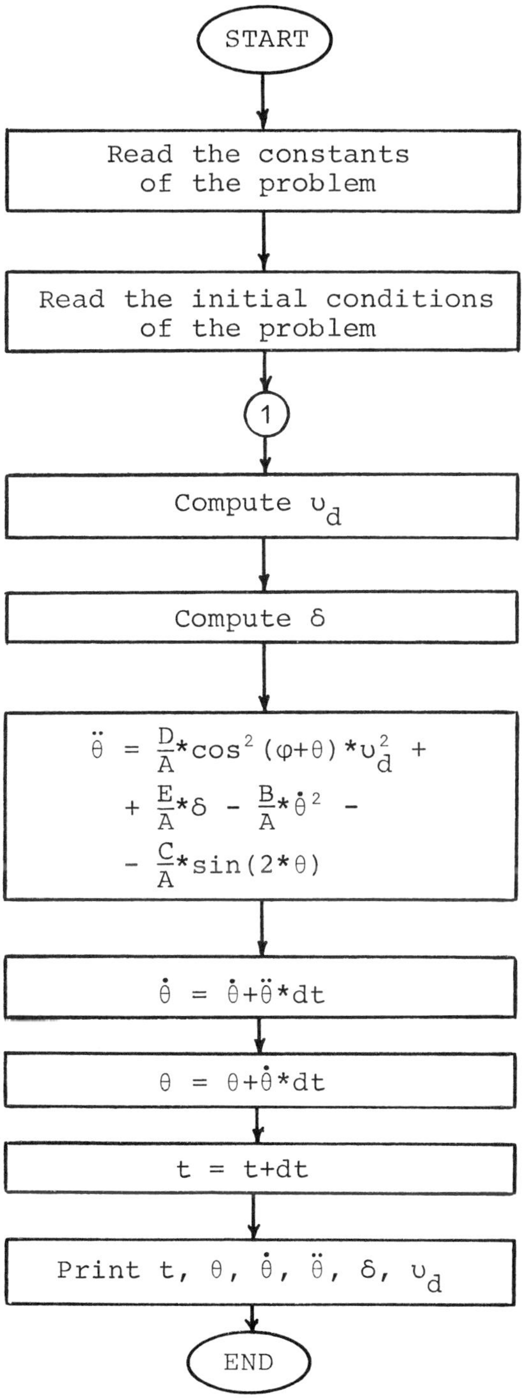

Fig.3. Flow-chart for the roll motion simulation.

Density of the Air (ρ) : 1.2256 (kg/m^3)
Gravity Acceleration (g) : 10 (m/sec^2)

The ailerons of the aircraft are supported by the wing in such a way, that the torque applied due to the lift is zero.

The step motor used (AIRPAX), has the following characteristics:

Step Motor	: Unipolar
Start-Without-Error Capability	: 100 (steps/sec)
Step Angle	: 7.5 (^{0}) (With the use of a 2:1 gear becomes 3.75^0)
Running Torque (Start-without-error curve) for the above stepping rate	: 50 (mNm)
Holding Torque	: 92 (mNm)
Rotor Moment of Inertia	: 6.0 10^{-3} (gm^2)
Power supply	: 5 (V)

The torque induced by this motor is adequate to drive the existing inertial and frictional load, during its movement, at the required stepping rate of 100 steps/sec (see Appendix B).

The response of the aircraft was tested for the following cases:

1. Aircraft free at an angular position of 60^0
2. " " " " " " " 30^0
3. " " " " " " " 10^0
4. Disturbance: Step. Wind velocity : 5 (m/s)
5. " " " " : 1 (m/s)
6. Disturbance: Ramp. Slope of wind velocity: 1 (m/s^2)
7. " " " " " " : 0,5 (m/s^2)
8. Disturbance: Sinusoidal. Ampl. 5 (m/s). Freq. 2 (Hz)
9. " " " " " " 1 (Hz)
10. " " " " " " 0.5 (Hz)
11. " " " " " " 0.4 (Hz)
12. " " " " " " 0.2 (Hz)
13. " " " " " " 0.1 (Hz)

For each case, all other initial conditions are zero.

The values of the coefficients, obtained for the above cases, are shown in Table I. Each set of coefficients was obtained for the worst conditions for the control. In particular, for the last set of Table I, the worst case is No. 12, since the frequency of the disturbance is very close to the resonance frequency of the aircraft.

It should be noted that utilization of the coefficients found for cases (6) and (7), yields satisfactory results for

TABLE I

Case	Values of coefficients				
	a_1	a_2	a_3	a_4	a_5
(1),(2),(3)	-15.02707	-8.53989	-0.94333	-10.86854	-0.64316
(4),(5)	-176.58483	-19.42268	-78.80837	-17.82152	-8.42693
(6),(7)	-200	-40	-2	-4	-2
(8),(9),(10)					
(11),(12),(13)	-150	-120	-8	-10.5	-11

all other cases and most of the time they are better than the dedicated so to speak coefficients. This implies that if the controller is realized with the coefficients found for cases (6) and (7), the system can easily handle almost any kind of disturbance.

Various plots were obtained for the cases investigated. Each plot comprised two parts: the first part showing the response of the aircraft with no control, while the second part showed the response to the same initial conditions and disturbances, using the controller described before. The coefficients used in the controller correspond to each individual case.

As it was seen from the plots of the response with no disturbance and no control, the aircraft is statically stable but dynamically unstable. The reason is, that, when the aircraft is oscillating at low amplitude, the angular velocity generates a torque, which eventually constitutes a negligible damping factor.

The non-linearity of the model is ascertained by calculating the resonant frequency, by dividing the number of oscillations completed in a certain duration, by this time interval. For different amplitudes of oscillation, the resonant frequencies were obtained and they were found to be around 0.2 Hz.

From the responses to the step and the ramp disturbances in the absence of the controller, one may think that, when the disturbance increases in amplitude, the dynamic stability of the aircraft increases too, approaching 90^0 of deflection. This is not true. Actually, because the disturbance is a wind current blowing upwards and affecting only one of the wings, (the aircraft is rotated by an angle such that this particular wing is vertical, that is by $90^0 - \varphi = 90^0 - 4^0 = 86^0$), the disturbance will not affect it at all. For a further rotation of the aircraft beyond 86^0, the wind will attack the wing in such a way, that the induced torque will tend to restore the aircraft 86^0. The simultaneous effect of this action, along with the torque due to the dihedral angle of the aircraft, define the final angular position, at which the aircraft will be stabilized. With this reasoning, angular deflections of the aircraft outside the range of -94^0 to $+86^0$ can be excluded.

Comparison of the responses for the cases with and without controller reveals the remarkable effectiveness of the controller proposed. From the plots of the flaps position, it was seen that, even when the aircraft had reached an acceptable deflection, the controller kept on moving the ailerons. This occurs because of the discrete positions the ailerons can take, which, in general, does not coincide with the exact torque required (quantization error). Actually, the controller is forced to move the ailerons, so that the average torque induced by the ailerons over time is that required.

For the ramp disturbance, it was seen that, after a certain time the controller was unable to control the aircraft, because of the limited torque induced by the ailerons, while the torque due to the disturbance increased indefinitely.

6. CONCLUSIONS

The use of a microprocessor to control the roll motion of a model aircraft has been described. The results of the computer simulation showed that the proposed controller is very effective in controlling the aircraft in various and severe atmospheric disturbances.

The microcomputer was able to execute the program, based on a rather complex algorithm in less than 10 msec, that is within the limits of acceptance. This gratifying result is due to the efficiency of the microcomputer, which can be used in demanding control schemes. This advantage, combined with the flexibility of the device itself, makes the microcomputer an invaluable tool in step motor control applications.

APPENDIX A

The equations discussed here represent an approximation to the roll motion of an aircraft, since an accurate model of the aircrafts' roll motion would be very difficult to simulate in a computer within a reasonable execution time interval. Actually, the objective of this study is not to introduce an exact model of the aircraft, but rather to employ a non-linear model, which will be controlled by the microcomputer. The following analysis is based on this philosophy.

Torque due to the aerodynamic resistance of a wing at roll, M_τ

Consider a wing rotating about an axis x with angular velocity $\dot{\vartheta}$, as in Fig.1, and let an infinitesimal part of the wing with dimensions C and $d\ell$ lie at a distance ℓ from the axis rotation. Due to the angular velocity $\dot{\vartheta}$, that part of the wing will have a linear velocity,

$$v_{\tau} = \dot{\vartheta}\ell \tag{23}$$

The aerodynamic force dF induced on this part is

$$dF_{\tau} = \frac{1}{2} C_D \rho \dot{\vartheta}^2 \ell^2 C d\ell \tag{24}$$

and the torque dM_{τ}, caused by this force with respect to the axis of rotation, is

$$dM_{\tau} = \ell dF_{\tau} = \frac{1}{2} C_D \rho C \dot{\vartheta}^2 \ell^3 d\ell \tag{25}$$

Integrating the last equation over the length of the wing, and considering that there are two wings contributing to this torque, we get

$$M_{\tau} = 2 \int_0^{\ell} dM_{\tau} = \frac{1}{4} C C_D \rho \dot{\vartheta}^2 \ell^4 \tag{26}$$

<u>Torque due to the angular displacement of the aircraft, M_{ϑ}</u>

The cause of this torque is the loss of height by the aircraft, that occurs when the aircraft is rolling, in conjunction with the dihedral angle of the wings. Although we deal with a one-degree-of-freedom model, we now introduce a second degree of freedom. We do this because of the great importance of this torque component, which is the only one to restore the aircraft to its equilibrium position. For modelling purposes though it will be assumed that the cause of this torque is quite different from the real one. In fact, the aircraft will be considered as a body, which can rotate about its longitudinal axis and on which a vertical force is applied through a bar, which has a length H, given by the formula:

$$H = \frac{\ell}{2} \sin\varphi \tag{27}$$

This length equals the distance of the point of application of the lift force from the longitudinal axis x of the aircraft.
The magnitude of the force is given by

$$F_{\vartheta} = m\ g\ \cos\vartheta \tag{28}$$

(This is the magnitude of the vertical coefficient of the lift force, acting on the aircraft).
Thus, the torque M_{ϑ}, applied to the aircraft is,

Thus, the torque M_ϑ, applied to the aircraft is,

$$M_\vartheta = \frac{1}{2} m\ g\ \ell \sin\varphi \cos\vartheta \sin\vartheta, \text{ or}$$

$$M_\vartheta = \frac{1}{4} m\ g\ \ell \sin\varphi \sin(2\vartheta) \tag{29}$$

Torque due to atmospheric disturbances, M_δ

The atmospheric disturbances are, in fact, currents of wind, which effect the distribution of pressure over the wing surfaces, resulting in a torque, which tends to deflect the aircraft from its equilibrium position.

Assuming that the wind current is vertical, affecting only one wing, while the aircraft is at an angular displacement ϑ and considering an infinitesimal part of the wing with dimensions C and $d\ell$, at a distance ℓ from the axis of rotation x, the aerodynamic force on it will be:

$$dF_\delta = \frac{1}{2} C_D \rho \left[v_d \cos(\varphi+\vartheta)\right]^2 C\ d\ \ell \tag{30}$$

The torque due to this force with respect to the axis x, is:

$$dM_\delta = \ell\ dF_\delta = \frac{1}{2} C_D \rho \left[v_d \cos(\varphi+\vartheta)\right]^2 C\ \ell\ \ d\ell \tag{31}$$

Integrating over the length of the wing and keeping in mind that only one wing is contributing to this torque, we obtain:

$$M_\delta = \int_0^\ell dM_\delta = \frac{1}{4} C_D \rho v_d^2 \ell^2 \cos^2(\varphi+\vartheta) C \tag{32}$$

Torque due to the controller, M_k

Only that part of the wing, which contains the aileron, will produce the torque M_k. An assumption made is that the lift force acting on the parts, which contain the ailerons, is applied in the middle of these parts.
The variation of the lift coefficient for the length ℓ_f of the wing, due to the deflection of the aileron by an angle δ from its idle position, is given as [12]:

$$\Delta C_\ell = 2\ \pi\ \left(1 - \frac{\varphi_1}{\pi} + \frac{\sin\varphi_1}{\pi}\right)\ \delta \tag{33}$$

In the case δ is negative, ΔC_ℓ would also be negative, since all other terms in this equation are positive.
The torque M_k is produced by deflecting the two ailerons by angles δ_1 and δ_2 respectively, with δ_1 and δ_2 being opposite in sign and usually having the same absolute value.
Denoting by A_1 and A_2 the lift forces acting on the two wings respectively, for a straight, undisturbed, steady

flight the equation

$$A_1 = A_2 \tag{34}$$

will be in effect.

By deflecting the ailerons by an angle δ, there will be a change in the lift forces given by

$$\Delta A = \frac{1}{2} \rho \Delta_{\ell} V_{\infty}^{2} C \ell_f \tag{35}$$

Since the signs of δ will be opposite on the two wings, the corresponding signs of ΔA would also be opposite. ΔA contributes to the torque M_k, which is given by

$$M_k = 2 \Delta A R_f \tag{36}$$

Substituting ΔA from Eqs. (35) and (33), finally yields,

$$M_k = 2 \pi R_f \ell_f C \rho V_{\infty}^{2} (1 - \frac{\varphi_1}{\pi} + \frac{\sin\varphi_1}{\pi}) \delta \tag{37}$$

APPENDIX B. Evaluation of the Step Motor

The load that the step motor should drive consists of an inertia (of the wings) and a friction.

The mass of the wings is approximately $m_w = 0{,}053$ Kg, assuming fiberglass material with surface density $\rho_s = 2$ Kg/m^2 and a total area of,

$$s_f = 2.1_f.c_f = 2.0.044 \text{ m}.0{,}30 \text{ m} = 0{,}0264 \text{ m}^2$$

Assuming that the wing mass is equally distributed over its surface, and that the axis of rotation coincides with the side of the wing, the moment of inertia is,

$$I_f = \frac{2}{3} .m_w.C_f^2 = \frac{2}{3}.0.053.(0.044)^2$$
$$= 0.0684 \text{ g.m}^2$$

The total moment of inertia I_T, is thus

$$I_T = I_f^* + I_R$$

where,

I_R = moment of inertia of the rotor of the motor

I_f^* = moment of inertia of the load "seen" by the motor.

For the step motor selected (Section 5), the rotor moment of inertia I_R is,

$$I_R = 0.006 \quad g\cdot m^2$$

The load moment of inertia, as "seen" by the motor, is one half of the actual, since a gear reducer 2:1 is being used. Thus,

$$I_T = \frac{1}{2}\, 0.0684 + 0.006 = 0.0402 \quad g\cdot m^2$$

The required torque is obtained by the formula,

$$T = I_T \cdot \frac{V^2}{2} \cdot \frac{2\pi}{\text{steps/rev}}$$

where,

I_T = total moment of inertia (load + rotor) "seen" by the motor

V = angular velocity of the motor

steps/rev = number of steps corresponding to a full revolution of the motor.

For the specific step motor selected we have,

$$I_T = 0{,}0402 \text{ g.m}^2$$

$$V = 100 \text{ steps/sec}$$

$$\text{steps/rev} = 360^o/7.5^o = 48$$

Therefore,

$$T = 26.31 \text{ mNm}$$

Taking the efficiency of the gear reducer to be 0.98 and that of the bearings 0.98, the required torque is,

$$T_r = \frac{T}{(0.98)^2} = 27.40 \text{ mNm}$$

The step motor proposed, at 100 steps/sec, supplies a torque of 50 mNm. There is, thus a safety coefficient equal to 1.8.

REFERENCES

[1] B.C.Kuo, Theory and Applications of Step Motors, West Publishing Co., 1974.

[2] G.S. Buja and P. Fiorini, "Microcomputer control of PWM inverters", IEEE Trans. Ind. Electron., Vol. IE - 29, p.p. 212 - 216, Aug. 1982.

[3] J.J. Hill and W.E. Alderson, "Design of a microprocessor-based digital wattmeter", IEEE Trans. Ind. Electron. Contr. Instrum. Vol. IECI - 28, pp. 180 - 184, Aug. 1981.

[4] G. Oliver, V.R. Stefanovic, and G.E. April, "Microprocessor controllers for a thyristor converter with an improved power factor", IEEE Trans. Ind. Electron. Contr. Instrum., Vol. IECI - 28, pp. 188-194, Aug. 1981.

[5] K. Rajaraman, "A microprocessor - based time sequence controller for process control application" , IEEE Trans. Ind. Electron. Contr. Instrum., Vol. IECI - 28 pp. 214-219, Aug. 1981.

[6] D.M. Alley and H.W. Mergler, "Digital control of a front-end loader linkage" , IEEE Trans. Ind. Electron., Vol. IE - 29, pp. 220-225, Aug. 1982.

[7] C.A. Klein and J.J. Maney, "Real - time control of a multiple - element mechanical linkage with a microcomputer", IEEE Trans. Ind. Electron. Contr. Instrum., Vol. IECI-26., No 4, Nov. 1979.

[8] C. Barrientos, "Development of a multiple microprocessor-

based Controller for a legged locomotion process using interactive design tools", M.S. thesis, Ohio State Univers., Columbus, OH, Dec. 1982.

[9] T.J. Flis, "The use of microprocessors for electronic engine control", IEEE Trans. Ind. Electron., Vol. IE-30, pp. 75 - 87, May 1983.

[10] J. Bereisa, "Applications of microcomputers in automotive electronics," IEEE Trans. Ind. Electron., Vol. IE-30, pp. 87-96, May 1983.

[11] N. J. Krikelis and S.D. Fassois, "Microprocessor implementation of PID Controllers and lead - lag compensators IEEE Trans. Ind Electron., Vol. IE-31, pp. 79 - 85, Feb. 1984.

[12] B. Etkin, Dynamics of Atmospheric Flight, John Wiley and Sons, Inc., 1972.

[13] MCS - 80/85 (TM) Family User's Manual, Intel Corp., CA, 1979.

[14] 8080/8085 Assembly Language Programming Manual, Intel Corp., C.A, 1979.

CHAPTER 10

A MICROPROCESSOR-BASED CONTROL SYSTEM FOR A GLASS FURNACE

Heikki N. Koivo, Ari Setälä
Tampere University of Technology
P.O. Box 527
SF-33101 Tampere
Finland

ABSTRACT: A microprocessor based control system for a glass furnace is presented. The basic hardware and software of this control system are described. The control system has three main purposes: a) to plan the furnace operation to meet given glass production specifications, b) to provide measurement and alarm information, and c) to control the furnace both during glass production and during start-up. In a) data collected from past furnace operation is used to estimate the required melting power. A dynamic model of a glass furnace is also given. This model is used to test various control strategies before their implementation in c).

1. INTRODUCTION

The increasing price of energy has forced the glass industry to pay more attention for developing the efficiency of the glass melting process. This has led to several proposals of using mini- and microcomputers in the glass industry: Finger (1978), Finger and Law (1978), Davies (1979), Gaige (1979), and Rawson (1981). The system descriptions are quite general and control philosophies are not spelled out. In this paper a detailed design of a microprocessor based control system is developed for a continuously operating, electric glass melting furnace introduced in an earlier paper (Setälä and Koivo, 1983). The design is based on modelling and simulation for finding the most useful control strategies.

A short description of the process is given in section 2. Main ideas of the modelling are summarized in section 3. In section 4 the control strategy for a melting furnace is discussed. A brief review of computer control is given in section 5. Section 6 presents the design of a microcomputer implementation of the proposed control system.

N. K. Sinha (ed.), Microprocessor-Based Control Systems, 205–222.

2. PROCESS DESCRIPTION

2.1. The furnace

The glass manufacturing process is displayed in Fig. 2.1. In this study only the glass melting is of interest. The melting occurs in a continuously operating, electric glass melting furnace. The structure of the furnace is shown in Fig. 2.2. The glass furnace can be divided into the following parts:

- melting tank, where glass is melted with electricity
- refiner, where the homogenization is continued and
- forehearths, wherefrom glass is taken out for shaping.

In addition the refiners have a common overflow, through which extra molten glass flows down.

The glass furnace is a continuous process. When blowers take molten glass from forehearths for production of articles, new raw material (soda, limestone, quartz sand, and some additional ingredients) is fed into the melting tank. There is always a layer of unmolten raw material in the melting tank. It is called a batch blanket and the furnace is therefore of cold crown type.

Glass is melted with electricity. The melting electrodes are built into the bottom refractories and the electrodes are connected to a 50 Hz AC voltage. The glass mass around the electrodes works as a medium for the electric current. The produced heat flow is inversely proportional to the glass resistivity, which is quite nonlinear -in room temperature glass is a good insulator, but at higher temperatures a conductor.

From the melting tank glass flows into the refiners, wherefrom glass flows further into the forehearths and overflow. In the refiners too quick a cooling of glass is prevented with radiation elements. These are built above the glass level and their electric power can be controlled.

In the forehearths the glass temperature is brought to the right level by burning a gas-air mixture in the forehearth crown.

The glass quality depends strongly on the operations of the furnace. The most important thing in the operating of a glass furnace is to keep the glass temperature in the forehearth at its setpoint in spite of disturbances. The viscosity of glass is a very nonlinearly behaving quantity when temperature varies. The glass shaping characteristics and the quality of the finished article depend mostly on viscosity.

2.2. Measurements

With the present instrumetation the voltages and the currents of the melting electrodes and the radiation elements, and the corresponding electric powers can be measured. Unfortunately no measurement of the gas flow fed to the forehearth is available.

The most important control variables are the temperatures in

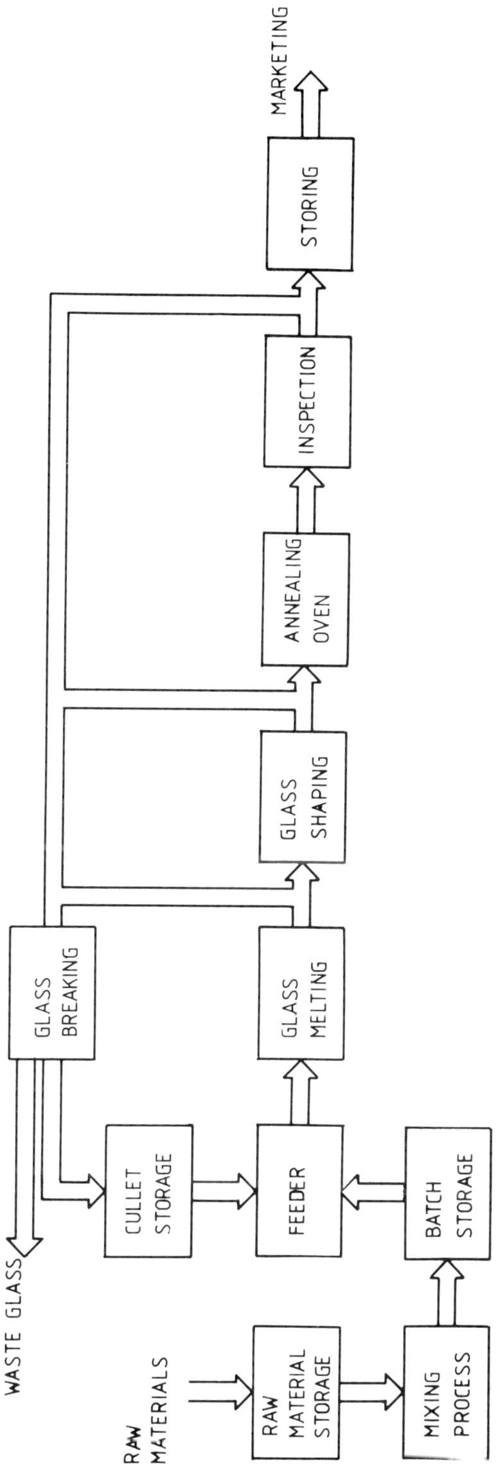

Fig. 2.1. Block diagram of glass manufacturing process.

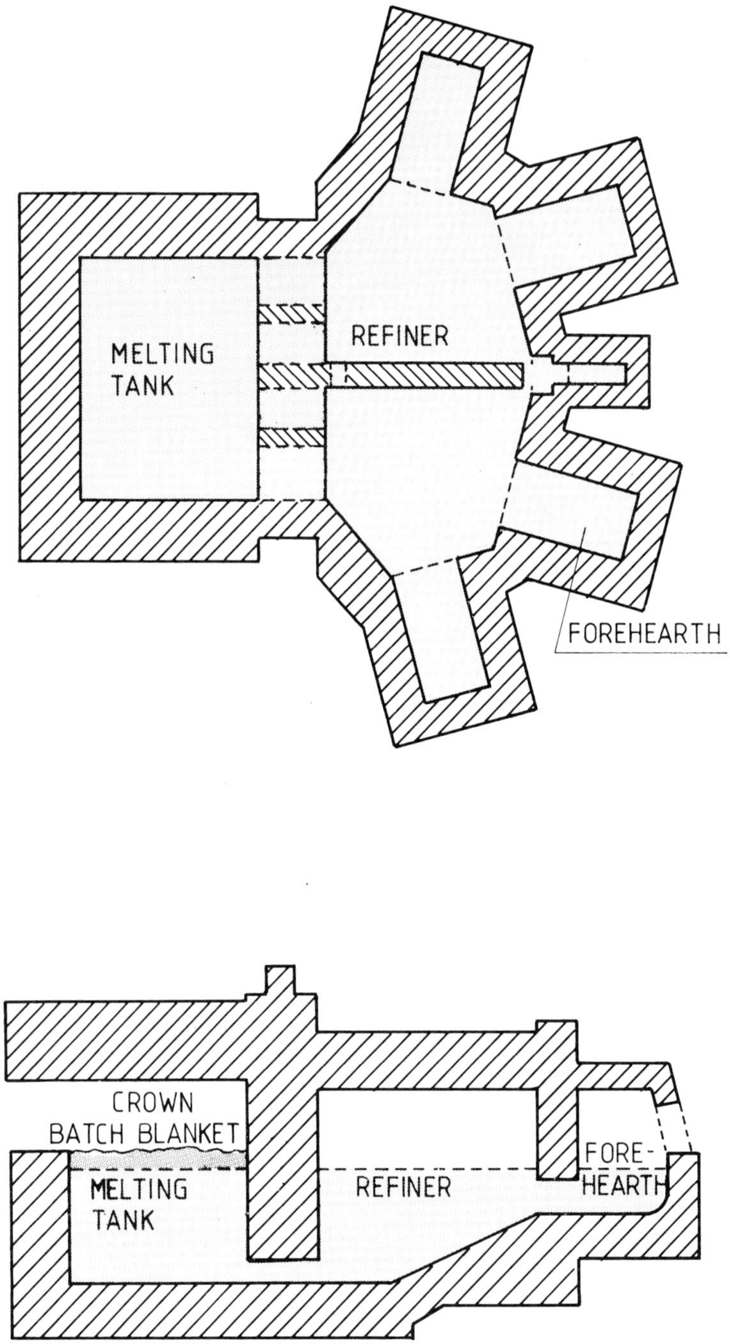

Fig. 2.2. Structure of the glass furnace from above (above) and from the side (below).

different parts of the furnace. From the melting tank we can measure the glass temperature and above the batch blanket, the crown temperature. From the refiners we can only measure the crown temperatures. In the forehearths we can measure both crown temperature and glass temperature, and we will use these both to control the glass temperature in the forehearth. Useful measurements that are not available are the glass-mass level measurement from the forehearth or from the refiner, or the measurement of the glass-mass flow to overflow.

3. MODELLING

Many kinds of physical and mathematical models have been built for glass furnaces (McConnell, 1976). In some of these the furnace is divided into differential volume elements, where outputs of one element are inputs to the next element. The construction of this kind of model is very tedious - and requires more measurement information about the process than is available.

Here the goal in building a mathematical model for the glass furnace was twofold. First to understand the dynamic behaviour of the furnace better and secondly to develop more efficient control strategies for both start-up and regulation purposes. The model has been constructed partly on the basis of physical laws and partly on the basis of experimental information. The overall model is pieced together using the models of different parts of the process.

From Fig. 3.1 the mass balance for the melting tank is

$$\dot{m}_m + \dot{m}_{rm} = \dot{m}_{mr} \quad , \tag{3.1}$$

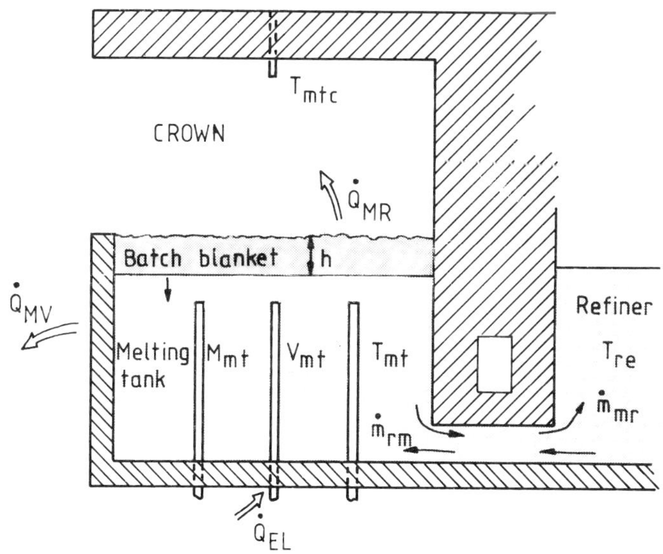

Fig. 3.1. Melting tank.

where

$\dot{m}_m$ = mass flow of molten glass from the batch blanket
$\dot{m}_{rm}$ = mass flow from the refiner to the melting tank
$\dot{m}_{mr}$ = mass flow from the melting tank to the refiner

Because there was no way to accurately measure the mass flow from the refiner to the melting tank, $\dot{m}_{rm}$ was assumed constant. The mass flow $\dot{m}_m$ of molten glass from the batch blanket is calculated from the glass temperature in the melting tank T_{mt}. The functional relationship between $\dot{m}_m$ and T_{mt} is obtained from measurements.

The energy balance for the melting tank is

$$c_g M_{mt} \frac{dT_{mt}}{dt} = 2c_g \dot{m}_{rm} T_{re} - 2c_g \dot{m}_{mr} T_{mt} - Q_{ME} \dot{m}_m + \dot{Q}_{EL} - \dot{Q}_{MW} - \dot{Q}_{MR} \tag{3.2}$$

where

c_g = heat capacity of glass
M_{mt} = mass of glass in the melting tank
T_{mt} = temperature of glass in the melting tank
T_{re} = temperature of glass in the refiner
Q_{ME} = heat needed to melt glass
$\dot{Q}_{EL}$ = heat flow produced by melting electrodes
$\dot{Q}_{MW}$ = waste heat flow through the walls
$\dot{Q}_{MR}$ = waste heat flow by radiation

The first and second terms on the right hand side of (3.2) represent the input heat flow caused by the cooler glass flowing from the refiners into the melting tank and the output heat flow due to glass flow the melting tank into the refiners, respectively. The third term, $Q_{ME}\dot{m}_m$, is the heat consumed to melt the batch to glass.

The heat flow $\dot{Q}_{EL}$, produced by the melting electrodes, is calculated from

$$\dot{Q}_{EL} = \eta_{EL} P_{EL} \tag{3.3}$$

where

η_{EL} = efficiency (constant)
P_{EL} = electric power of electrodes

The heat conduction $\dot{Q}_{MW}$ through the brick walls of the furnace is given by

$$\dot{Q}_{MW} = h_W A(T_{mt} - T_{out}) \tag{3.4}$$

where

h_W = heat transfer coefficient
A = heat transfer area
T_{out} = ambient temperature

Heat flow $\dot{Q}_{MR}$ due to radiation

$$\dot{Q}_{MR} = \varepsilon\sigma A(T_{mt}^4 - T_{out}^4) \tag{3.5}$$

where

ε = the average emissivity of the surface
σ = Stefan - Boltzmann constant

Mass and energy balance equations for the refiner and the forehearth are derived in a similar fashion (Setälä and Koivo, 1983).

The heat transferred to the glass from burning gas, Q_{GAS}, is proportional to the temperature of the forehearth crown. It influences the dynamics of the temperature measurement more than the dynamics of the crown temperature, which is quite fast. Based on experimental data the following transfer function between the crown temperature and the control signal of the gas valve were obtained

$$\frac{T_{cr}(s)}{\dot{m}_{gas}(s)} = \frac{3.8}{1+8.5s}, \text{ forehearth shutter closed,} \tag{3.6}$$

$$\frac{T_{cr}(s)}{\dot{m}_{gas}(s)} = \frac{4.9}{1+29.2s}, \text{ forehearth shutter open.} \tag{3.7}$$

The process dynamics can now be summarized as follows

$$\dot{T}_{mt} = \frac{-2\dot{m}_{mr}T_{mt} + 2\dot{m}_{rm}T_{re}}{M_{mt}} + \frac{\dot{Q}_{EL} - Q_{ME}\dot{m}_{m} - \dot{Q}_{MW} - \dot{Q}_{MR}}{M_{mt}c_g} \tag{3.8}$$

$$\dot{T}_{re} = \frac{\dot{m}_{mr}T_{mt} - (\dot{m}_{rm} + m_{rd} - 2\dot{m}_{rf})T_{re} + 2\dot{m}_{fr}T_{fo}}{M_{re}} + \frac{\dot{Q}_{RE} - \dot{Q}_{RW} - \dot{Q}_{RR}}{M_{re}c_g} \tag{3.9}$$

$$\dot{T}_{fo} = \frac{\dot{m}_{rf}T_{re} - (\dot{m}_{fr} - \dot{m}_{p})T_{fo}}{M_{fo}} + \frac{\dot{Q}_{GAS} - \dot{Q}_{FW} - \dot{Q}_{FR}}{M_{fo}c_g} \tag{3.10}$$

$$\dot{T}_{cr} = \frac{1}{\tau_{cr}} \cdot T_{cr} + k \cdot \dot{m}_{gas}$$

where in addition to the previously defined symbols we have

M_{re} = mass of glass in the refiner
$\dot{T}_{fo}$ = temperature of glass in the forehearth
$\dot{Q}_{RE}$ = heat flow produced by radiation elements
$\dot{Q}_{RW}$ = waste heat flow through the walls
$\dot{Q}_{RR}$ = waste heat flow by radiation
M_{fo} = mass of glass in the forehearth
$\dot{T}_{fo}$ = temperature of glass in the forehearth
$\dot{Q}_{GAS}$ = heat flow produced by burning of gas in the forehearth crown
$\dot{Q}_{FW}$ = waste heat flow through the walls
$\dot{Q}_{FR}$ = waste heat flow by radiation
τ_{cr} = time constant; 8.5 sec in (3.6) and 29.2 in (3.7)
k = constant; 3.8 in (3.6) and 4.9 in (3.7)

In addition to the dynamical equations there are algebraic equations analogous to (3.3)-(3.5).

The parameters in (3.8) to (3.11) were determined by fitting the model to experimental step-response data from a furnace at Nuutajärvi glass factory.

The model was used to build a computer simulator for the glass furnace to test various control strategies.

4. CONTROL OF THE MELTING FURNACE

The furnace has been mainly controlled manually up till now. A new control strategy was developed for the furnace in an earlier paper (Setälä and Koivo, 1983). The basis for the proposed control strategy is that the glass quantity needed in the production is known. The products which will be produced in every forehearth, are known at least one week before. This list of products is typed into the memory of the control computer to a production file as code numbers. Another file, contains information about the weights of the products, the production rates and the amount of waste glass. Using these we can now calculate the common amount of glass needed in the production.

The amount of molten glass produced by the furnace depends almost totally on the electric power of the melting electrodes. The electric melting power can be calculated, because we know the electric power needed to produce a certain amount of molten glass from process experiments. The principle of determining the electric melting power with the aid of the previous knowledge about the production is shown in Fig. 4.1. Fig. 4.2 gives melting power as a function melting capacity.

When the glass melting capacity of the furnace must be changed, for example after a weekend or when the grade of articles is changed, it must be done before the moment when more glass is needed. This anticipated time depends both on the amount of the change and on the initial capacity level.

Several test runs with a computer simulator have been performed by Setälä and Koivo (1983) to determine, how fast it is possible to

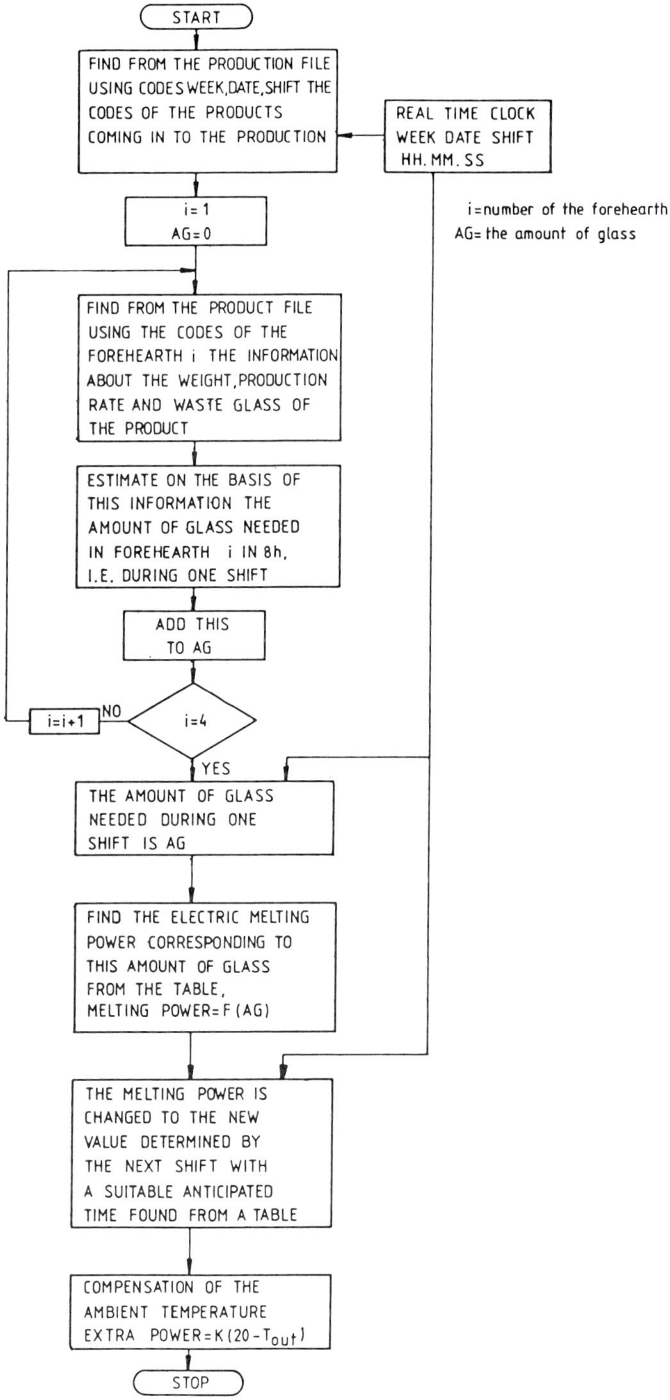

Fig. 4.1. Block diagram of estimating the amount of glass, melting power, and the anticipated time.

MELTING CAPACITY [kg/24h] AT THE END	MELTING POWER [kW] AT THE END	ANTICIPATED TIME [h] — MELTING CAPACITY [kg/24h] AT THE BEGINNING 1200; MELTING POWER [kW] AT THE BEGINNING 207	2400; 265	3600; 305
1350	215	5.67		
1500	224	8.42		
1650	231	9.08		
1800	238	10.17		
1950	245	10.33		
2250	259	11.42		
2550	271		4.42	
2700	277		6.75	
2850	282	13.33	7.67	
3000	287		9.00	
3150	292		10.50	
3450	301	13.75	11.08	
3750	309			1.83
3900	312	14.25	11.67	5.17
4050	315			5.58
4350	321	14.33		6.67
5100	332	14.75	12.33	8.75

Fig. 4.2. Relationship between melting capacity, melting power, and the anticipated time to change the glass melting capacity from one operating point to another.

change the glass melting capacity from one operating point to another. The results, which are the anticipated times, are shown in Fig. 4.2. We can see, that it takes more time to change the melting capacity at lower melting power levels than to make the same change at higher power levels. Therefore the furnace is planned to be operated at the top power levels where it is more efficient.

The electric power of the radiation elements should be controlled during start-ups according to the proposed time optimal strategy introduced in Setälä and Koivo (1983). The algorithm sets the power of the radiation elements at first to the maximum value. When the glass temperature in the refiner rises close enough to its new set-point, we change to a PI-control. This kind of strategy speeds up the glass heating in the refiner but keeps the overshoot in bounds. The above control strategy has been tested in simulation runs and was found quite effective compared to other strategies.

The main objective of the forehearth control is to keep the glass temperature at the set point and when it is necessary to transfer the temperature as fast as possible to the new set point. A cascade loop is proposed for that purpose. In the new control strategy the set point is given to the glass temperature and not to the crown temperature as has been done before. The structure of the control loop is shown in Fig. 4.3.

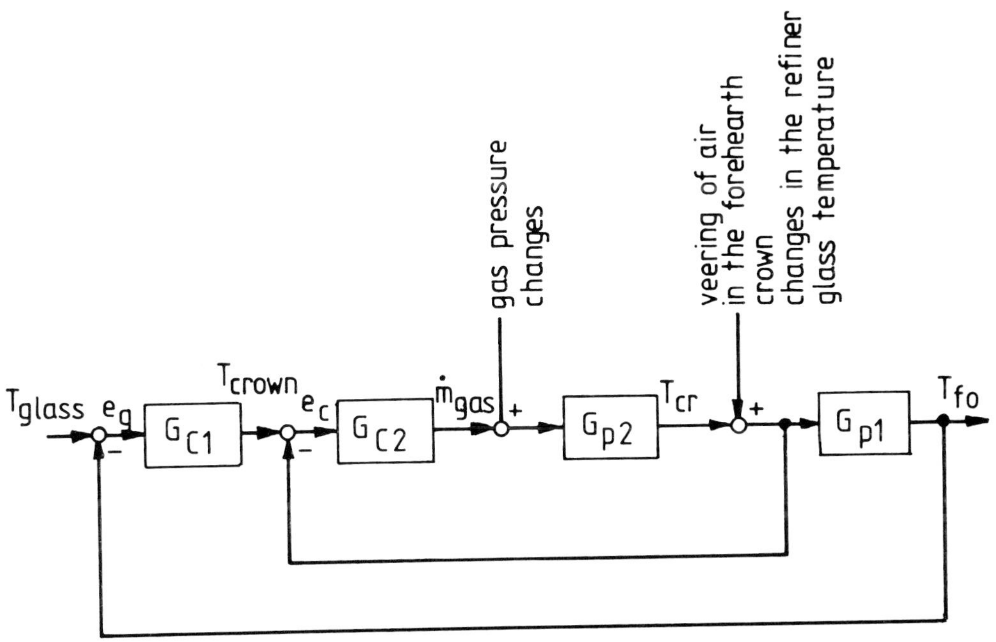

Fig. 4.3. Cascade control strategy for glass temperature control in the forehearth.

The output of the primary process is the forehearth glass temperature and the input the forehearth crown temperature. The output of the secondary process is the forehearth crown temperature and the input the gas flow. Both the primary and secondary controllers are PID-controllers.

The cascade control compensates efficiently the disturbances appearing in the secondary process before they have had effect on the primary process. Such disturbances are e.g. changes in gas pressure and in gas-to-air ratio. It also eliminates disturbances that are produced by setting the forehearth shutter on or off. In this situation the waste heat flowrates vary a lot. Because the forehearth glass temperature is a much more important than the crown temperature, the cascade control loop is clearly a better solution to control forehearth temperatures. The cascade control loop eliminates also the change of the glass temperature in the forehearth caused by the change of glass temperature flowing from the refiner.

The cascade control strategy is based on the fact that the dynamics of the secondary loop is much faster than that of the primary loop. The control period for the primary loop is about 10 min and for the secondary loop about 3 min.

5. COMPUTER CONTROL SYSTEMS IN THE GLASS INDUSTRY

The first energy crisis in 1970's forced the glass industry to pay more attention to energy consumption in the process. To achieve energy savings in glass melting a better understanding of the furnace operation is needed to more intelligent control strategies. These can be realized with computers. At first the computers were used only as supervising and reporting devices in the glass industry. They were used to collect measurement data from the process and to give important information about the direction and the rate of change of the process variables. The control variables were then set on the basis of this information.

At the next stage the computer also gave the set point to an analog controller based on the measurement information. The analog controller then controlled the corresponding outputs to the given set point.

At present direct digital control is used, where the computer computes directly the value of the control signal using measurement data. It then sends the control signal to the actuator. In the latest, more extensive control applications in glass industry modern distributed control systems have been used. The tasks have been divided to microcomputers and to process stations operating at different levels.

In the glass industry computers are applied among other things in the following areas:

- storing, transferring, mixing, and feeding of batch and cullet
- control of melter, refiner, and forehearth
- control of shaping machines

- control of annealing oven and
- counting of glass articles and inspection.

Depending on the extent of the application a microprocessor based control system, a microcomputer, or a bigger process computer is installed. In the application considered here a microprocessor based control system is suitable. The control system includes both direct digital control and supervisory control tasks.

6. A MICROCOMPUTER SYSTEM FOR A GLASS FURNACE

6.1 Hardware

The control system has to control the process according to the requirements of the production schedule and to take care of the shutdowns and start-ups at weekends in additions to the regular control tasks. The control system must therefore be able to handle tables, make calculations, operate in real time according to the state of the process, and make decisions and sound alarms on the basis of the state. A microprosessor control system is well-suited for this purpose. The basic structure of the control system is shown in Fig. 6.1.

The system includes a control unit, a terminal, a floppy disc station and a printer. The control unit contains a central processing unit, a clock, enough memory space, A/D- and D/A-converters, binary input and binary output boards, serial interface boards, a power source and other necessary boards. The control unit performs the measurements and calculations, sends the control signals to the process, and regulates how the other devices function. With the help of a terminal and a keyboard the production program is fed into the memory of the computer, and the necessary changes are carried out. The state of the process can be followed in numerical form and with trend curves. When a disturbance occurs the necessary control actions are performed. Both the product file, which contains information about the weight, production speed etc. and the production file, which contains the weekly production program, are recorded on the floppy discs. Process information for trend displays and reports are also recorded there. Most of the programs are located on the floppy discs and they are only loaded into the central memory to be run when needed. The printer output includes alarms, reports and the product and production files.

If it is necessary to expand the usage of the control system to other parts of the process such as the annealing oven or press and blow machines, and the software is developed according to use and experience, it is justified to think about linking a program development system to the control system. The structure of such a control system is shown in Fig. 6.2. The hardware is built around the program development system. The disc station, the terminal and the printer are connected to the program development system. A separate control unit communicates with the program development system through a serial interface bus. The control unit is able to operate fully independently when some disturbance has occured in the program development system.

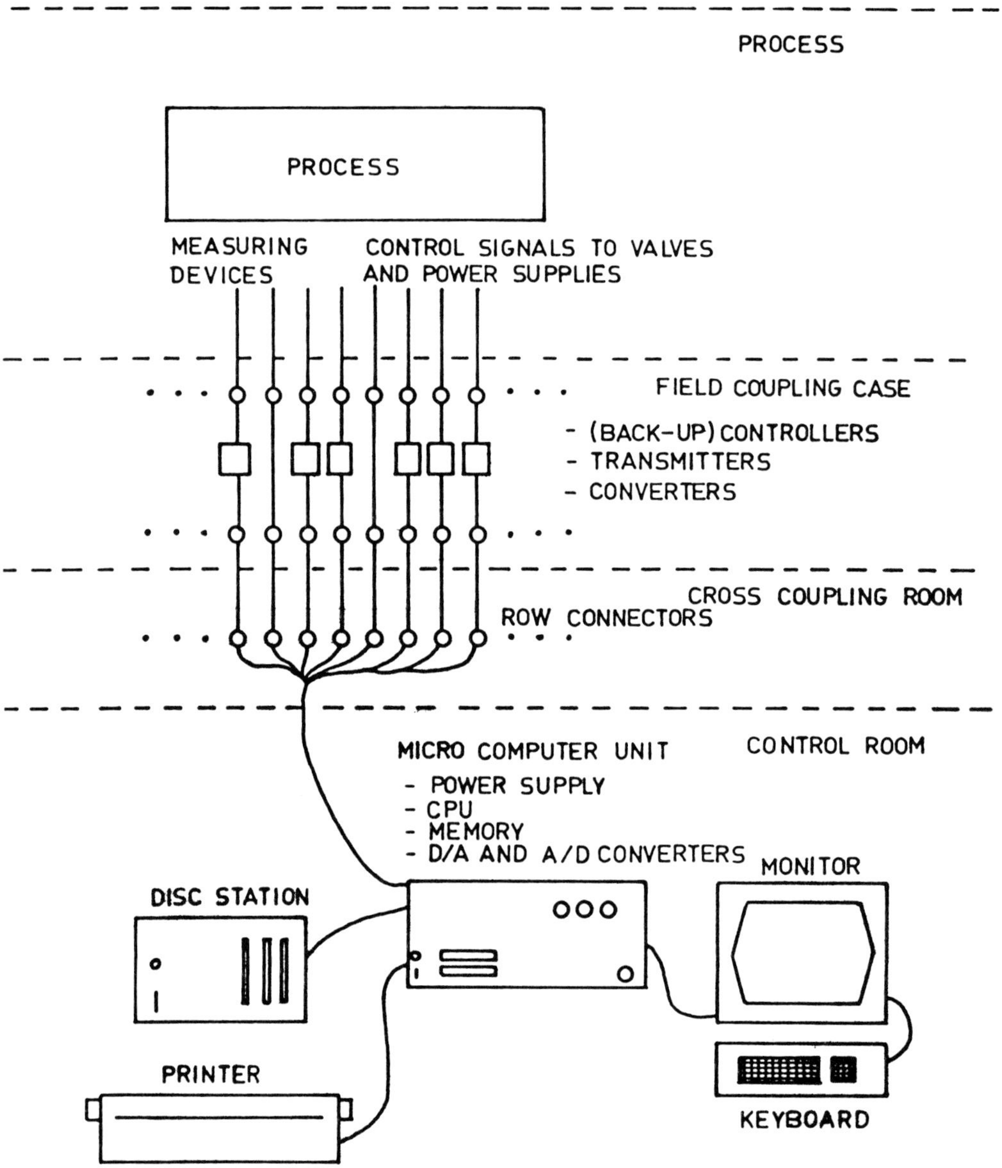

Fig. 6.1. Basic structure of the control system.

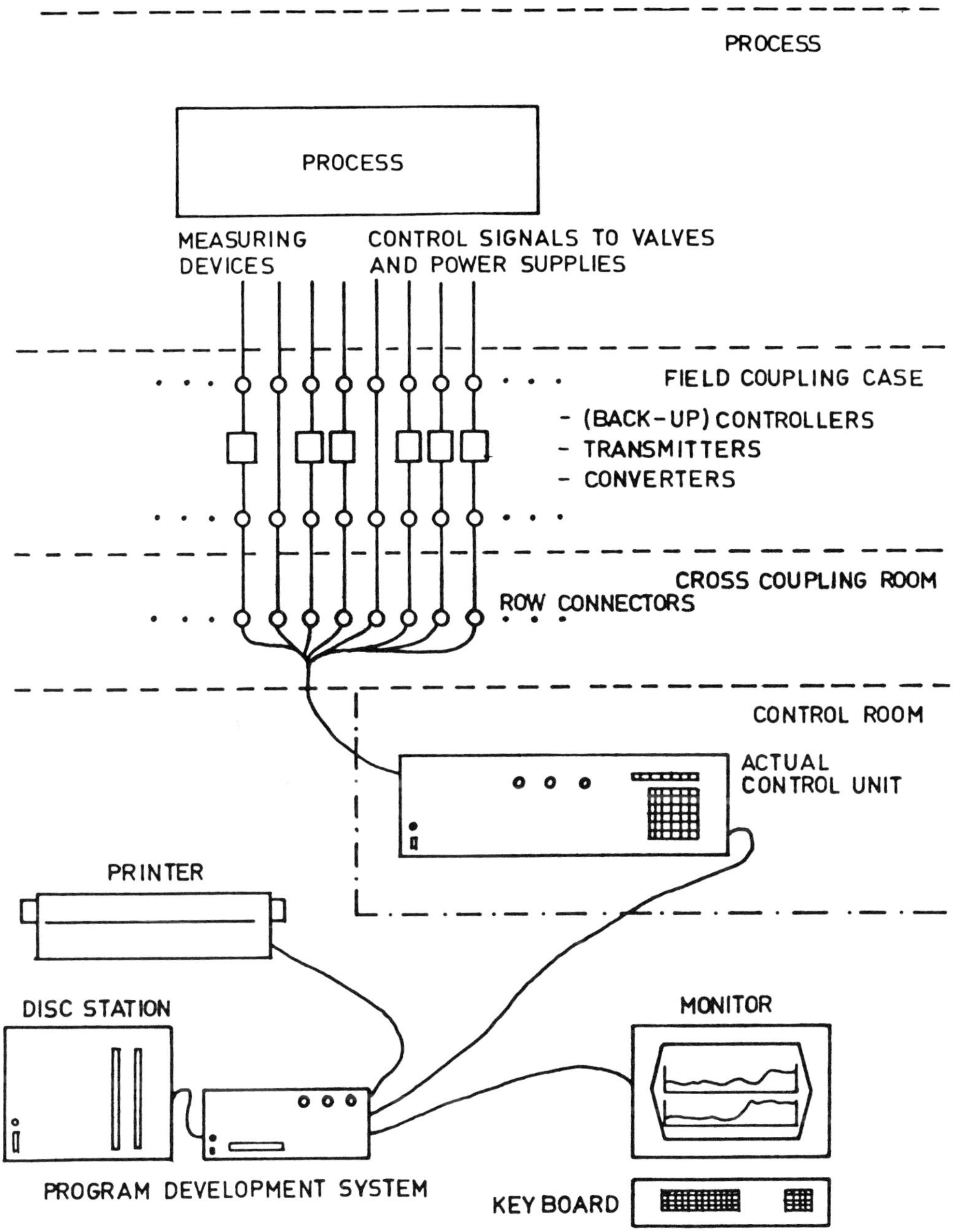

Fig. 6.2. Alternative structure for the control system.

The sharing of the tasks is arranged in the following way. A separate control unit takes care of the measurements and the calculations of the control signals according to the setpoint given by the program development system. The control unit sends the control signals to the process and measurement information, alarms and other important messages to the program development system. In case of failure at the development system the process can be operated with a minikeyboard of the control unit. The program development system records and saves the information coming from the control unit on the floppy disc. The data can be displayed on the terminal or printer if such a request is made. The program development system is also required to take care of maintaining the production and product files, and of the updating of control parameters and setpoints in real time according to the production program. In addition to these it has to handle alarms and write reports. Furthermore, it is possible to develop and change the system software.

6.2 Application software

The software of the proposed microcomputer control system can be divided into two different entities: a real time operating system and application software. The operating system controls the system resources. It forms the interface between the system hardware and application tasks. The application software can be divided into three groups:

- estimation software for the glass melting capacity and for the melting power
- measurement tasks and
- control tasks.

A block diagram of the application software is shown in Fig. 6.3. The first group includes software for processing of production and product files introduced in section 3. In addition it contains software for estimating the amount of glass needed in the production and further for estimating the electric melting power and the anticipated time of the power change. The second group includes the usual measurement and filtering tasks, software for processing of measurement information e.g. to remove bias of thermoelements, display software and software for changing and updating parameters. Alarm and reporting tasks are included here, too. The third group contains the control tasks: control of the electric power of the melting electrodes and the radiation elements, and the cascade control of the gas flow. The control of the raw material feeder is easily linked to this, too.

CONCLUSIONS

A detailed design of a microcomputer -based control system for a continuously operating, electric glass furnace is presented. It is based

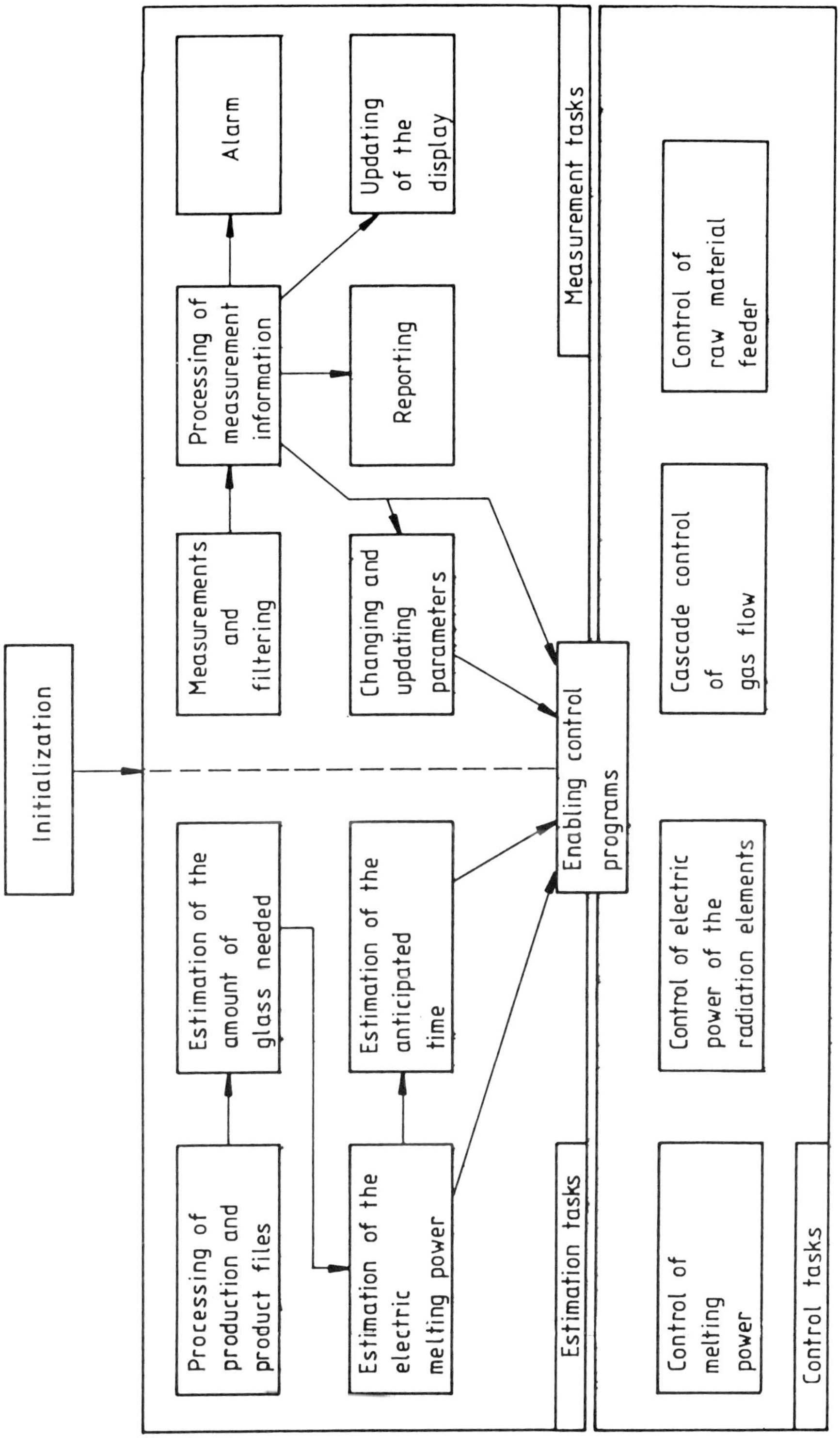

Fig. 6.3. Block diagram of the software.

on being able to determine the melting power needed to produce a certain amount of molten glass. This, on the other hand, is determined by the desired production level.

The paper gives the structure of the control system hardware and software. The application software is designed using TOP-DOWN method by dividing the tasks into modules. The main application tasks are: (1) estimation software for the glass melting capacity and for the melting power, (2) measurement tasks, and (3) control tasks.

REFERENCES

Ashfield, A.W.F. (1980). 'Furnace control strategies'. *Journal of Non-Crystalline Solids,* 38 & 39.

Davies, R.B. (1979). 'Microprocessor based control systems in the glass industry'. *Glass,* 56 508-509.

Edgington, J.H. (1979). 'Controls in the glass industry and future automation: Part 1'. *Measurement and Control.* Vol. 12, No. 7.

Edgington, J.H. (1979). 'Controls in the glass industry and future automation: Part 2'. *Measurement and Control.* Vol. 12, No. 8.

Finger, T.H. (1978). 'Digital control- mini and microcomputers in glass melting'. *Proc. of the IEEE-IAS 13th Annual Meeting.* Toronto, 975-977.

Finger, T.H. and J.M. Law (1978). 'Direct digital control of a glass melter, refiner and forehearths using small computers'. *IEEE Trans on industry applications,* IA-14, 62-67.

Gaige, R.A. (1979). 'The installation of a digital control system for a glass container furnace'. *Proc. of the IEEE-IAS 14th Annual Meeting, Cleveland,* 665-667.

McConnell, R.R. (1976). 'Modeling glass furnace, refiner and forehearth design and operation for energy efficiency and improved control'. Ph.D. Thesis, Purdue Univeristy, Indiana, USA.

Pincus, A.G. and G.M. Diken (1976). *Electric Melting in the Glass Industry.* New York.

Rawson, E. (1981). 'A microprocessor based system for process control applications'. *Glass Technology* 22, 108-112.

Setälä, A. and H.N. Koivo (1983). 'Modeling and control of a class melting furnace'. *Proc. of the First IAESTED Int. Symposium on applied Control and Identification*, Copenhagen, Denmark.

Stanek, J. (1977). 'Electric melting of glass' in *Glass science and technology,* vol. 1, Elsevier, New York.

CHAPTER 11

MULTIPROCESSOR ROBOT CONTROL SYSTEM USING MICROPROGRAMMING

Professor Milan Jelšina
Head of Department of Technical Cybernetics,
Technical University,
Švermova 9/B
041 20 Košice
Czechoslovakia

ABSTRACT. This contribution is oriented to some problems of microprogramme means synthesis intended for the robot control system. The need of the microprogramme control in the executive level of the control system is motivated in the above contribution. Simultaneously the systematic approach of the microprogramme processor design and some of its fundamental components are suggested. In the conclusion are presented some microprogrammes fragments of the process control in the robotic system and their application in the suggested control system.

1. CONTROL OF ROBOTIC SYSTEM

The robotic system is a programmable heteregenous automaton intended for operational activity of a technical feature, which is performed as a result of its purposeful interactions in the working space. From the system point of view it is reasonable to decompose the robotic system into operational and control part. These parts have a quite different physical structure in this case. The operational part of the robot system is represented by the robot, the effectors and sensors of which mediate the informational and mechanical interaction with the working space. The control part of the robot system represents an electronic control system, the computer means of which make it possible to programme, control and correct the execution of the required activity by the operational part. The control of the robot system is oriented to provide for the desired robot movement and consequently we distinguish [1], [2], [3]:

1. point-to-point control (successive reaching of the stationarily defined points),
2. path control (movement along defined space trajectory),
3. force control (control of the time layout of the applied force and (or torque).

The path control can be regarded as a generalized case of the point-to-point control and the dual case of the force control. For this reason we have taken choice of the path control as a foundation

N. K. Sinha (ed.), Microprocessor-Based Control Systems, 223–247.

for the illustration of the requested computational rate which is one of the factors influencing the structure of the control system's computer means.

Let $x(t)$ be the generalized vector to characterize the position and orientation of the robot's end effector in the Cartesian space and $q(t)$ - vector of the joint variables to characterize the relative position of the individual of the robot mechanism links. Then the equation

$$\bar{x}(t) = F(\bar{q}(t)) \quad (1)$$

indicates the mathematical formulation of the direct problem of kinematics which defines a movement of the robot's end point according to the established joint variables. For the robot path control system the solution of the opposite task has a greater importance : to compute the values of the joint variables so that the robot's end point may move along the defined trajectory. The solution is represented by the inverse problem of kinematics in the form

$$\bar{q}_{id} = F^{-1}(\bar{x}_{id}) \quad (2)$$

where $\bar{x}_{id}$ is the vector of the discrete points of the required space trajectory, and q_{id} the vector of the relevant generalized coordinates in the joint variables' space.

The computed values q_{id} represent the control statement acting at the input of the joint mechanisms servoes.

There are many methods for the solution of the problems stated above. One possible approach suggests that in order to be able to control we need to have the knowledge of the speed values of the individual joints. For this reason it is necessary to distribute the movement speeds in the Cartesian space according to the equation

$$\dot{\bar{q}} = J^{-1}(\bar{q}) * \dot{\bar{X}} \quad (3)$$

where $J(\bar{q})$ is Jacobi matrix with elements

$$J_{ij} = \partial x_i / \partial q_j$$

The relation (3) follows from the modification of the linear transformation in the form

$$dx = J(q) * dq \quad (4)$$

which expresses the equation (1) for the case of the path distribution of the robot's end point to the increments dx [5]. The computation of the increments requires to perform complicated goniometric dependences and/or nonlinear arithmetic operations.

The relationship (3) may be solved on the basis of the minimization of the actual position deviation x of the robot's end point from the required position x_d which stems from the application of the gradient method in the form

$$\dot{\bar{x}} = -\lambda \left[\frac{\partial Q}{\partial x}\right]^T \quad (5)$$

where Q is a cost function

$$Q = \frac{1}{2} (\bar{x}_d - \bar{x})^T * (\bar{x}_d - \bar{x}) \tag{6}$$

After substituting the relationships (6) and (5) into (3) and (1) we can obtain

$$\dot{\bar{q}} = \lambda \, J^{-1}(\bar{q}) * (\bar{x}_d - F(\bar{q}))$$

from where the final solution will be obtained

$$\bar{q} = \lambda \int J^{-1}(\bar{q}) * (\bar{x}_d - F(\bar{q}))dt \tag{7}$$

The organization of the computation of the joint variables to the performance of the robot's end point movement along the determined path in the Cartesian space according to relationship (7) is illustrated in Fig. 1a.

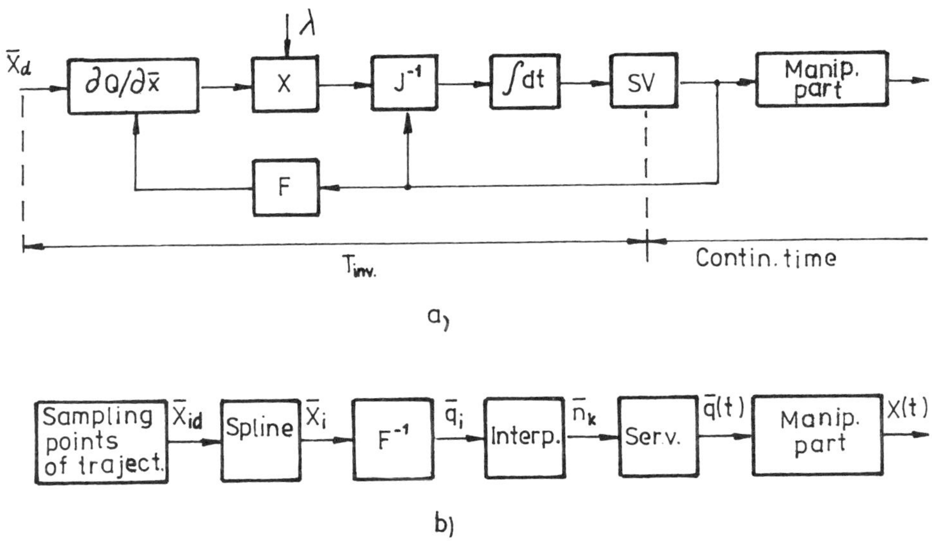

Figure 1. Two principles of robot control (a-by the close loop, b- by the open loop).

The design of the robot control described above will be included in the class of the computation with the closed loop. The particular class is represented by the computations with the open loop in the framework of which the relationship (2) may be solved directly [5] on the basis of the restored points of the primarily performed trajectory of robot movements.

In this case the procedure of the process control is as follows:

1. through the discrete points of the space trajectory the approximating spline function will be put and the interpolated values of the discrete points will be computed within the sampling interval,

2. the inverse problem of kinematics will be solved,

3. the linear interpolation on the sections with the computed values q_i will be solved,

4. the increment of the point variables values for the indivi - dual servoes of the joint mechanisms will be represented by the above obtained values, and

5. after completing the adequate transient process which is expressed by the equation

$$\dot{\bar{q}}(t) = f(\bar{q}(t), \bar{q}_j) \tag{8}$$

the position of the robot's end point will coincide with the desired value x_{id}.

Such organization of the control is illustrated in Fig. 1b. To execute the required computations in real time it is necessary to design a reasonable structure of the control system which in general has also to provide the control mechanism for the next activity :

1. interface with the sensor system for the creation of the feed-back,

2. interface with the primary performance system for the teach-in process, and

3. interface with the language means for the programming of the robot activity.

In following the attention is concentrated to the structured organization of the computing means of the robot control system from the view of the primary performance and/or use of the microprogramme means for the control mechanisms.

2. MULTILEVEL STRUCTURE OF ROBOT CONTROL SYSTEM

In general the robot control system can be structured into three levels: strategic, tactic, and executive. The strategic level determines the main objective of the control of the robot movement and operation, it designs the optimal layout of its activity and plans its driving trajectories. The tactic level solves and optimizes the individual tasks which are defined on the strategic level and provides the necessary interface with the surroundings for the fulfilment of those activi - ties. The executive level performs the direct interface with the discrete environment of the control including the relevant data, statement and state communications.

In following the control system of the robot (CSR) is proposed in which the strategic and tactic part is not presented explicitly. With regard to it we will assume two levels in that CSR: coordinative and executive. The architecture of the CSR is illustrated in Fig. 2 [6] . In this architecture stress is laid upon the means for the primary performance and/or programming of the robot movement.

The display (DP) is intended for the communication of the operator with CSR in the programme control mode. The conversation is performed through the technological language that makes it possible to programme two individual partial activities as well as to create technological programmes. The manual control board (MCB) is intended for the mode of the robot manual control. In that mode the operator can execute the primary performance of the driving trajectory in the teach-in process. The primary performance can be accomplished in the joint mode (by the individual driving of the simple joint mechanism), in the coordinate mode (by the movement of the robot's end point in the direction of the chosen coordinate or its rotation, and the leading - through mode (by the movement of the robot's end point along an arbitrary space trajectory by the joystick J). Moreover the MCB makes it possible to define the various types of the production of the primarily performed movement (cyclic, with return into the last point, with tracing, etc.) and further discrete statements (robot movement stop, interval set of the robot movement stop, effector driving etc.).

To the proposed functional structure of the CSR corresponds the adequate architecture of the computer means.

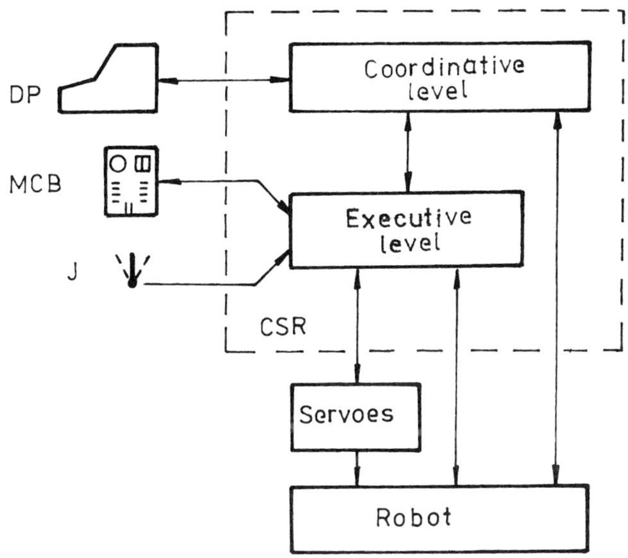

Figure 2. Two-level structure of robot control system

3. ARCHITECTURE OF MULTIPROCESSOR ROBOT CONTROL SYSTEM

The architecture of the CSR computer means is dependent on some factors from which in the first place appears :

1. the computation of the increment of the joint variable values for the path control in real time,
2. the configuration of the CSR's interface with the robot and its work environment, and
3. the approach to the programming and primary performance of the robot activity.

It is known [1] ,[5] that in manipulator path control in the Cartesian space the sampling rate of the driving trajectory and joint angle increment generation frequency must be chosen to satisfy [6] [7]

$$f_v < f_r < f_g \tag{9}$$

where f_v is the sampling rate, f_r - the resonant frequency of the mechanical structure of the robot, and f_g - the increment generation frequency for the servoes of the robot joint mechanism.

For the characteristic values of the digital control of the robot servoes ($f_r \approx 10$ Hz, $f_g \approx 10^5$ Hz) the unequation (9) is always fulfilled (some problem can arise during the optimal design of f_r from the point of view of the memory capacity needed for the restoring of the points of the primarily performed movement trajectory).

At the digital control of the servoes are by the incremental pulses with the frequency f_g generated N pulses ($N = T.f_g$) in the course of a chosen time interval T. These pulses represent the precomputed increment of the joint variable adequate to this interval.

If the period T_p required for the computation of the joint variable increment during the interval T by the monoprocessor CSR is not satisfied then it is not possible to fulfil the unequation

$$T_p < T < T_v \tag{10}$$

where $T_v = 1/f_v$, and then considering the restriction of the value T (as a rule to be chosen [7] $T = T_r = 1/f_r$) the only solution is to bring down the value T by the design of a more efficient RCS's computer means. This approach leads to the multiprocessor CSR architecture.

For the interface organization with the robot's components and its environment (the sensor system, the servoes of the joint mechanisms, manual control board etc.) the CSR is to be supplied with individual microcomputers which are parts of the multiprocessor system as shown above. Because for the above mentioned activities no general purpose microcomputers are required, it is reasonable to design specialized microcomputers with microprogrammable architecture which allows to suit perfectly the interface control mechanism.

In Fig. 3 is shown the basic structure of the two-level CSR from which follows its functional organization (see chap. 2). The coordinative level of CSR includes :

1. the processor subsystem (μP1, μP2) on the basis of two 16 bit microprocessors,

2. the memory subsystem which is represented by two 64 kbyte operational memory (M1,M2) and by 512 kbyte external tape memory (EM) intended for the technological programme and the system software, and

3. the sensor subsystem (SN) for obtaining the information about the robot's word by which it can achieve features of artificial intelligence.

The computer architecture is created on the principle of two buses (B1,B2) which are mutually connected by the bus adapter (BA).

The main component of the executive level of the CSR is a specialized microprogrammable microprocessor (EP) which is intended for the transformation of the control and move statements generated on the coordinative level into signals directly driving the servoes of the joint mechanisms. Moreover it also provides for the necessary technological commands and the primary performance of the driving trajectory at the teach-in process by the joystick (J) or the manual control board (MCB) and their specialized electronics denoted EB and EJ, respectively.

The function of the individual components at the path control is illustrated in Fig. 4 in which the symbols are applied with this meaning :

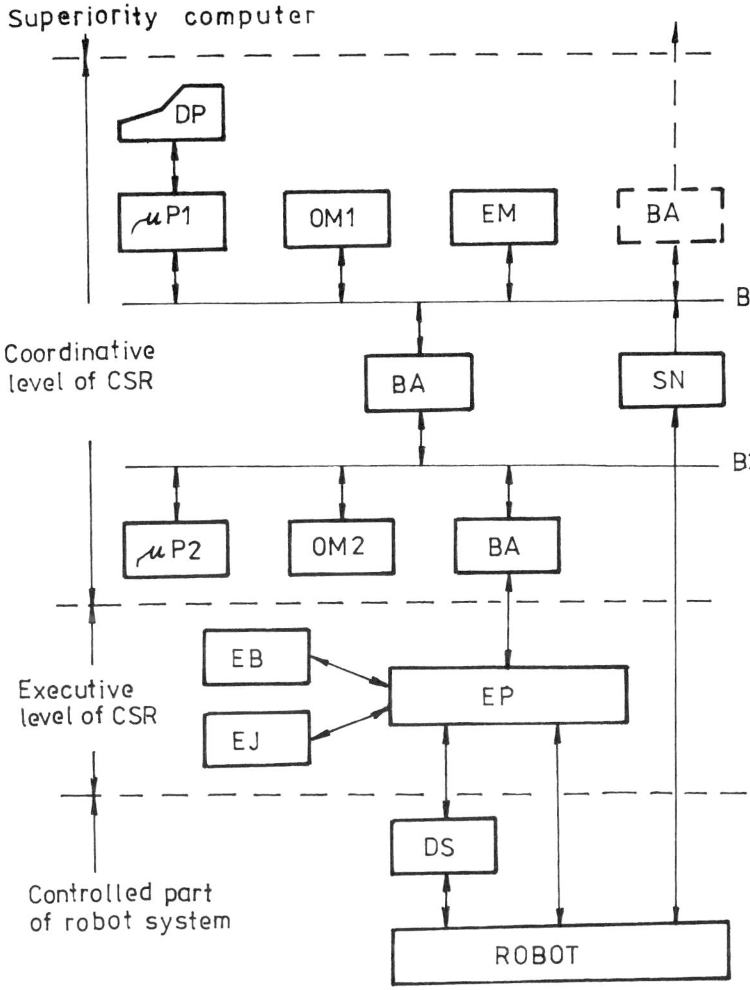

Figure 3. Multiprocessor architecture of robot control system

$\bar{X}$ - the driving trajectory of the robot's end point movement in the Cartesian space ($\bar{X}_{id}$ - a vector of the stored discrete points, $\bar{X}_i$ - a vector of the discrete points after the approximation by the spline function),

$\bar{q}$ - the driving trajectory in the point space ($\bar{q}_i$ - a vector of precomputed discrete points following from the solution of the inverse task of kinematics),

$\bar{n}$ - the reproduced form of the linearly interpolated trajectory along the sections T, and

$\bar{n}_k$ - the vector of the pulsed function acting on the servoes inputs.

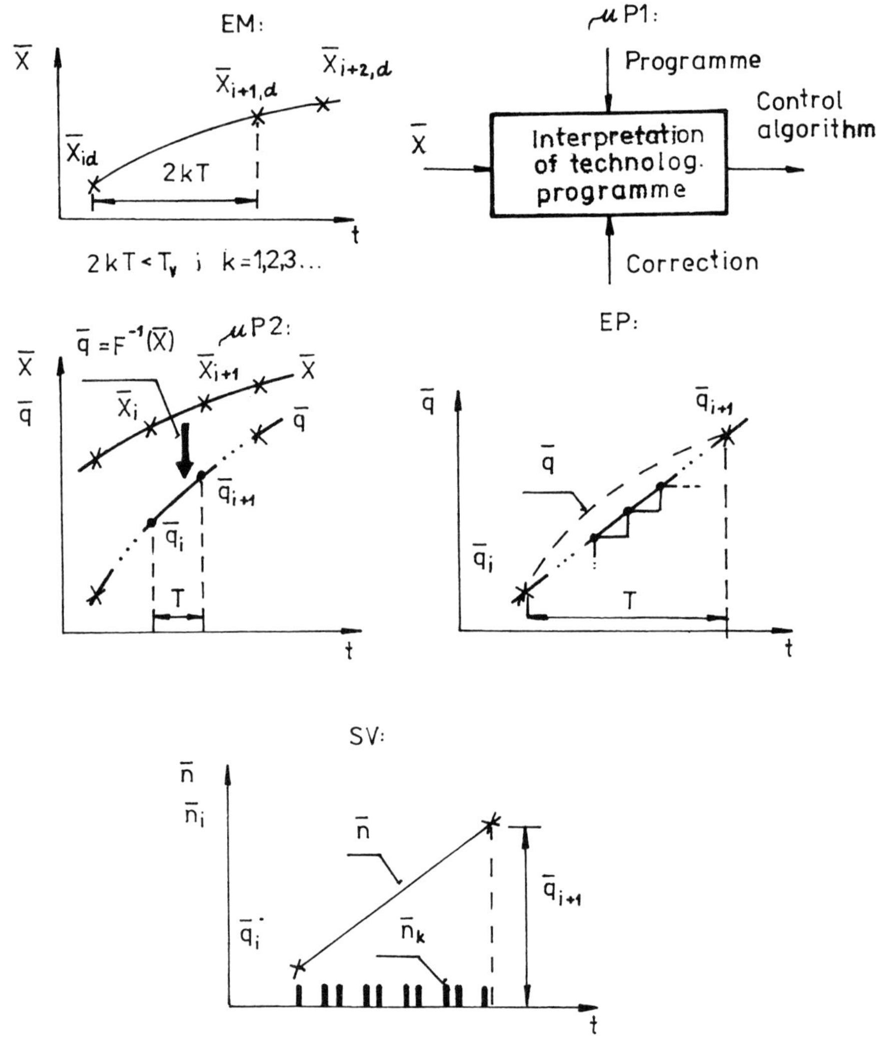

Figure 4. The individual activities of multiprocessor RCS's components

4. MICROPROGRAMMABLE ARCHITECTURE OF EXECUTIVE LEVEL OF ROBOT CONTROL SYSTEM

4.1. The Micrcprogramme Control Principles

4.1.1. The Microprogramme Control Unit. The robot control systems (CSR) represent a specialized class of digital systems intended for discrete processing of data (DPD). In general the DPD can always be decomposed into two basic parts : the control part (CP) and the operational part (OP) which communicate by their input and output channels with a controlled discrete environment (DE). The channels represented in the decomposition (Fig. 5) have the following meaning :

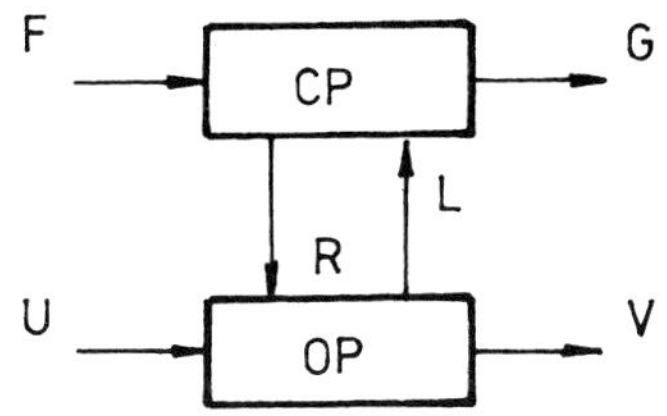

Figure 5. The decomposition of discrete processing system to control (CP) and operational (OP) part

U is input data channel, V - output data channel, R - internal control statements channel, L - internal logic conditions (predicates) channel, F - input control statement channel, G - output control statement channel.

In following the j-th vector of the input or output word will be denoted

$$\mathfrak{A}^j = \mathfrak{A}_{j_1}\mathfrak{A}_{j_2}\ldots\mathfrak{A}_{j_a}, \quad A \in \{U,V,R,L,F,G\} \tag{11}$$

and the input or output word

$$A^j = A_{j_1}A_{j_2}\ldots A_{j_{a_j}}, \quad a_j \leqslant a, \; j = 1,2,\ldots,2^a \tag{12}$$

in which for the binary variables $A_{j_k} \in A = \{A_1, A_2, \ldots, A_a\}$ it holds

$$\bigcap A_{j_k} = 1, \; j_k \in \{1,2,\ldots,a\}$$

The control part of the DPD which appears in a relation to the OP as the autonomous special-purpose or universal complex of the control means is called the control system (CS). The control system which can be repeatedly decomposed into the control and operational part will be called the data control system (DCS). The characteristic feature of the DCS is that it controls the OP which in this case is the external operational part (EOP), on the basis of processing data in the internal operational part (IOP) which is the component of the DCS (Fig. 6a). The external control statements $R_i \in RE$ need not be dependent on the processing data during the whole control interval. On the other hand the IOP can directly define the control statement for the EOP (see Chap. 4.2). The control system in which it is not possible to make a decomposition as seen above, is called the ordered control system.

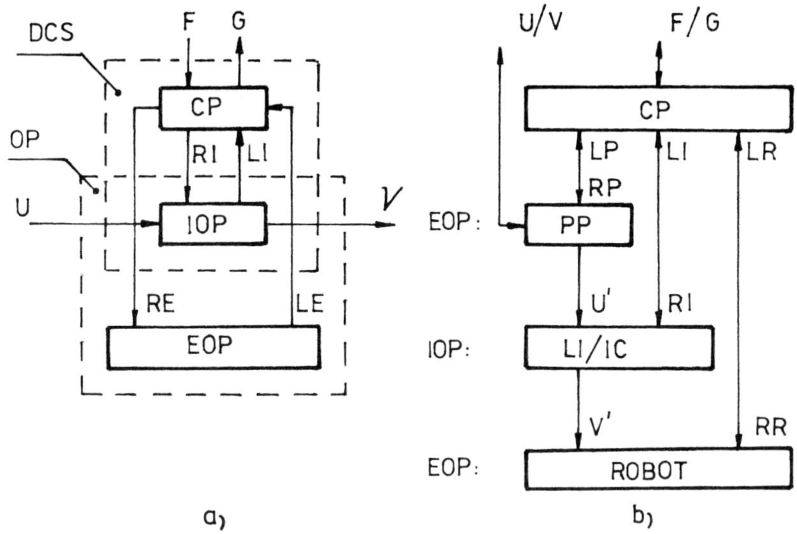

Figure 6. The data control system (a- in general, b- aplication to executive processor of CSR

The hardware design of the CS or its control part can be performed as a control unit with fixed or programmable logic. For CRSs are more significant the control units with a programmable logic which from the point of view of flexibility, control, general-purpose and systematism of design have a lot of advantages.

In the following our attention will be devoted to the microprogrammable control units which in comparison to programmable ones have a special-purpose character. The feasibility of those units makes it possible to obtain a substantially greater operational speed for the digital control. The latter property is especially important in the control of the robots which require to solve a great extent of computations and to execute a complicated heterogenous operation in real time.

The microprogrammable control unit introduced above is, as for its function, in fact a microprogrammes' control unit (MCU). The substantial role in a MCU is played by the microprogrammes stored in the microprogrammes' memory (MM), in form of the microinstruction sequence in which are coded the control statements for the control system. In the MCU's cycle the microinstructions are fetched from the MM to be written into the microinstruction register (RGM), on the output of which they will explicitly define the internal R^j and the external G^j control words as well as the information about the mode of the creation of the next microinstruction address. This address is generated by the address and control logic network (ACL) according to the direct information from the microinstruction, external control word F^j

and the internal status word L^j.

4.1.2. Process of microprogramme control. The process of micropro - gramme control follows from the mutual interaction of the MCU and OP by the control (R^j) and the status (L^j) words operating in the relevant internal channels R and L.

The microinstruction initiating the microoperation M on the OP will be denoted as a general statement in the form [8]

$$n_i : M/\alpha_1 : n_1, \alpha_2 : n_2, \ldots, \alpha_k : n_k \tag{13}$$

where $n_i, n_1, \ldots, n_k$ are labels, $\alpha_1, \alpha_2, \ldots, \alpha_k$ - predicates defined by the OP for which it holds :

$$\alpha_i \wedge \alpha_j = \emptyset, \ i \neq j, \quad \bigvee_{i=1}^{k} \alpha_i = 1$$

From the general commands we can derive the following form of the commands :

1. operational statement : for $\alpha_j = 1$ (α_j is a constant predi - cate) in (13) it holds : $n_i : M/\alpha_j : n_j$ or simply

$$n_i : M/n_j \tag{14}$$

2. control statement : for M = NOP(M is an empty microoperation) in (13) it holds : $n_i : NOP/\alpha_1 : n_1, \ldots, \alpha_k : n_k$ or simply

$$n_i : \alpha_1 : n_1, \ldots, \alpha_k : n_k \tag{15}$$

3. STOP statement : for $\alpha_i = 1$ and M = NOP in (13) it holds : $n_i : NOP/\alpha_i : n_i$ or simply

$$n_i : STOP \tag{16}$$

4. residual statement : for the necessity of holding the oxecu - ted microoperation to a longer period than is the time interval of the microinstruction cycle the commands are proposed in the form :

$$\begin{aligned} n_i &: M/n_j, \ M\Uparrow \\ n_k &: M/n_m, \ M\Downarrow \end{aligned} \tag{17}$$

where $M\Uparrow$ and $M\Downarrow$ denotes the beginning and the end of the microoperation holding respectively. (Analogically we can also apply the conception of the residual predicate).

The family of those microinstruction types can be extended in accordance with the requirements of the microprogramme control (see Chap. 4.4.).

The semantical interpretation of the microinstruction types as seen above can be presented by the flow chart fragments in Fig. 7.

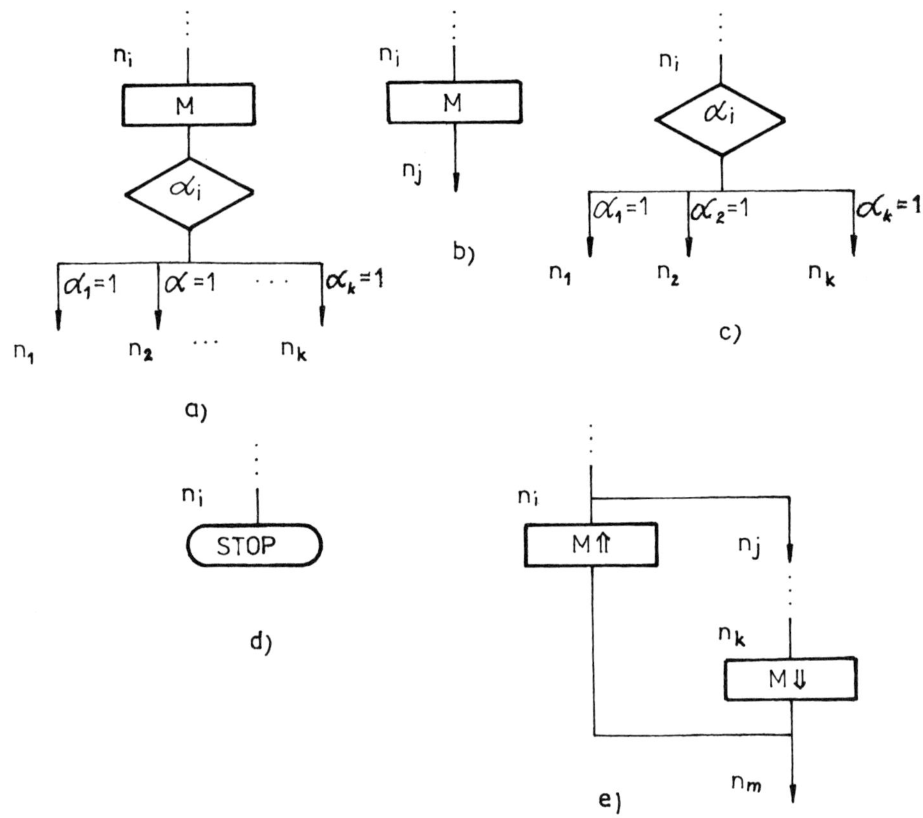

Figure 7. The flow chart fragments to express the elementary types of microprogram coded microinstructions by adequate commands (a- general, b- operational, c- control, d- STOP, e- residual)

The fact that the microoperation M_i is initiated by the statement $R_i = 1$ can be denoted in the form [6]

$$R_i)M_i \tag{18}$$

where R_i is control variable. In general the control word R^j initiates, with regard to the formulas (11) and (12), exactly r_j local microoperations which can be expressed in the form :

$$R^j)M_{j_1}, M_{j_2}, \ldots, M_{j_{r_j}} \tag{19}$$

Those microoperations are always executed simultaneously during one step of the microprogramme control.

The fact that the information medium about the fulfilment of the condition (predicate) α_i is state variable L_j which is set according to $L_j = 1$, if the condition is fulfilled, it will be formally described as :

$$L_j)\ \alpha_j \tag{20}$$

In general we can define a composite predicate

$$\alpha^j = \alpha_{j_1} \wedge \alpha_{j_2} \wedge \quad \cdots \wedge \alpha_{j_{\ell_j}}$$

the truth value of which we will express according to formulas (14) and (17) in the form :

$$L^j)\ \alpha_{j_1} \wedge \alpha_{j_2} \wedge \quad \cdots \quad \wedge \alpha_{j_{\ell_j}} \tag{21}$$

The application of some microinstruction types introduced in the definitions is illustrated in the creation of the linear interpolation microprogramme in the Chap. 4.5.

4.1.3. <u>Interpretation of microinstructions in the microprogramme control unit</u>. From the substance of the microprogramme control follows that the microprogramme to be stored in the microprogramme's memory will be executed by the interpretation of its microinstructions sequence in the MCU. The interpretation of individual microinstructions rests on the generation of the control words vector $\tilde{R}$ as well as of the addresses of the next microinstructions defined by the adequate labels on the basis of the logical conditions fulfilments defined by the status words vectors $\tilde{L}$.

From the definition of microinstruction follows that the format of its binary standing consists of two basic parts : the command and the address. (It stands also for the additional information). It is as well in accordance with the layout of the fundamental structure and function of the MCU which is responsible for the interpretation of the microprogramme microinstructions. Then, in general, the code KMI of the microinstruction M can be expressed in the form of the Boolean vectors chain to represent the individual segments of the microinstruction format as follows :

$$KMI = K_1(M).K_2(R^i).K_3(L^j).K_4(n_j).K_5(Z) \tag{22}$$

where $K_1(M)$ is the segment of the microinstruction type responsible for the internal control of the MCU, $K_2(R^i)$ - the segment of control words R^j responsible for the generation of the control commands , $K_3(L^j)$ - the segment of status word L^j selection intended for choosing the condition by which are generated the addresses of the next microinstructions, $K_4(N_j)$ - the segment of the direct addresses of the next microinstructions, and $K_5(Z)$ - the segment of the addresses sources of the next microinstructions.

The binary chain (22) represents a general format of the microinstruction in which $K_1(M)$, $K_2(R^i)$ and $K_3(L^j)$, $K_4(n_j)$, $K_5(Z)$ stand for its command and address part, respectively. The main component of the MCU is the address and the control logic (ACL) the structure of which follows directly from the microinstruction format and from the given set of the address sources. The segment $K_1(M)$ of the micro - instruction to be written into the microinstruction register (RGM) defines the internal control commands (IR) for the MCU's components. The generation of the commands $R_k = 1$, $k \in \{1,2,\ldots,r\}$ is provided by the control commands generator (GC) through the transformation $K_2(R^i) \to \bar{R}^i$.

The next microinstruction address to be generated in the framework of the operational order execution of the type (14) with unconditional jump, or also of the type (15) with conditional jump after fulfilling the adequate branch condition, can be directly set in the microinstruction by the segment $K_4(n_j)$, or in some of the address sources defined by the segment $K_5(Z)$. The selection of the logical conditions, by which will be executed the branching in the microprogramme, is defined by the segment $K_3(L^j)$ and it is performed in the condition logic (CL). The finite address of the next microinstruction is generated by the microprogramme processor (MP) which includes two elementary components : address logic (AL) and source logic (SL). AL is responsible for the setting of the relevant branching address into the microprogramme on the basis of the selected logical condition values in the CL. The components CL and MP (AL,SL) create the structure of the ACL. The example of the ACL design and the microinstruction interpolation from the point of view of the next address generation and address sources in the microprogramme branching is introduced in Chap. 4.4.

4.2. The microprogramme control of the executive level of the robot control system

The task of the CSR's executive level is :

1. to generate precomputed increments of the values of the joint mechanisms angle variables,
2. to generate control commands for the robot and its environment,
3. to receive the state signals about a robot status and its environment,
4. to propose the communication of the control system compo - nents with the primary performance subsystem, and
5. to propose the own communication with the coordinative level of the control system.

The execution of these activities is formulated as a control process which can be described by a control microprogramme. In con - sequence of the requirements on the processing of a great number of data, heterogeneous control commands and state signals in the real time, the executive processor (EP) with the microprogrammable architecture will become the optimal hardware solution.

In the proposed architecture of the CSR, the executive processor (EP) is designed as a data control system according to Fig. 6a the individual components of which are represented by the following structure (see Fig. 6b) :

CP - microprogramme control unit (MCU),

IOP - system of linear interpolators and increment counters (LI/IC),

EOP - subsystem primary performance (PP), servoes of joint mechanisms (SM) and discrete surroundings of robot (DR),

DE - coordinative level of control system (CLCS).

The communication EP with CLCS is performed by the logic channels F, G,U,V which mediate the transfer of the following information :

F - executive instructions (represent a flag instruction, operational code, number of the next data and complementary statement information),

G - status words (represent information about the modes of manual control and primary performance, the alarm status and the error reporting),

U - input data (represent the precomputed values of joint variables increments), and

V - output data (represent the primary performed values of joint or coordinative variables increments).

The MCU communicates with its operational part by the control commands $R_i \in R = \{RPURIURR\}$ and state variable $L_i \in L = \{LPULIULR\}$ which are defined by the microinstruction of the microprogramme. (The symbols P, I and R in the set defined above stand for the primary performance, interpolator and robot, respectively.) The processing data in IOP is distinguished from the current one in which $V = F(U)$ so that $\widetilde{V} = f(\widetilde{U}')$ and $\widetilde{V}' = g(\widetilde{U})$, where f is the function following from the microprogramme control by commands $R_i \in \{RIURP\}$, g - function following from $R_i \in \{RI\}$ and U',V' - discrete control commands (they have not a data character of increments).

4.3 The microprogramme processor of executive level of the control system

In Fig. 8 is introduced the structure of the robot control system executive processor. The CP of this processor communicates with the coordinative level of the control system (CLCS) by the output and the input data registers : RGOUT and RGIN, respectively. The RGOUT is intended for writing the status words from the status register (RGST) and/or the data (values of primarily performed joint variables) from the DO - output of the OP. The RGIN is intended for writing the executive instruction (EI) and the precomputed increments (PI) from the CLCS.

The main components of the EP's control part is a microprogramme processor (MP), mapping memory (MAP), microprogramme memory (MM) and condition logic (CL) which together with the OP represent the MCU. The communication of the MCU with the MCB is performed by two status registers : the hand control (RGHC) and the control acknowledgement (RGCA). The role of the control signale generator is ensured by the decoder (DC) and the joint distribution (JD) for the generation of the pulses $R_I \approx 1$ to the adequate servoes in accordance with the performance of the microoperation. R.JD[I]) R_I:=1, where JD[I] stands for the set of the JD to the I-th joint mechanism.

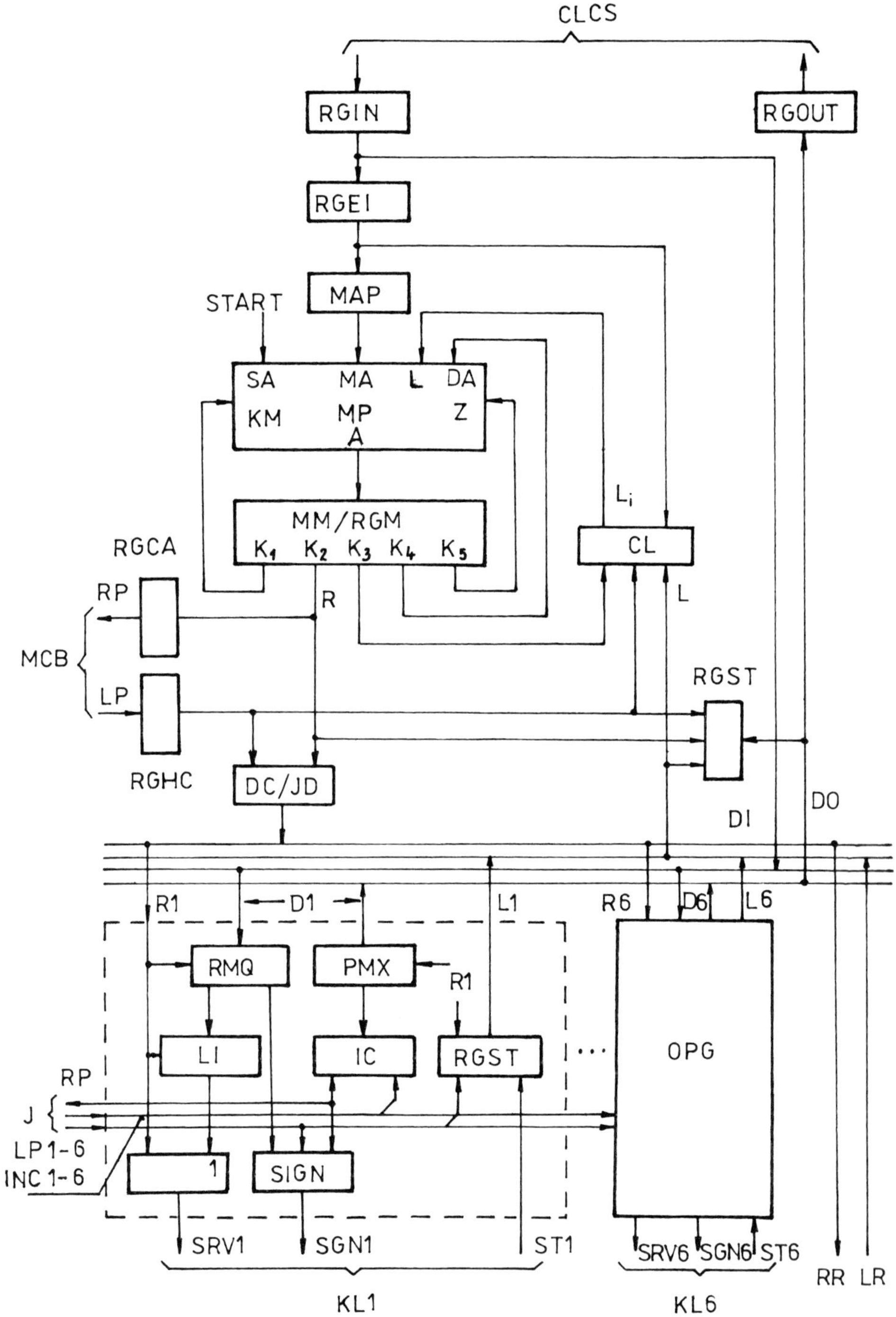

Figure 8. The structure of executive processor

The address and the control logic (ACL) defines the address of the next microinstruction by two of its types :

$$n_i \; : \; R^j/L_k \; : [Z] \; , \; L_m \; : \; n_m \tag{23}$$

and

$$n_i \; : \; R^j/[Z] \tag{24}$$

where Z is one from the following sources : direct address (DA), microprogramme starting address (SA), address of mapping (MA), and internal address (IA) becoming accessible after the fulfilment of the selected logical condition.

The individual segments $K_1, K_2, \ldots, K_5$ of the microinstruction fetched from the MM to the microinstruction register (RGM) stand for parts of its format according to the description (22).

The operational part is divided to six sections from which each is intended for the autonomous control of one of the servoes of the joint mechanisms. The section dealing with this includes two regis - ters : the input (PMQ) and the output (PMX) which are necessary for writing the increment values that the linear interpolator (LI) treats and distributes to the input's servoes (by outputs SRV_i and SGN_i), and for the accumulation of the increment pulses obtained by means of the increment counter (IC) in course of the chosen modes of the primary performance, respectively.

For the mode of primary performance by the joystick (J) are the sections of the OP directly connected with its interface electronics, where RP, INC1-6 and LP1-6 denote the collective control command for all axes, and state signals, respectively. The microprogramme architecture proposed above enables to obtain the required high speed of the movement control of the joint mechanisms in real time. In Chap.4.4, 4.5. and 5. are given some details of the microprogramme control design.

4.4. A synthesis of the main components of the executive processor control part : the address logic

The analysis of the microinstructions types, their formats, and address logic, according to Chap. 4.1.3., enables a systematic design of the microprogramme control and adequate logic networks [9], [10]. There are five steps of this design which are illustrated by examples as follows :

1. The control process is expressed by a microprogramme.
2. On the basis of the microprogramme structure analysis we de - fine the types of statements (instructions).

Further on we suppose the microinstructions of (23), (24) types in the form :

$$\text{Type 1: } IR_k . R^j/[Z] \;\; ; i \in \{1,2,\ldots,8\}$$

$$\text{Type 2: } IR_k . R^j/L_k : [Z] \; , \; L_m : n_m ; \; k,m \in \{1,2,3,4\}$$

where IR_ks are variables of internal commands, and Z is one of next sources : mapping memory (MAP), stack memory (SM), microprogramme counter (MPC), and direct address (DA).

3. In accordance with the chosen types of microinstructions the codes of segments $K_1(M)$ for the internal control can be defined (e.g. the rewriting of MPC into the stack memory, the decrementing of stack pointer, etc.). The segment $K_2(R)$ is intended for direct generation of commands $R_i = 1$.

4. The codes of segments $K_3(L)$, $K4(n)$ and $K_5(Z)$ are chosen so as they may be able, from the defined logical conditions L_1, L_2, L_3, L_4, to set all the possible couples $L_1 L_2, L_1 L_3, \ldots, L_3 L_4$, to generate the direct address (DA), and to define one of four sources (Z) of the address (DA,EA,MA,IA), respectively. For example, by the interpretation of the microinstruction of type 2 is executed the microoperation $R^j)M^j$ as well as generated the next address (A) of microinstruction in accordance with the values of the conditions L_k and L_m ($k,m \in \{1,2,\ldots,\ell\}$) and selected the address source to be defined in the segment $K_5(Z)$, $Z \in \{DA,EA,MA,IA\}$ and as a result it holds :

$$A = \begin{cases} DA \Leftarrow L_m = 1 \\ DA \Leftarrow L_k = 1, \ K_5(Z) = 00 \\ EA \Leftarrow L_k = 1, \ K_5(Z) = 01 \\ MA \Leftarrow L_k = 1, \ K_5(Z) = 10 \\ IA \Leftarrow L_k = 1, \ K_5(Z) = 11 \end{cases}$$

The external address EA is represented by the starting address (SA) and the internal address (IA) by one of the stack memory (SM) or the microprogramme counter (MPC).

5. On the basis of the codes defined above we will create the hardware design of the condition (CL), source (SL), and the address (AL) logic of the MCU.

For the considered case the structure ACL is shown in Fig. 9. The design of AL is as a rule performed by standard components that possess the required elements (SM,MPC,etc.), e.g. microsets KP 1800 [11], Am 2910 [12] , I 3001 [13] .

4.5. A synthesis of main components of the executive processor operational part : interpolators

The main component of the proposed executive processor is the linear interpolator. For the aim of the interpolation design at the executive level it is started from the modification of the programme inter - polator [14] intended for the discrete control of the movement trajectory in the state space. For the needs of the proposed EP the inter - polation is designed to the movement control in the time space [15] which characterises the condition of the physical performability in the form $N \leqslant T$, where N is number of pulses (an increment value) to

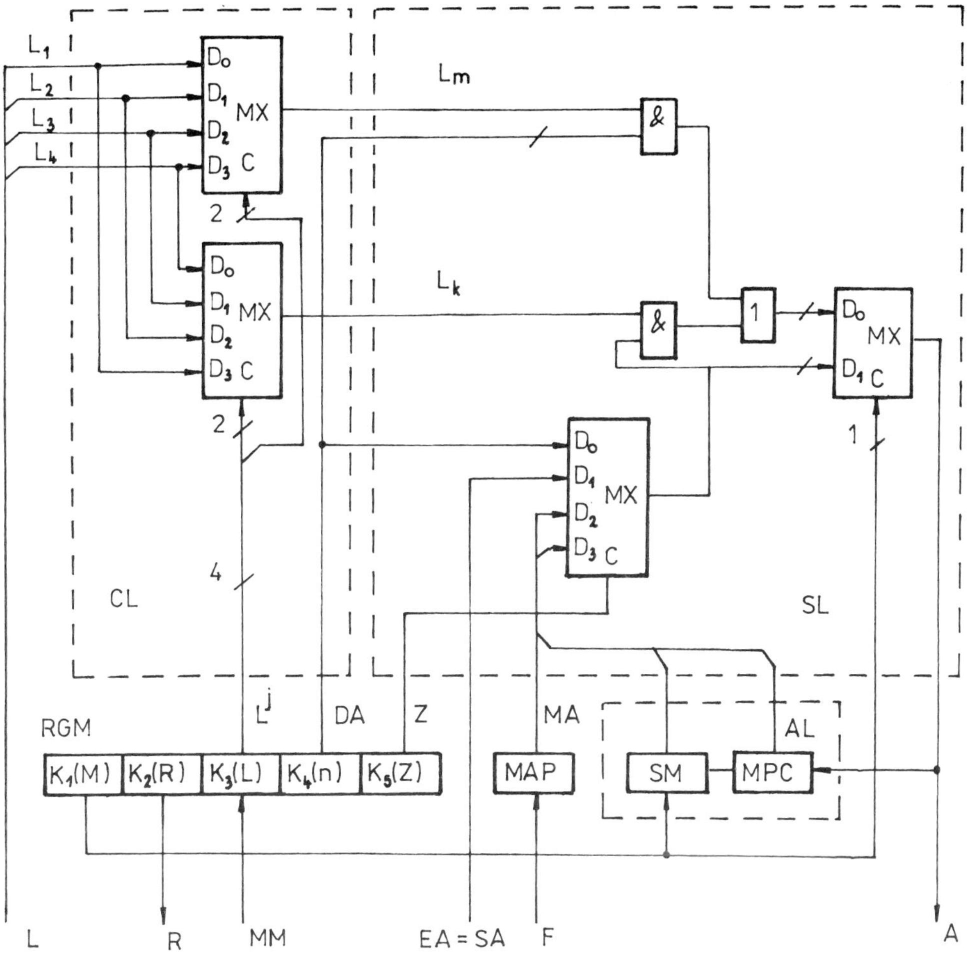

Figure 9. The example of address and control logic

be generated on the interpolation interval T. This interval T expresses the pulses number which can be generated by the interpolator during this T.

In accordance with [15] under the linear interpolation we will understand the generation of the discrete function n_i for which during the interval T holds :

$$\Delta n_i = \begin{cases} 1, \text{ if } D'_i < 0 \Longrightarrow D_{i+1} = D'_i + T \\ 0, \text{ if } D'_i \geqslant 0 \Longrightarrow D_{i+1} = D_i \end{cases} \tag{25}$$

where $D_i' = D_i - N$.
The variable D_i in (25) stands for a finite value of the discriminant ($D_i = n_i T - N t_i$, $t_i = 0,1,2,\ldots$) on that interpolation interval the calculation of which follows from the recurent formula :

$$D_{i+1} = \begin{cases} (1+n_i)T - N(t_i+1), & \text{if } D_i' < 0 \\ n_i T - N(t_i+1), & \text{if } D_i' \geqslant 0 \end{cases}$$

The hardware realization of the linear interpolator can be designed as an automaton with the programmable or fixed logic. In Fig.10a is shown the flow chart of the proposed linear interpolation which can be executed by the microprogramme or technical means. The adequate microprogramme is expressed by the following description :

$n_1 : R^1/n_2$

$n_2 : R_1/L_2 : n_6,\ \bar{L}_2 : n_3$

$n_3 : R_2/L_1 : n_4,\ \bar{L}_1 : n_5$

$n_4 : R^2/n_5$

$n_5 : R_3/n_2$

$n_6 :$ STOP

in which are used these microoperations and predicates :

$R^1)$ IC : = 0, D : = 0	$L_1)$ $D < 0$
$R_1)$ NOP	$L_2)$ IC = T
$R_2)$ D : = D-N	
$R_3)$ IC : = IC+1	
$R^2)$ D : = D+T, n : = 1	

where n stands for the variable of the discrete increments of the interpolated function n_i.

The hardware design of this interpolator is based on the above presented modified microprogramme in the form :

$R_o)$ PC : = 0, D : = 0

$R_1)$ D : = D-N

$$R_1)\ IC := IC+1, \qquad D := \begin{cases} D+T,\ n:=1 \Leftarrow L_1 = 1 \\ D \qquad\qquad\ \Leftarrow L_1 = 0 \end{cases}$$

The proposed linear interpolator is shown in Fig. 10 b,c.

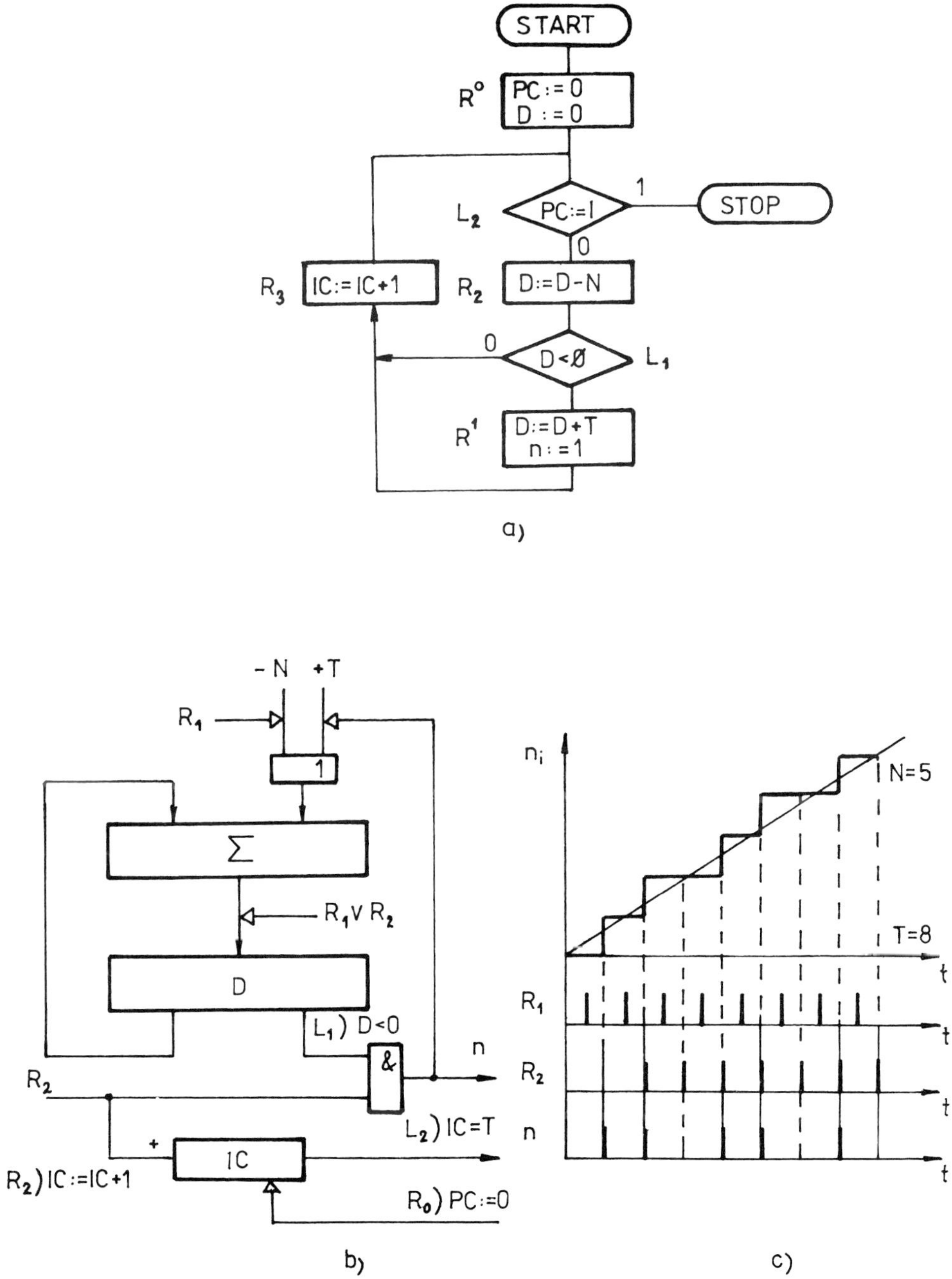

Figure 10. The linear interpolation (a- microprogramme flow chart, b- hardware design, c- result of interpolation for N = 5 and T = 8)

5. THE ALGORITHMS OF THE MICROPROGRAMME CONTROL OF THE ROBOT CONTROL SYSTEM EXECUTIVE LEVEL

The structure of controlling microprogrammes in the proposed EP is organized in the following mutually interconnected parts :

1. microprogrammes of control organization,
2. microprogrammes of executive instructions, and
3. microprogrammes of primary performance and hand control.

The microprogrammes of the process control organization are intended for the initiation of the adequate control microprogrammes performance. This choice is defined by the status statement with particular meaning, e.g.: the setting of some modes of control (programme, hand), the setting of some modes of the movement repetition (cyclic, one-shot, etc.), the calibration, the temporary movement holding, etc.

The microprogrammes of the executive instructions provide the performance of the statements to be generated in the framework of the programme control by the interpretation of the technological programme instructions or in the framework of some modes of the hand control.

The microprogrammes of hand control make it possible to execute the primary performance of the robot movement and to define the technological instructions that will be used in the movement reproduction.

As an example we will introduce the microprogramme of the point-to-point control at which the end point of the robot is to be set to the required position. In Fig. 11 is presented the simplified flow-chart of the primary performance microprogramme at the joint mode of hand control and in Fig. 12 is presented the basic microprogrammes structure, as well, in which are indicated their types and mutual connections.

At the joint coordinative mode of the robot hand control there are preset by the MCB the adequate state variables of modes and chosen the I-th joint. The movement along the direction of the I-th joint or Cartesian space coordinate begins by pressing the forward (F) or backward (B) direction button in consequence of which in the next microprogramme cycle is respectively incremented or decremented, i.e. the increment counter (IC) of the chosen joint executes one of the two next microinstructions

$$R_2)\, IC[I] := IC[I]+1$$

$$R_3)\, IC[I] := IC[I]-1$$

At the joint mode ($J = 1$) by the microoperation $R_4)\, SRV[I] := 1$ is generated the increment pulse to the input of the I-th joint mechanism servoes. The microprogramme loop, in the course of which this pulse has been generated, is completed by means of testing the time interval T, and after its elapsing ($T = 1$) the increment of joint variable value, represented by the accumulated number of pulses in IC[I] during the time interval T, is transmitted into the coordinative level. Moreover the generation of these pulses can be finished within the interval T by the data validity acknowledgement (ACK). The fre -

quency of the increment pulses generation, dependent upon the run of the microprogramme loop, is adapted to the needs of the real servoes inputs by the frequency brake (FB). This FB enables the transmission through the microprogramme loop on the preset time interval.

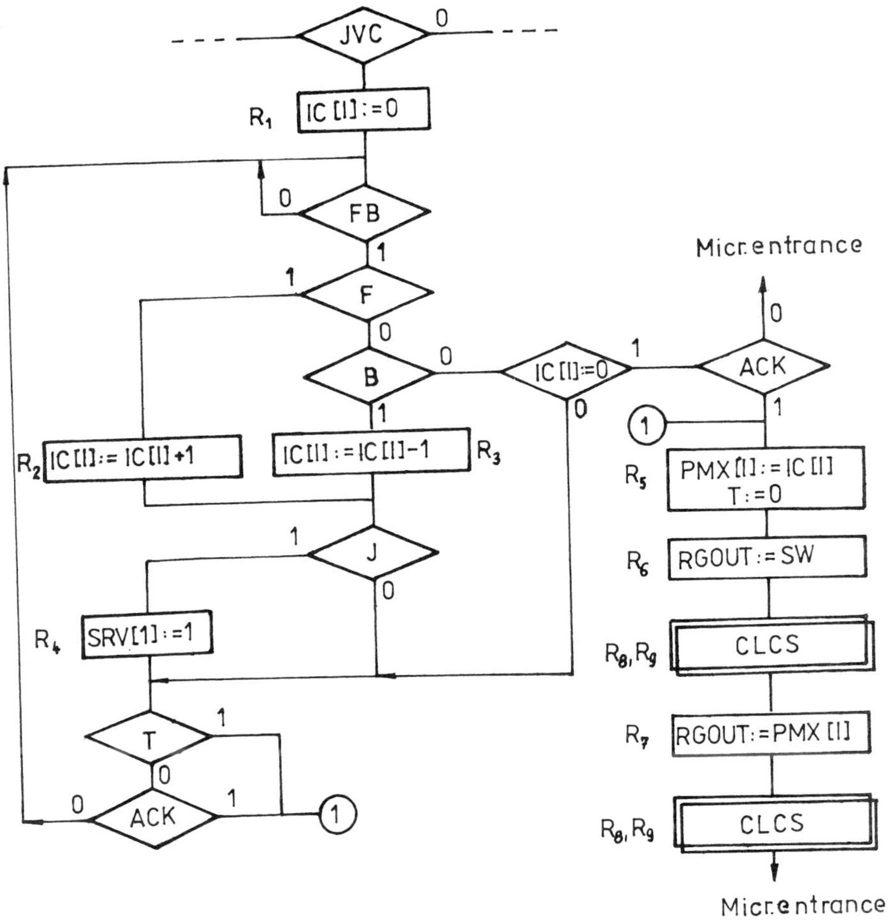

Figure 11. The simplified flow chart of primary performance microprogramme in joint/coordinate mode of hand control

After elapsing of the interval T the content of the IC [I] is transmitted into the coordinative level by the short communication microprogramme

R_8)RGOUT : = PMX [I]

R_9)CLCS : = RGOUT, $R_9\Downarrow$

before which has been transmitted the status word SW that is to carry out tho adequate state information about a control mode, joint chosen, etc.

If the above presented microprogramme represents the primary performance in the coordinative mode (C = 1), the microoperation R_4)SRV [I] : = 1 will be left out from the microprogramme loop.

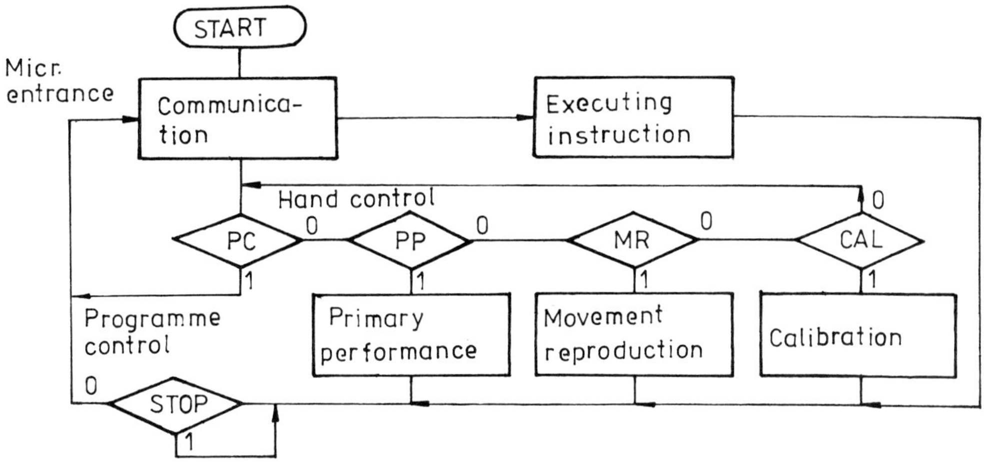

Figure 12. The structure of basic microprogrammes

The transmitted values of the content IC [I] in this case represent the movement increments in the direction of the Cartesian coordinate I. In the coordinative level they are precomputed in real time to the values of joint variables (inverse problem of kinematics), and by the executive instruction transmitted to the adequate linear interpolation input in the executive processor of the CSR. Some further details about the primary performance microprogrammes are given in [16] .

6. CONCLUSION

The submitted work represents a contribution to the research in the field of microprogramming for process control in the robot control system. In the introductionary part are indicated some elementary problems concerning the control of robot as well as the function and computer structure of its control system.

For the control system interface with the discrete surrounding we have tried to motivate the suitability of the specialized control units with a microprogramme architecture, the design of which represents the substance of the submitted work. We have started from the systematic approaches of the description of the individual microprogramming elements, on the basis of which it is possible to design the main components of the microprogramme control units intended for control of robotic systems. The work deals with a method of synthesis of these components which follows from the given structure of micro - instructions and their formats. Some of them intended for the main component of the executive processor (address and control logic,linear interpolators, and hand control microprogrammes) are proposed more in detail in the work.

The suggested approach and interpretation of control system microprogramming creates the space for obtaining higher operation speeds and for flexible adaptability of the control mechanism of its interface with the surrounding. The above presented approach has a general character and it is applicable in any control systems.

7. REFERENCES

[1] Paul, R.P.C.: 'Manipulator Cartesian Path Control.' IEEE Transaction on SMC, 9, 1979, 6, p. 449-455

[2] Luh, J.M.S.: 'An Anatomy of Industrial Robots and Their Controls.' IEEE Transaction on AC, 28, 1983, 2, p. 133-152

[3] Milenkovič, V. - Huang, B.: 'Kinematics of Major Robot Linkage.' In: [4], p. 16.31 - 16.47

[4] 13th International Symposium on Industrial Robots and Robots Application Worldwide. Chicago 1983, vol. 2

[5] Kováč, J.: Control Systems For Robots Path Control (in Slovak). PhD thesis, Department of Technical Cybernetics, Faculty of Electrical Engineering, Technical University Košice, 1984

[6] Jelšina, M. and oth.: Microprogramme Architecture of Executive Processor in Structure MUDRS (in Slovak). Research report, Department of Technical Cybernetics, Faculty of Electrical Engineering, Technical University Košice, 1984

[7] Jelšina, M. - Kováč, J.: 'A Microprocessor System with Parallel Architecture for Control of Robots.' In : Preprints IFAC - 9th World Congress. Budapest, 1983, vol. 2, p. 241-246

[8] Sabo, P. - Frištacký, M.: 'Updated Theory of Microprogram Control and Microparallelism for Microprogram Compaction.' Computer and Artificial Intelligence, 3, 1984, 5, p. 445-467

[9] Frištacký, M.: 'A Contribution to microprogramm Control Design.' 27. Intern. Wiss. Kolloquium, TH Ilmenau 1982, Vortragsreihe Entwurf, Analyse und Einsatz von informations-verarbeitenden Geräten und Systemen, p. 43-47

[10] Frištacký, M. - Kotočová, M.: 'A Systematic Approach to Microprogram Control Unit Design.' Electrotechnical journal, 34, 1983, 10-11, p. 816-831

[11] Zlotnik, E.M.: Bit-slice microprocesors (in Russian). Nauka i technika Inc., Minsk, 1984

[12] The Am 2900 Family Data Book Related Support Circuits. AMD Inc., Sunnyvale, 1979

[13] Series 3000 Reference Manual. Intel. Corp. USA, 1976

[14] Lee, B.J., Nho, T.S.: 'Interpolation for Two-dimensional Contouring Control.' In : [4] , p. 13.121-13.126

[15] Jelšina, M.: 'The Linear Interpolator of Discrete Signals In Digital Control Systems' (in Slovak). Electrotechnical journal, 36, 1985 in printing.

[16] Jelšina, M.: 'Microprogrammable Computer Means for a Primary Performance of Robot Movement by Joystick,' submitted to publication.

CHAPTER 12

COMPUTER CONTROL SYSTEM FOR THE TRAINING SIMULATOR

Katsuhisa Furuta
Faculty of Control Engineering, Tokyo Institute of Technology
Oh-Okayama, Meguro-ku, Yokyo, JAPAN
Yasuhiro Ohyama
DK Laboratory, 202 Asahi ishikawadai mansion, 1-26
2-chome Ishikawa-cho, Ohta-ku, Tokyo, JAPAN

ABSTRACT. This chapter is concerned with the computer control of a mechanical system to behave like a mathematical model to the command. This is applied to the control of the training simulator of Radio Control Helicopter. The design of control system is discussed first and the real time computer control system developed is also presented. The experiment shows that the proposed control is satisfactory.

1. INTRODUCTION

For the training simulator to a certain system, like airplane, the output should be controlled to behave similar to the output of the mathematical model to the command. Such model following control problem has been considered by many authors. Tyler[1] and Kreindler[2] solved the above problem using the linear optimal control theory. However, their design method do not give a satisfactory results in that steady state offset of the plant cannot be eliminated. Furthermore, the controller designed by this method is not robust to the disturbances and small variations of plant parameters. On the other hand, the Davison[3]'s robust controller asymptotically tracks a given reference signal in the presence of disturbances and small parameters variations in the steady state, though the transient response is no paid attention.

Furuta et al.[4] gave a synthesis method of a controller called "the model following servo controller" which gives control law similar to Kreindler's method and at the same time includes the servo compensator for the elimination of the plant offset as well as the effect of constant disturbances and/or small plant parameter's variations. In this report, an implementation of the model following servo controller and experiment are done for a training simulator of RC helicopter.

The simulator itself is desirable to have similar physical configuration of the objective system for the training . In this paper, we use real RC electric helicopter model with 5 degree of freedon as a simulator for training general RC helicopter operation. The important aspects of the training simulator are that the motions of the simulator represent those of real one and the trainee can imagine himself as if he

N. K. Sinha (ed.), Microprocessor-Based Control Systems, 249–262.

is flying a real helicopter.

This simulator used is composed of one arm, gimbals and a small RC helicopter as shown Photo 1 and Fig.1. The gimbals have three degrees

Photo 1. Configuration of gimbal-arm system

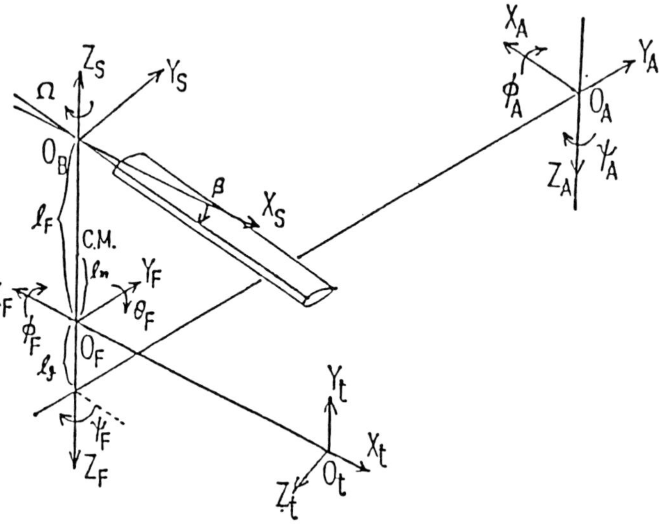

Fig.1 Axes of gimbal-arm system

of freedom and the arm has two degrees of freedom. In particular, the sound and rotor rotation are same as those of real RC helicopter and the effect of disturbance due to the vibration or backlash of the linkage, which is not included in the linear mathematical reference model, can be effectively reproduced, because the rotor thrust driven by electric motor of the helicopter is used to control the simulator.

The control system of this simulator is designed to follow the motions of real hovering RC helicopter flown manually. Since all variables of this simulator can be measured, it can be used to evaluate both the mathematical model and the control system designed.

This simulator is found to be able to reproduce the motion of the hovering helicopter and to be very useful for training the operation. The real time computer control system developed is also presented.

2. DESIGN OF MODEL FOLLOWING SERVO SYSTEM

The motions of the simulator have to follow precisely those of the reference model. For this kind of problem, Model Following Servo technique is proposed and is proved to be effective [4,6]. This design method is represented as follows.

Let the plant equation (simulator dynamics) and the reference model as follows.

PLANT

$$\dot{x} = A_p x + B_p u \qquad y = C_p x \tag{1}$$

REFERENCE MODEL

$$\dot{z} = A_M z + B_M v \qquad y_M = C_M z \tag{2}$$

By digitalizing these equations, the augmented system

$$\begin{pmatrix} x(k+1)-x(k) \\ e(k+1) \\ z(k+1)-z(k) \end{pmatrix} = \begin{pmatrix} A_1 & A_2 \\ 0 & A_3 \end{pmatrix} \begin{pmatrix} x(k)-x(k-1) \\ e(k) \\ z(k)-z(k-1) \end{pmatrix} + \begin{pmatrix} B_1 \\ 0 \\ 0 \end{pmatrix} [u(k)-u(k-1)] \tag{3}$$

where

$$A_1 = \begin{pmatrix} e^{A_p T} & 0 \\ C_p & I \end{pmatrix} \qquad A_2 = \begin{pmatrix} 0 \\ -C_M \end{pmatrix} \qquad A_3 = [e^{A_M T}] \qquad B_1 = \begin{pmatrix} \int_0^T e^{A_p T} dt B_p \\ 0 \end{pmatrix}$$

is derived. The optimal control law is designed to minimize the criterion function,

$$J = \frac{1}{2}\sum_{k=1}^{\infty}\left\{ \begin{pmatrix} x(k)-x(k-1) \\ e(k) \\ z(k)-z(k-1) \end{pmatrix}' \begin{pmatrix} Q_1 & Q_2 \\ Q_2 & Q_3 \end{pmatrix} \begin{pmatrix} x(k)-x(k-1) \\ e(k) \\ z(k)-z(k-1) \end{pmatrix} \right.$$
$$+ 2 \begin{pmatrix} x(k)-x(k-1) \\ e(k) \\ z(k)-z(k-1) \end{pmatrix}' \begin{pmatrix} S_1 \\ S_2 \end{pmatrix} [u(k)-u(k-1)]$$
$$\left. + [u(k)-u(k-1)]' R_1 [u(k)-u(k-1)] \right\} \tag{4}$$

The feedback law is given as follows.

$$u(k) = -F_1 x(k) - F_2 \sum_k e(k) - F_3 z(k) \tag{5}$$

where

$$[F_1, F_2] = -(R_1 + B_1' P_1 B_1)^{-1} (B_1' P_1 A_1 + S_1)' \tag{6}$$

$$F_3 = -(R_1+B_1'P_1B_1)^{-1}(B_1'P_1A_2+B_1'P_2A_3+S_2') \tag{7}$$

P_1 and P_2 are solutions of following equations.

$$P_1 = Q_1 + A_1'P_1A_1 - (A_1'P_1B_1+S_1)(R_1+B_1'P_1B_1)^{-1}(A_1'P_1B_1+S_1)' \tag{8}$$

$$P_2 = Q_2 + A_1'P_1A_2 + A_1'P_2A_3 - (A_1'P_1B_1+S_1)(R_1+B_1'P_1B_1)^{-1}((A_2'P_1+A_3'P_2)B_1+S_2)' \tag{9}$$

Note that the feedback parameter F_1 and F_2 are determined independent of the reference model.

3. DYNAMICS OF SIMULATOR

The configuration of the simulator is shown in Photo.1 and Photo 2.

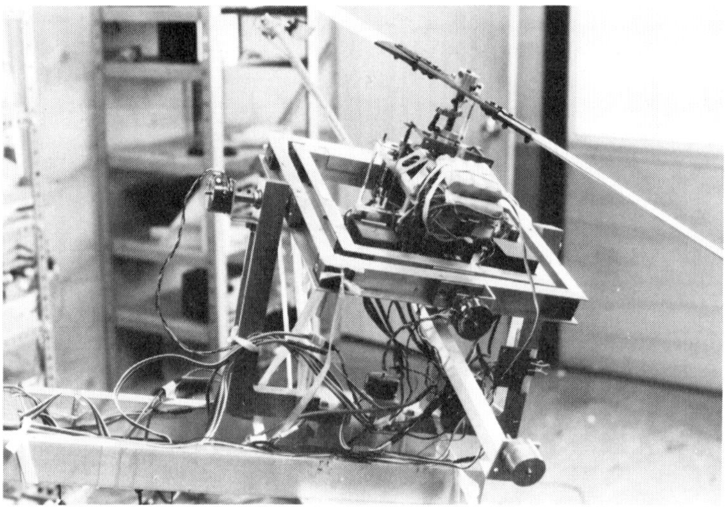

Photo.2 Gimbal and potentiometers (roll, pitch)

The small RC helicopter whose rotor is driven by electric motors is mounted on the gimbals with three degrees of freedom and these are attached at the end of the arm with two degrees of freedom. A counter mass is mounted at the other side of the arm because the rotor thrust is small. The angles of five joints are measured by potentiometers, and the inputs of this system are pitch angles of mainrotor and tailrotor and the rate of rotor rotation.

In order to design a control system, a mathematical model of this system has to be derived. Arm axes, fuselage axes and main rotor axes depicted in Fig.1 is introduced. The model is linearized at the two neutral points ($\psi_{Fo}=0$ or $\psi_{Fo}=90$, where $\psi_F=\psi_{Fo}+\psi_{Fd}$), because the fuselage logitudinal or lateral degree of freedom has to be restricted.

4. EXPERIMENT

4.1. Dynamics of simulator and reference model

Substituting the data in Appendix A into the simulator model in Sec.2, the model of this simulator is derived.

The yawing motion is relatively decoupled from the other motions and can be regulated well by using the rategyo in the real RC helicopter, so that the independent servo control loop is used for the yawing control.

The dynamics of the flapping angles are considered to be very fast compaired with the other dynamics, so that these state variables are reduced [1].

The parameters of the simulator are shown in Table 1 and the eigenvalues are shown in Table 2. θ_s, θ, and θ_0 are representing pitch angles of the blade approximated by

$$\theta = \theta_0 + \theta_s \sin\psi + \theta_c \cos\psi \tag{10}$$

where θ_0: corrective pitch angle, Ω_d is the variation of the main rotor angular velocity Ω from the steady state velocity Ω_0. The eigenvalues with positive real part show that the system is unstable.

On the other hand, the parameters of the hovering RC helicopter is shown in Table 3 and Table 4, whose main rotor radius is 1.4m and weight is 4.5kg.

4.2. Real time computer control

The real time control is done by using a micro computer (DG MP200). The over all computer system is depicted in Fig.2. The real time

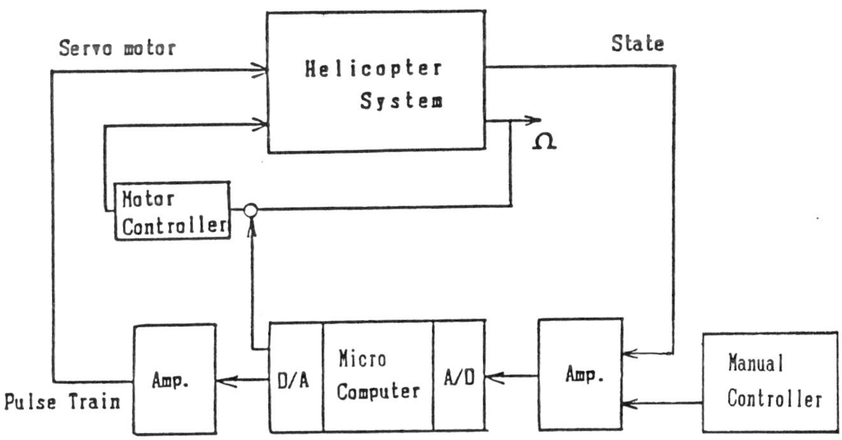

Fig.2 Configuration of control system

computer control is done under RCOS(Real Time Control Operating System) developed by authors, which runs under RDOS of the DG machine. RCOS has two modes in operation. One is the real time control mode and the off control mode. Once the control mode starts, the handling of input and output data (AD, DA) and the computation of the control input are done in the sampling period. In the period after those tasks are finished, if the time is still left, several other tasks can be done according to the command from the console. Therefore the real time control is done interactively. Those commands are HELP, TYPE, CPARA, DUMP, LOAD, SAVE, FUNC, PRINT, COMMAND. HELP is to type all possible command names. TYPE is to type parameter values. CPARA is to change parameter values. DUMP is to dump all parameters except logging data to disk from CPU memories. LOAD is to load the parameter data from the disk. SAVE is to store the logging data to disk. FUNC denotes the functions given by the users. COMMAND is to change the reference values. Clock is to set the sampling period. The configuration of RCOS is shown in Fig.3. An example of the use of RCOS is shown in Fig.4.

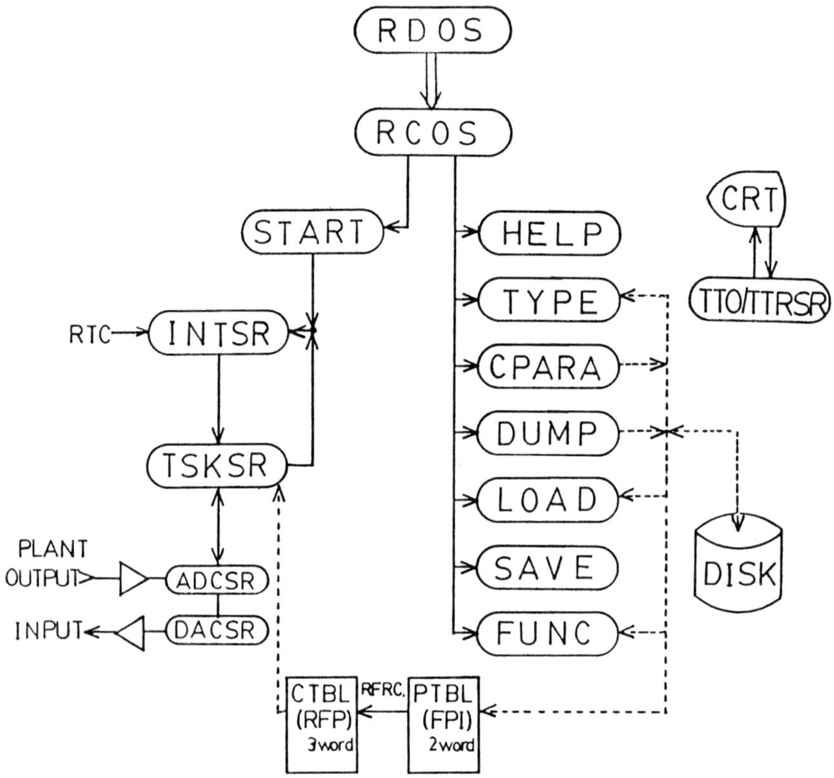

Fig.3 Real Time Control Operating System RCOS

4.3. Design

The lateral degree of freedom is restricted, and the outputs of the simulator and model is decided as follows.

$$\begin{aligned} y &= [1_a\psi_A,\ \phi_F+\phi_A,\ 1_a\phi_A] \\ y_M &= [\ X\ ,\ \phi_F\ ,\ Z\ ] \end{aligned} \tag{11}$$

The block diagram of M.F.S. is shown in Fig.5 and the control loop is devided into two parts, servo control loop and the model.

```
R
RTS4

REAL TIME CONTROL OPERATIING SYSTEM ON MP/200 DOS
( FOR EH3 TRAINING SIMULATOR )

/HELP
  /DUMP
  /LOAD
  /CLOCK
  /SAVE
  /START
  /PRINT
  /BYE
  *HELP
  *CPARA
  *TYPE
  *FUNC
  *COMMAND

  ESC     : STOP CONTROL
  CTRL L  : START LOGGING

/LOAD
FILE NAME :  RTS4D1.PR
COMMENT :  DATA NO.1 (1/4 MODEL)

/CLOCK
CLOCK PERIOD =  +00020
CHANGE TO

/TYPE
PARAMETER NAME :  F1.11
HOW MANY PARAMETERS ? :  5
F1.11      +.7947395E-01
F1.12      +.4546483E+00
F1.13      +.6480665E-01
F1.14      +.4369900E+00
F1.15      +.4673771E-02

/START
CLOCK PERIOD =  +00020

*COMMAND

WHICH VARIABLE   : YR.1
YR.1  :  FROM  +.0000000E+00  TO  0.1

WHICH VARIABLE   : YR.1
YR.1  :  FROM  +.9999853E-01  TO

*$
STOP

/BYE
R
```

Fig.4 Operation of RCOS

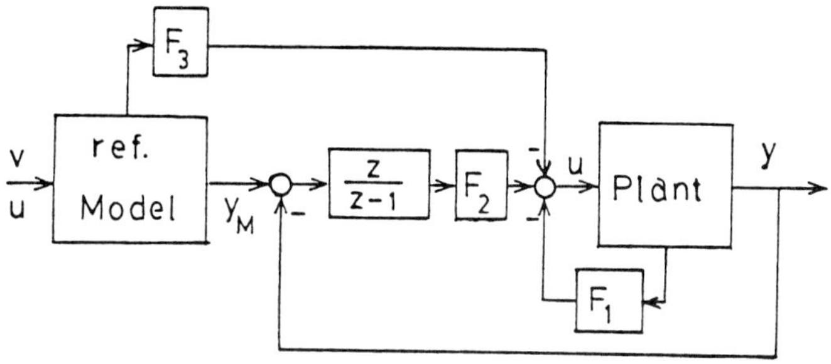

Fig.5 Configulation of M.F.S.

The sampling time is usually determined by the computer computation ability, because shorter period gives the better closed loop characteristics. However the sampling time can not be determined before running the program with designed parameters and is sometimes changed due to program change, for example, adding a filter routine or another check routine. So some design technique which does not change the closed loop characteristic when the sampling time is varies is necessary. For designing the control system by the optimal control, the criterion function for the continuous model can be chosen easily, since its weight matrices have the meaning in the engineering view point. Therefore in this paper, the weighting matrices of the discrete criterion function which is equivalent to the given continuous criterion function are used to design the discrete control system (Appendix A.7). The appropriate criterion function is obtained heuristically from the simulation results. The feedback parameter F1 and F2 depend only on Q and R, and F3 is determined using the servo control loop design parameters, so that once the F1 and F2 are determined to stabilize the simulator, F3 is determined to several reference model without changing Q and R.

In this simulator, the sampling interval of 20msec is enough for stabilization, however the computation of the model needs much time because of many multiplications, so it takes more than 35msec and this period is not enough for stabilization.

Considering the separation of the servo controller and model in M.F.S. block diagram, the servo controller can be controlled in 20msec and model can be computed within 40msec. The characteristics of the controlled system is a little different from the designed one except the servo control loop. The difference caused by the feedforward term F3z and C_Mz can be considered to be samll if the change of z is so small.

The reference model whose time scale is changed is made by digitalizing the continuous model in different sampling time. For example, if the model is digitalized in 10msec and designed in 20msec, the time is twice of the usual so that the speed of motions is more slowly without changing the modes of flying helicopter and the system becomes easy to control.

The designed parameters for 1/4 scale model are shown in Table 5 and the experimental results are shown in Fig.6, Fig.7 and Fig.8. The simulator is stabilized by the servo controller firstly, so that the training can be started from the stable position. The outputs of the simulator follow to those of the reference model.

5. CONCLUSION

A training simulator for the hovering RC helicopter is proposed and its control system is designed and implemented by a mini-computer.

This simulator with 5 degrees of freedom is driven by the rotor thrust of the small RC helicopter mounted on the gimbals, attached to the end of long arm, so that the sound and the rotor rotation are same as those of real one and the effects due to the disturbances by the vibration and backlash of the linkage, which can't be included in the linear mathematical reference model, can be considered in the simulator.

The control system is designed by using Discrete Model Following Servo thechnique based on criterion function in continuous time, which is very useful because the closed loop characteristics is equivalent even if the sampling time is changed.

The reference models whose time scale is changed and whose degrees of freedom are restricted are used according to the trainee's ability.

The experimental results show that the motions of the simulator follow those of the reference model and of the real one. Moreover, using the above reference models, the trainee could train the operation effectively.

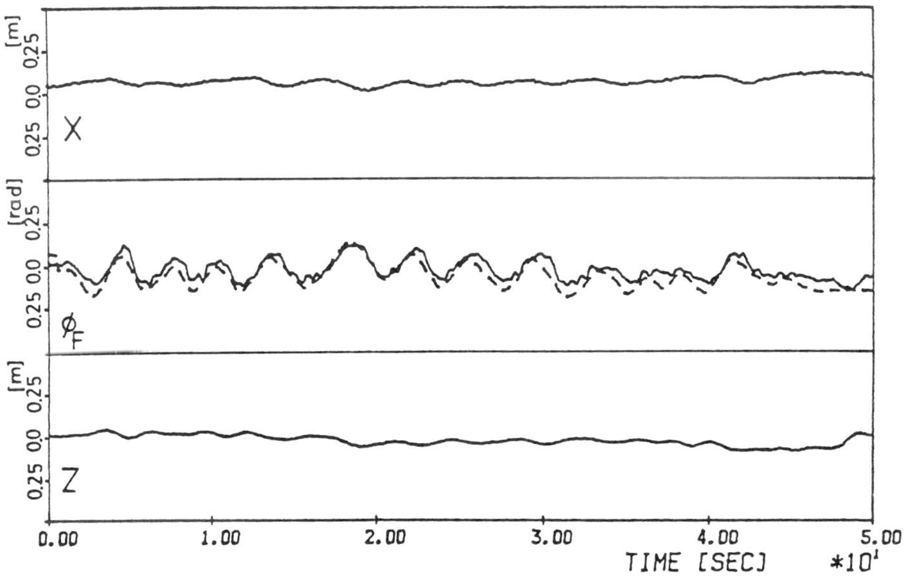

Fig.6 Experimental Result:1 degree of following 1/4 scale model

——— Simulator output
------ Model output

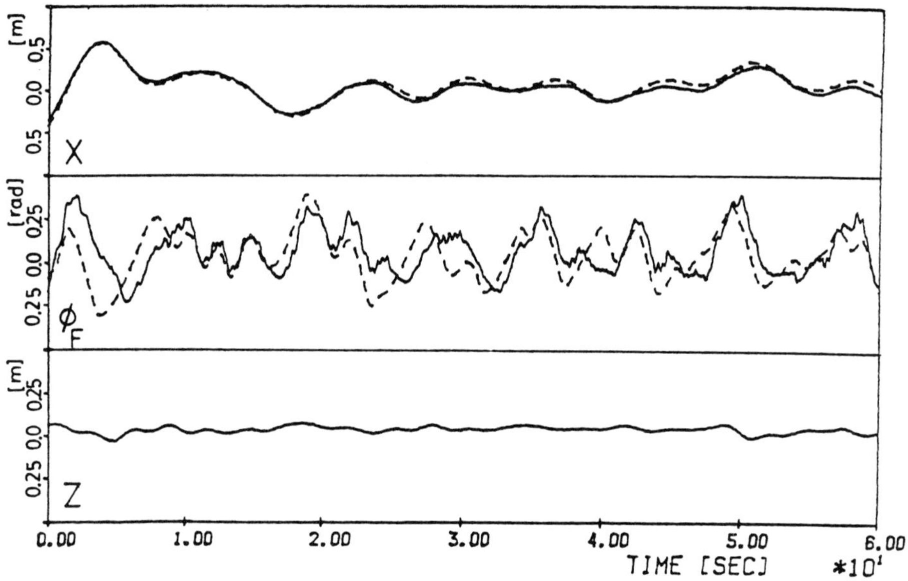

Fig.7 Experimental Result: 2 degree of following 1/4 scale model

——— Simulator output

------- Model output

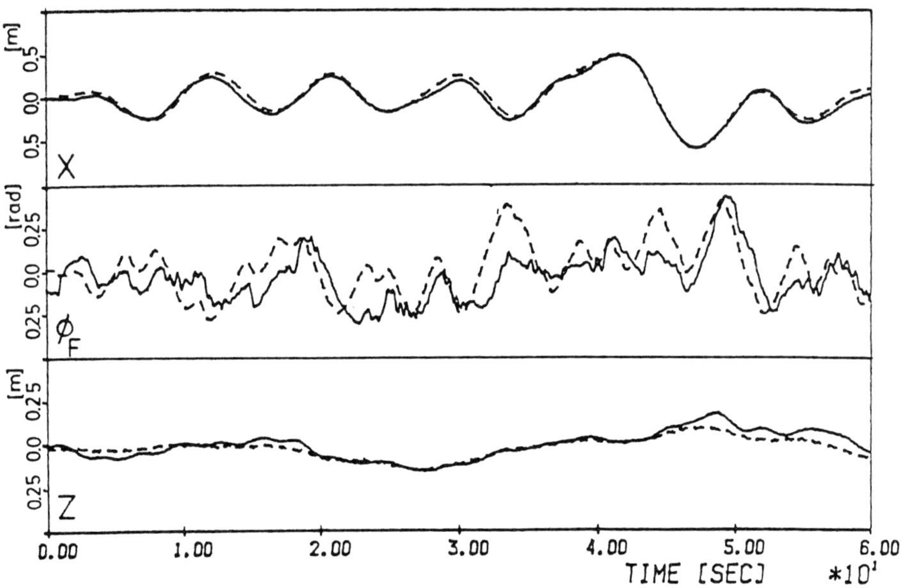

Fig.8 Experimental Result: 3 degree of following 1/4 scale model

——— Simulator output

------- Model output

$$\dot{x} = A\,x + B\,u$$

$$x = [\,\dot{\phi}_F \;\; \dot{\theta}_F \;\; \dot{\phi}_A \;\; \dot{\psi}_A \;\; \phi_F \;\; \theta_F \;\; \phi_A \;\; \psi_A\,]'$$

$$u = [\,\theta_s \;\; \theta_c \;\; \Omega_d\,]'$$

$$A = \begin{bmatrix} 0.00 & 0.00 & 0.00 & 0.00 & 1.00 & 0.00 & 0.00 & 0.00 \\ 0.00 & 0.00 & 0.00 & 0.00 & 0.00 & 1.00 & 0.00 & 0.00 \\ 0.00 & 0.00 & 0.00 & 0.00 & 0.00 & 0.00 & 1.00 & 0.00 \\ 0.00 & 0.00 & 0.00 & 0.00 & 0.00 & 0.00 & 0.00 & 1.00 \\ 51.0 & 3.02 & 0.00 & 0.00 & -37.6 & -1.66 & -40{,}0 & 1.99 \\ -2.82 & 10.1 & 0.00 & 0.00 & 1.99 & -7.01 & 2.18 & -1.27 \\ 0.004 & -0.003 & 0.00 & 0.00 & 0.03 & 0.001 & -0.26 & -0.03 \\ -0.04 & 0.67 & 0.00 & 0.00 & 0.03 & -0.10 & 0.03 & -0.02 \end{bmatrix} \qquad B = \begin{bmatrix} 0.00 & 0.00 & 0.00 \\ 0.00 & 0.00 & 0.00 \\ 0.00 & 0.00 & 0.00 \\ 0.00 & 0.00 & 0.00 \\ 99.9 & 391.0 & 0.023 \\ 68.4 & -36.2 & 0.00 \\ -0.09 & -0.35 & -0.023 \\ 1.05 & -5.54 & 0.00 \end{bmatrix}$$

TABLE 1 Parameters of simulator model

(1) 1.27
(2) 1.18
(3) 0.057
(4) 0.00
(5) 0.00
(6) -0.267
(7) -8.37
(8) -38.7

TABLE 2 Poles of simulator model

$$\dot{x} = A\,x + B\,u$$

$$x = [\,\dot{\phi}_F \;\; \dot{\theta}_F \;\; \phi_F \;\; \theta_F \;\; \dot{X} \;\; \dot{Y} \;\; \dot{Z} \;\; X \;\; Z\,]'$$

$$u = [\,\theta_s \;\; \theta_c \;\; \theta_0\,]'$$

ϕ_F : Fuselage roll angle
θ_F : Fuselage pitch angle
X : longitudinal position
Y : lateral position
Z : vertical position

$$A = \begin{bmatrix} 0.00 & 0.00 & 1.00 & 0.00 & 0.00 & 0.00 & 0.00 & 0.00 & 0.00 \\ 0.00 & 0.00 & 0.00 & 1.00 & 0.00 & 0.00 & 0.00 & 0.00 & 0.00 \\ 0.00 & 0.00 & -16.8 & 1.33 & -0.03 & -0.36 & 0.00 & 0.00 & 0.00 \\ 0.00 & 0.00 & -0.15 & -5.69 & 0.12 & -0.003 & 0.00 & 0.00 & 0.00 \\ 0.00 & -9.80 & 0.011 & 0.41 & -0.009 & 0.0002 & 0.0 & 0.00 & 0.00 \\ 9.80 & 0.00 & -0.41 & 0.03 & -0.0007 & -0.009 & 0.0 & 0.00 & 0.00 \\ 0.00 & 0.00 & 0.00 & 0.00 & 0.00 & 0.00 & -1.84 & 0.00 & 0.00 \\ 0.00 & 0.00 & 0.00 & 0.00 & 1.00 & 0.00 & 0.00 & 0.00 & 0.00 \\ 0.00 & 0.00 & 0.00 & 0.00 & 0.00 & 0.00 & 1.00 & 0.00 & 0.00 \end{bmatrix} \qquad B = \begin{bmatrix} 0.00 & 0.00 & 0.00 \\ 0.00 & 0.00 & 0.00 \\ 59.1 & 394. & 0.00 \\ 131. & -26.7 & 0.00 \\ -9.47 & 1.93 & 0.00 \\ 1.42 & 9.47 & 0.00 \\ 0.00 & 0.00 & -100. \\ 0.00 & 0.00 & 0.00 \\ 0.00 & 0.00 & 0.00 \end{bmatrix}$$

TABLE 3 Parameters of hovering helicopter

(1) 0.0179 -0.454j
(2) 0.0062 -0.456j
(3) 0.00
(4) 0.00
(5) -1.84
(6) -5.75
(7) -16.9

TABLE 4 Poles of helicopter model

$$Q = \mathrm{diag}\,(\;0, I_3\;) \qquad R = \mathrm{diag}\,(\,100,\;1000,\;10)$$

$$F1 = \begin{bmatrix} 7.95E-2 & 4.55E-1 & 6.48E-2 & 4.37E-1 & 4.67E-3 & 3.99E-2 & 1.06E-1 & 6.85E-1 \\ 2.75E-1 & -6.99E-2 & 2.44E-1 & -1.19E-2 & 6.81E-3 & -9.04E-3 & 3.11E-1 & -3.57E-2 \\ 7.40E-2 & -1.91E-2 & 2.57E0 & 1.38E-1 & 7.39E-3 & -1.13E-3 & -6.34E0 & -6.59E-2 \end{bmatrix}$$

$$F2 = \begin{bmatrix} 1.91E-3 & 1.84E-4 & 3.44E-5 \\ -5.67E-5 & 5.33E-4 & 2.99E-4 \\ 1.97E-4 & -3.08E-3 & 5.55E-3 \end{bmatrix}$$

$$F3 = \begin{bmatrix} -2.77E-2 & 2.51E-1 & -1.73E-3 & 1.23E-2 & -1.27E-1 & 1.21E-3 & 1.77E-3 & -3.12E-1 & 6.99E-3 \\ -9.64E-2 & -9.76E-2 & -5.33E-3 & -1.72E-2 & 3.65E-3 & 2.77E-3 & 1{,}97E-2 & 8.52E-3 & 6.70E-2 \\ 8.31E-1 & 1.52E0 & 4.76E-2 & 2.42E-2 & -1.06E-1 & -4.84E-2 & 1.46E0 & -9.84E-2 & 3.27E0 \end{bmatrix}$$

Table 5 Designed parameters

REFERENCES

[1] J.S.Tyler,Jr., 'The Characteristics of Model-Following System as Synthesized by Optimal Control', IEEE Trans.Aut.Contr., AC-9, pp. 485-498, 1964

[2] E.Kleindler, 'On the linear optimal servo problem', Int.J.Control, 9, pp.465-472, 1969

[3] E.J.Davison, A.Goldenberg, 'Robust Control of a General Servomechanism Problem, The Servo Compensator', Automatica, 11 , pp.461-471, 1975

[4] K.Furuta, K.Komiya, 'Design of Model-Following Servo Controller', IEEE Trans.Aut.Contr., 27, pp.725-727, 1982

[5] R.D.Murphy, K.S.Narendra, 'Design of Helicopter Stabilization System Using Optimal Control Theory', J.Aircraft, 6 , pp.129-136, 1969

[6] K.Furuta, Y.Ohyama, O.Yamano, 'The Modelling and Control of RC Helicopter', IMACS Symposium Preprints, pp.231-236, May 1983

[7] T.Shiotsuki, K.Furuta and M.Okazaki, 'Design of Training Simulator of R/C Helicopter Using MFS', Trans.of SICE, 19, pp.56-63, 1983

[8] A.H.Levis et al., 'On the behaviour of optimal linear sampled-data regulators', Int.J.Control, 13, pp.343-361, 1971

APPENDIX A

Conversion of Criterion Function

The discrete criterion function based on the continuous one is derived as follows.

Digitalizing (1) and (2),

$$\begin{cases} x(k+1) = A_p x(k) + B_p u(k) \\ y(k) = C_p x(k) \end{cases} \tag{A-1}$$

$$\begin{cases} z(k+1) = A_M z(k) + B_M v(k) \\ y_M(k) = C_M z(k) \end{cases} \tag{A-2}$$

$$e(k) = y(k) - y_M(k) \tag{A-3}$$

the augmented system (3) is derived. The criterion function in continuous time is,

$$J = \frac{1}{2}\int_0^\infty \{ \begin{pmatrix} \dot{x} \\ e \\ \dot{z} \end{pmatrix} \begin{pmatrix} Q & 0 \\ 0 & 0 \end{pmatrix} \begin{pmatrix} \dot{x} \\ e \\ \dot{z} \end{pmatrix} + \dot{u}' \, R \, \dot{u} \} dt \tag{A-4}$$

$$= \frac{1}{2}\sum_{k=1}^{\infty}\int_0^T \{ \begin{pmatrix} \dot{x}(k+t) \\ e(k+t) \\ \dot{z}(k+t) \end{pmatrix}' \begin{pmatrix} Q & 0 \\ 0 & 0 \end{pmatrix} \begin{pmatrix} \dot{x}(k+t) \\ e(k+t) \\ \dot{z}(k+t) \end{pmatrix} + \dot{u}(k+t)' R \, \dot{u}(k+t) \} dt \tag{A-5}$$

The augmented system in continuous time is,

$$\begin{pmatrix} \ddot{x} \\ \dot{e} \\ \ddot{z} \end{pmatrix} = \begin{pmatrix} Ap & 0 & 0 \\ Cp & 0 & -C_M \\ 0 & 0 & A_M \end{pmatrix} \begin{pmatrix} \dot{x} \\ e \\ \dot{z} \end{pmatrix} + \begin{pmatrix} B \\ 0 \\ 0 \end{pmatrix} \dot{u} \tag{A-6}$$

However,

$$\begin{pmatrix}\dot{x}(k+t)\\ e(k+t)\\ \dot{z}(k+t)\end{pmatrix} = \exp(\begin{pmatrix}Ap & 0 & 0\\ Cp & 0 & -C_M\\ 0 & 0 & A_M\end{pmatrix} t)\begin{pmatrix}\dot{x}(k)\\ e(k)\\ \dot{z}(k)\end{pmatrix}$$

$$+\int_0^t \exp(\begin{pmatrix}Ap & 0 & 0\\ Cp & 0 & -C_M\\ 0 & 0 & A_M\end{pmatrix} t)dt\begin{pmatrix}Bp\\ 0\\ 0\end{pmatrix}\dot{u}(k) \qquad \text{(A-7)}$$

and,

$$\dot{x}(k) = \frac{1}{T}(x(k)-x(k-1)) \qquad \dot{z}(k) = \frac{1}{T}(z(k)-z(k-1))$$

$$\dot{u}(k) = \frac{1}{T}(u(k)-u(k-1)) \qquad \text{(A-8)}$$

Then,

$$\begin{pmatrix}\dot{x}(k+t)\\ e(k+t)\\ \dot{z}(k+t)\end{pmatrix} = \exp(\begin{pmatrix}Ap & 0 & 0\\ Cp & 0 & -C_M\\ 0 & 0 & A_M\end{pmatrix} t)\begin{pmatrix}\frac{1}{T}(x(k)-x(k-1))\\ e(k)\\ \frac{1}{T}(z(k)-z(k-1))\end{pmatrix}$$

$$+\frac{1}{T}\int_0^t \exp(\begin{pmatrix}Ap & 0 & 0\\ C_p & 0 & -C_M\\ 0 & 0 & A_M\end{pmatrix} t)dt\begin{pmatrix}Bp\\ 0\\ 0\end{pmatrix}[u(k)-u(k-1)] \qquad \text{(A-9)}$$

Therefore the criterion function (A-5) is

$$J = \frac{1}{2}\sum_{k=1}^{\infty}\begin{pmatrix}x(k)-x(k-1)\\ e(k)\\ z(k)-z(k-1)\end{pmatrix}'\begin{pmatrix}Q_1 & Q_2\\ Q_2 & Q_3\end{pmatrix}\begin{pmatrix}x(k)-x(k-1)\\ e(k)\\ z(k)-z(k-1)\end{pmatrix}$$

$$+ 2\begin{pmatrix}x(k)-x(k-1)\\ e(k)\\ z(k)-z(k-1)\end{pmatrix}'\begin{pmatrix}S_1\\ S_2\end{pmatrix}[u(k)-u(k-1)]$$

$$+ [u(k)-u(k-1)]'R_1[u(k)-u(k-1)] \qquad \text{(A-10)}$$

where

$$Q_1 = \begin{pmatrix}\frac{1}{T}I & 0\\ 0 & I\end{pmatrix}\int_0^T g(t)'Qg(t)dt\begin{pmatrix}\frac{1}{T}I & 0\\ 0 & 0\end{pmatrix}$$

$$Q_2 = \begin{pmatrix}\frac{1}{T}I & 0\\ 0 & I\end{pmatrix}\int_0^T g(t)'Qh_M(t)dt\cdot\frac{1}{T} \qquad Q_3 = 0$$

$$S_1 = \begin{pmatrix}\frac{1}{T}I & 0\\ 0 & I\end{pmatrix}\int_0^T g(t)'Qg(t)dt\cdot\frac{1}{T} \qquad S_2 = 0$$

$$R_1 = \frac{1}{T^2}\int_0^T \{R+h(t)'Qh(t)\}dt$$

$$g(t) = \exp\{\begin{pmatrix}Ap & 0\\ Cp & 0\end{pmatrix}t\} \qquad h(t) = \int_0^t \exp\{\begin{pmatrix}Ap & 0\\ Cp & 0\end{pmatrix}t\}dt\begin{pmatrix}Bp\\ 0\end{pmatrix}$$

$$h_M(t) = \begin{pmatrix} 0 \\ -C_M \end{pmatrix} \int_0^t \exp(A_M t)\, dt$$

CHAPTER 13

MICROCONTROLLER–BASED DIRECT NUMERICAL CONTROL

R. Kitai
McMaster University
Hamilton, Ontario, Canada

T.P. Taylor
Atlantis Flight Research Inc.
Woodbridge, Ontario, Canada

ABSTRACT. A direct numerical control system based on an Intel 8031 microcontroller has been designed and implemented. The hardware and software required to perform the basic direct numerical control functions are outlined. The system can download prepared numerical control part programs from a host computer and then transfer the numerical control data to a wide variety of numerically controlled machines. The system that has been implemented is intended for use in the metal machining industry. The fundamental principles of numerical control systems are first described. The operations of conventional numerical control, computer numerical control, and direct numerical control are also reviewed.

1. INTRODUCTION

Conventional numerical control (NC without computers) had its beginnings in the late nineteen forties when John T. Parsons proposed the control of a machine tool by the use of punched cards which directed the machine tool to move in specified directions to produce the desired part. Further research was conducted at the servomechanisms laboratory of MIT, resulting in a prototype NC machine that could manufacture different parts by having its input program changed [1]. After the usefulness of the prototype machine was demonstrated, machine tool manufacturers began producing their own NC machines to satisfy particular requirements. No standard control structure was proposed, so that today there are many types of NC machines, all slightly different, and all using different input program formats.

The basic components of an NC machine are shown in Fig. 1 and are discussed below. The input program consists of symbols, encoded on some form of medium, which instruct the machine as to what movements are to be made, and thus what part is to be made. Input media such as punched cards and magnetic tapes have been used, but the most common is one inch wide paper tape.

N. K. Sinha (ed.), Microprocessor-Based Control Systems, 263–278.

Numerical Control Systems

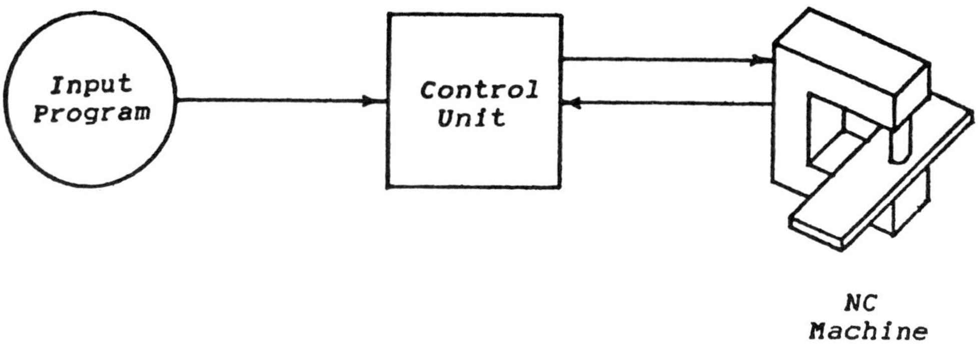

Figure 1 Basic Components of an NC Space

Paper tape is well suited to a machine shop environment whereas magnetic tape may be accidentally erased or contaminated by foreign substances. Punched cards are rugged, but the cards can get out of order and the use of an unsorted deck may cause damage to the work part or machine.

Paper tape has many disadvantages, the most notable being that a paper tape that is incorrectly punched cannot be used at all. Even a small error may necessitate the repunching of an entire paper tape. Other disadvantages of paper tape are as follows:

- Holes may be mispunched or misread.
- The tape can break.
- The reels of tape must be stored; therefore they can be lost or misplaced.
- Paper tape cannot be erased, therefore there is the continuing expense for new tape.
- Tapes eventually wear out.

These and other disadvantages mean that the paper tape has become an unreliable component of an NC system [1].

The control unit reads and interprets the input program and controls the machine's servomechanisms to produce the required movements. The control unit has evolved from its beginnings as a hardware logic circuit used in conventional NC to today's programmable minicomputers used in CNC machines.

The machine which performs the actual work may be a relatively simple drill press which is directed to move to a certain x,y position and drill a hole, or it may be as complicated as a 5 axis milling machine on which complex airfoil shapes may be cut.

Numerical control has found many applications such as assembly, drafting, automatic inspection, welding, and sheet metal punching. A major use of NC, and the one that is stressed in this chapter is metal machining.

Most machined parts today are made in small quantities, even one off, and NC is well suited to this end. When a new part is required, a new NC program is written and run on the same NC machine that made the last part, the two parts being very likely substantially different. Thus no changes to the machine (except of course the cutting tools) are required; this is the major benefit of NC, its flexibility. Many types of parts can be made on one machine simply by changing the NC program and the cutting tools.

2. PRINCIPLES OF NUMERICAL CONTROL

2.1 Part Production using NC Technology

The flow diagram in Fig. 2 shows how a part is manufactured using conventional NC techniques. First an engineering drawing or blueprint is created which precisely defines the part to be made. The information necessary to make the part is then entered into a computer using Computer Aided Design (CAD) techniques which today consist largely of interactive graphics. Once the part is completely defined in the computer, the Cutter Location (CL) file can be produced. This is a file which contains the information of the cutting tool path on a generalized machine tool. For example, if a part is to be made on a drill press, the CL file would not contain information about the dimensions or the control statements of a specific drill press. The CL file would contain position information relative to an origin (usually on the work part) specified by the designer.

The CL file is then fed to a post processor which converts the general information into specific commands to control a specified NC machine. Each NC machine has its own associated post processor which creates the NC program for that NC machine. The NC program contains all the information required to make the part. This program is then punched onto paper tape and passed to the machinist along with a process plan so that the machinist can follow what the machine is doing and make cutting tool changes when required. The paper tape is mounted on the NC controller's paper tape reader; then the controller reads and interprets the tape, and controls the machine. Older controllers read one block at a time and carry out the instructions before reading the next block, (eg. a block may instruct the NC machine to drill a hole at a specified location), while more recent controllers read the entire tape into internal storage before starting any machine actions.

If a tape has been tested and is deemed to be correct, then the actual metal part can be made; otherwise the tape must be proved by making a test part, usually out of hard foam. If an error is found in the paper tape during proving, the machinist informs the

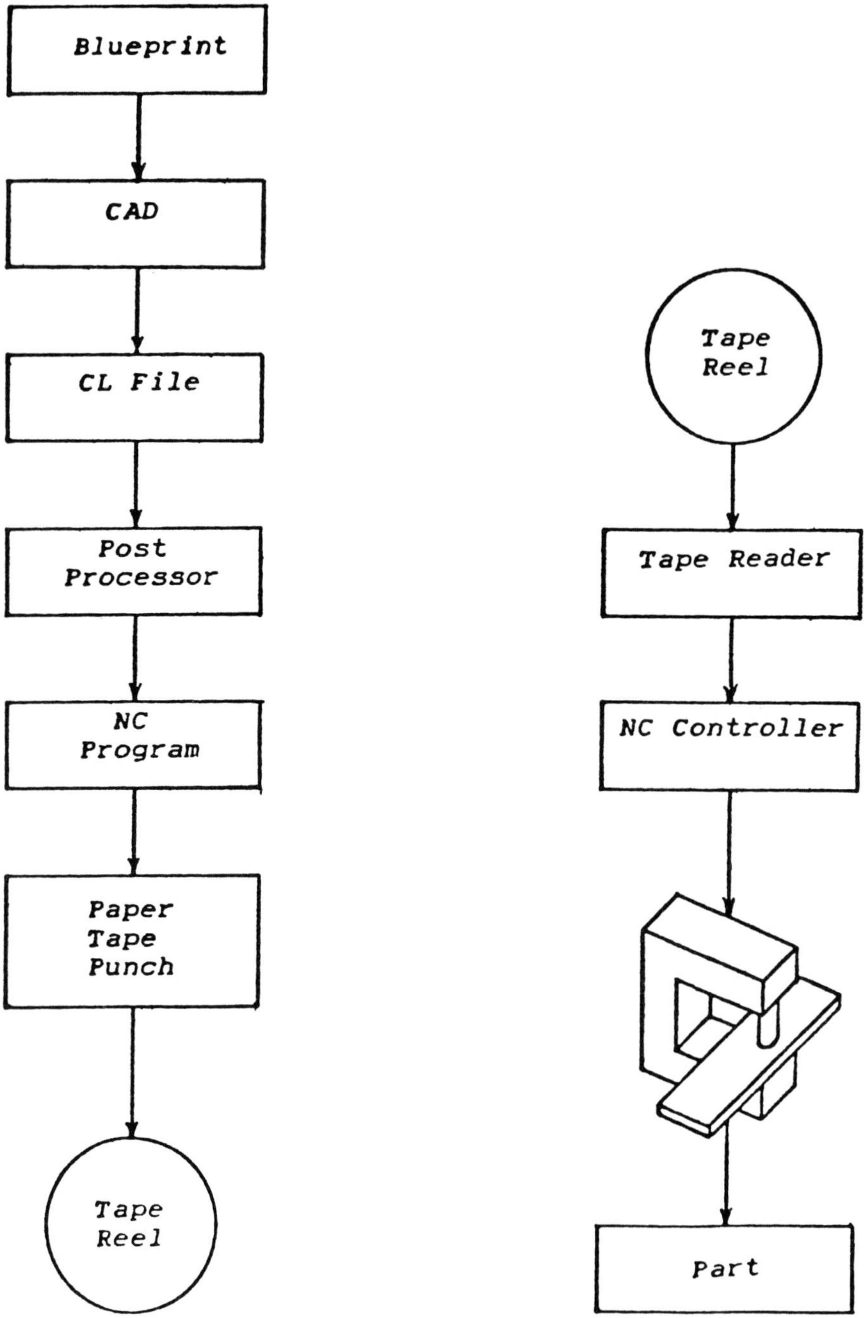

Figure 2 Part Production on an NC System

designer who then corrects the error and punches a new tape. The new tape must also be tested. A significant amount of time may be allocated to tape proving as upwards of three trials may be necessary to create a correct tape [1].

Numerical control has advantages beyond its great flexibility: a large number of different parts can be made on an NC machine without modifying it, which results in higher productivity; a part can be duplicated simply by rerunning the same NC part program; very complex parts, not produceable by hand, can be made; and very tight tolerances can be maintained. While NC has many advantages,it also possesses some disadvantages such as very high purchase price. Since the NC machine is more complicated then its manual counterparts, its maintenance and operation costs are much higher. To offset this, the NC machine must be as highly utilized as possible. CAD facilities must also be available. Some firms lease CAD computer time while others purchase CAD computers. If purchasing the facilities, they must also be maintained; if the CAD computer is down for any length of time, all the NC machines are idle after they finish their current undertakings, resulting in high costs due to lost production time.

2.2 Computer Numerical Control.

A recent development, (commencing about 10 years ago), in NC technology is Computer Numerical Control (CNC). CNC is distinguished by the use of a minicomputer as the NC controller. Instead of having a hardwired machine control circuit which cannot easily be modified, CNC uses a computer as a programmable control circuit. The NC part program is not changed, but the way in which the machine will execute the part program commands can be modified by changing the controller's program. A typical CNC system is shown in Fig. 3. CNC has many advantages which are listed in [1]. One advantage is that a paper tape is read only once, no matter how many of the parts are to be made. This is possible because the tape is read into the internal storage of the controller. Thus many of the problems associated with repeated use of paper tape are eliminated, but the basic problems of breakage, storage, etc. still exist.

2.3 Direct Numerical Control

Direct Numerical Control (DNC) has been defined by the Electronic Industries Association as [5]:

> A system connecting a set of numerically controlled machines to a common memory for part program storage with provisions for on-demand distribution of data to machines.

Thus a defining characteristic of DNC is the elimination of the input media used by the NC controller. This is accomplished by the

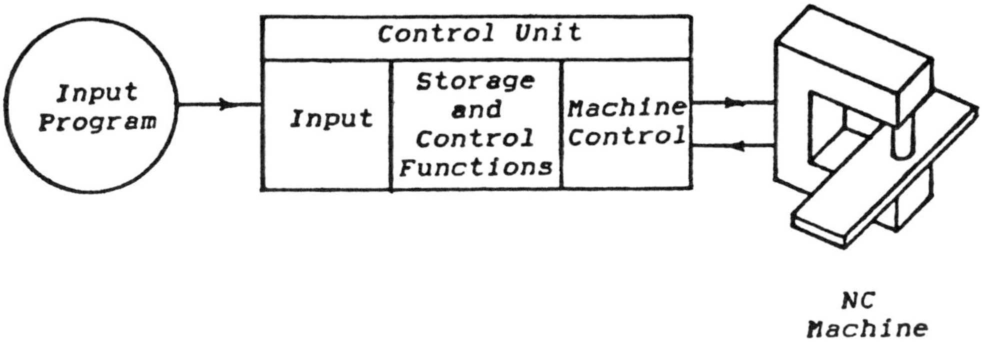

Figure 3 Basic CNC System

electronic transmission of the NC program from a storage computer to the NC controller.

The DNC systems described in the literature [1],[2],[5] use a large computer to control several machines as shown in Fig. 4. When DNC was first conceived, it was the only feasible implementation because of the high cost of computers. The main computer has to be able to satisfy the machines requests for data in a real time fashion since delay in data transfer cannot be tolerated by the NC machines. If many machines are to be controlled, a single computer may not be able to satisfy the demands, and so a system of satellite computers is used as shown in Fig. 5.

Utilizing the new microprocessor technology, it is now possible to implement a cost effective DNC system that has a computer for each NC machine. The basic components of such a DNC system are shown in Fig. 6. The host computer stores NC programs that are ready to be used on a machine. When the DNC coupler requests that a specific NC program be downloaded, the host finds the program and transmits it to the coupler. The host may be the CAD computer or a separate computer used to relieve the loading on the CAD facilities.

The transmission lines connect the host and the DNC coupler. The coupler is located with its NC machine on the shop floor while the host is usually located in a special room away from the shop environment.

The DNC coupler, which is a microprocessor system, has the ability to store one or more NC programs in its local memory. Under operator control the coupler requests that an NC program be downloaded from the host and as the program is received it is stored in memory. When the operator has the machine set up, the coupler is instructed to send the program to the machine through the machine interface, with no further interaction from the host.

The DNC coupler not only eliminates the paper tape, but can also upgrade a conventional NC machine to a CNC-like machine by offering such functions as local editing of the part program and manual entry of a program. Another function of the coupler is the gathering of machine operating statistics. Job billing times can be accurately tabulated and other statistics such as machine maintenance times can be gathered for use by management when scheduling machine operations.

The machine interface has been of one of two forms, depending on the NC controller of the machine. A conventional NC controller that has no built-in computer interface is accessed via a Behind the Tape Reader (BTR) interface. Figure 7 shows a typical BTR interface. The DNC coupler takes the place of the paper tape reader and the NC controller receives data as if a paper tape were being used. The paper tape reader is still usable so that in the event of a failure in the DNC system, tapes can be used until the system is repaired. The BTR interface is very common, since most conventional NC machines have an integral tape reader. The problem with the BTR interface is that at the connection point, each machine's tape reader will likely have different voltage levels; therefore a custom interface is usually needed for each machine.

If a CNC machine is being connected to, it may have a built-in computer port, in which case the DNC coupler can interface with the controller through the port. Most shops will require both types of interfaces.

2.4 Part Production Using DNC

Figure 8 shows a flow diagram of how a part is made using a DNC system. The steps are the same as for. conventional NC up to and including the production of the CL file or the NC program.

The CL file can be transferred to the host computer where it is postprocessed to create the NC program, or the NC program can be created on the CAD computer and then transferred to the host. When the part is to be made, the NC program is transferred to the DNC coupler. The operator can now inspect the program by displaying it on a CRT, which is located at the machine, and when ready, he can start the NC machine. If the program is being proved and the operator discovers an error that he knows how to correct, he can do so by using the local editor of the coupler. Thus tape proving time and paper tape expenses are saved. Once the program is corrected, it can be transferred back to the host for storage.

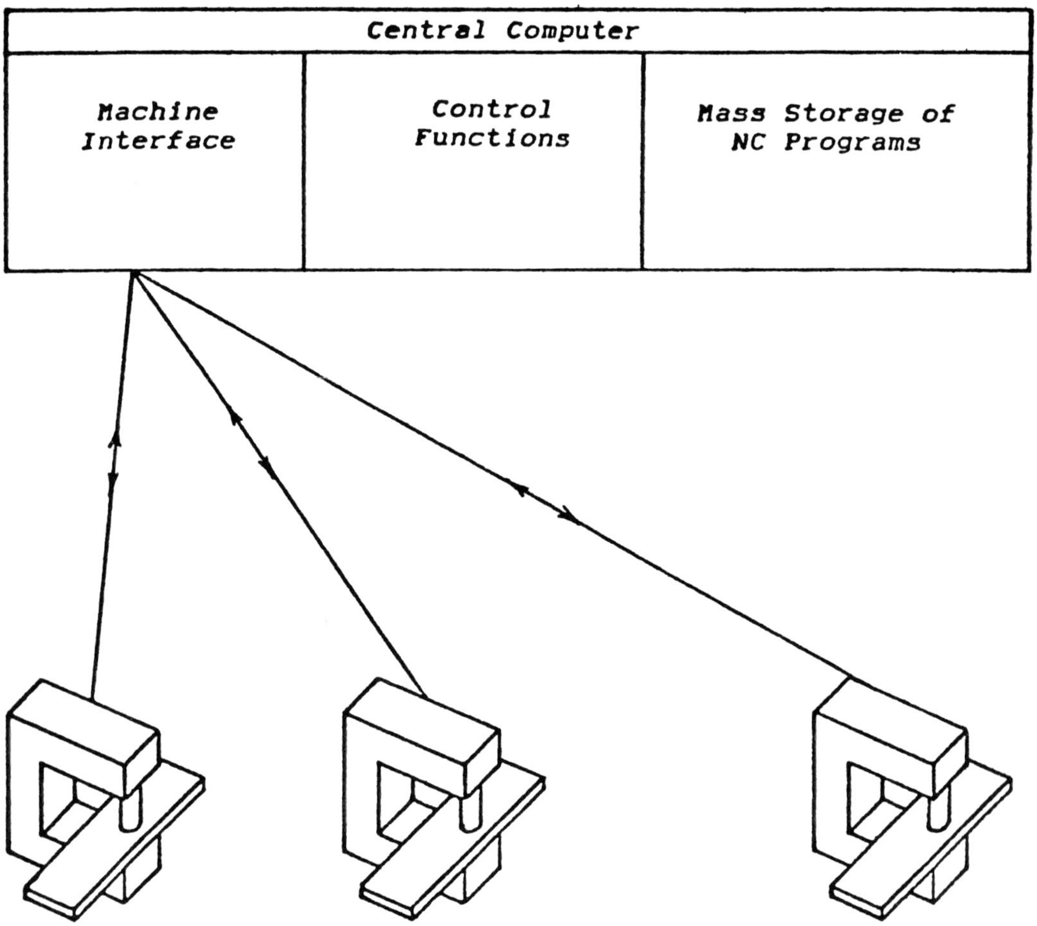

Figure 4 Central Computer DNC System

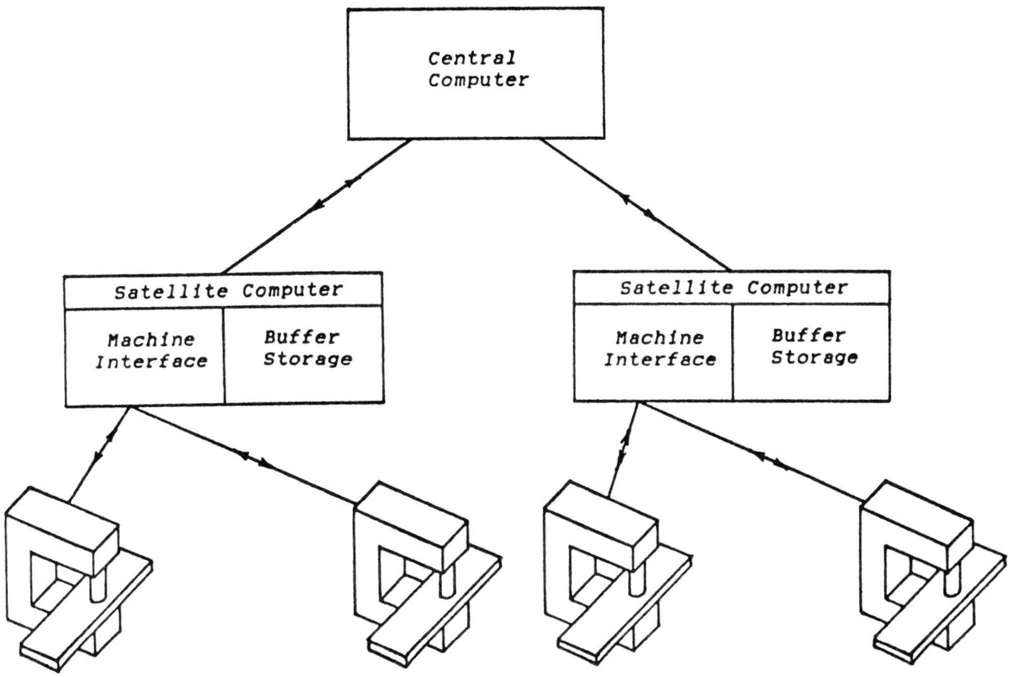

Figure 5 Satelllite Computer DNC System

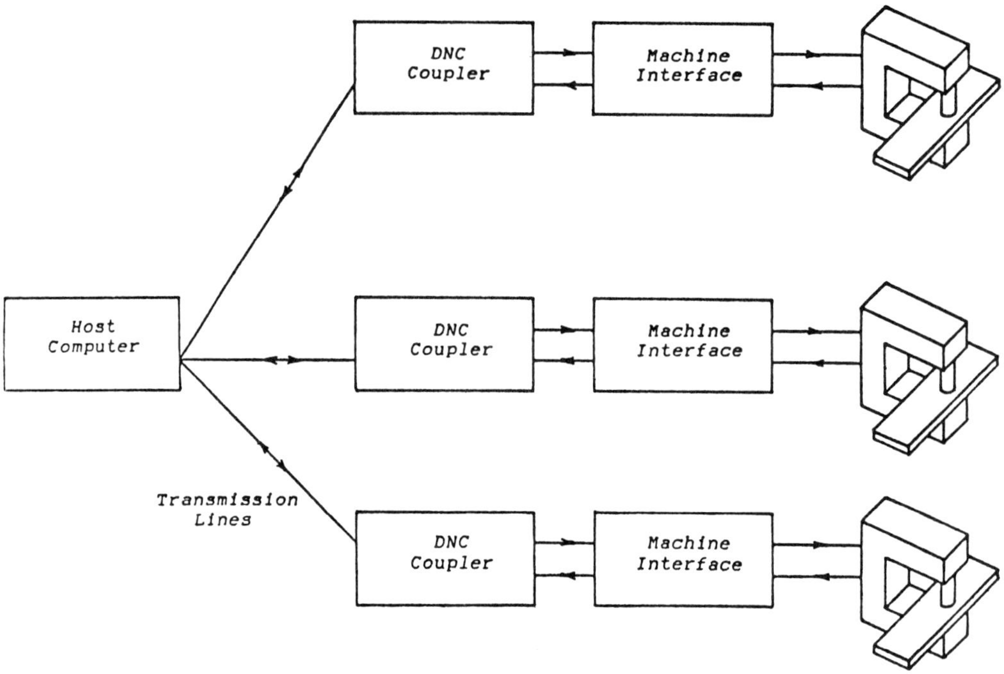

Figure 6 System Utilizing One Computer Per NC Machine

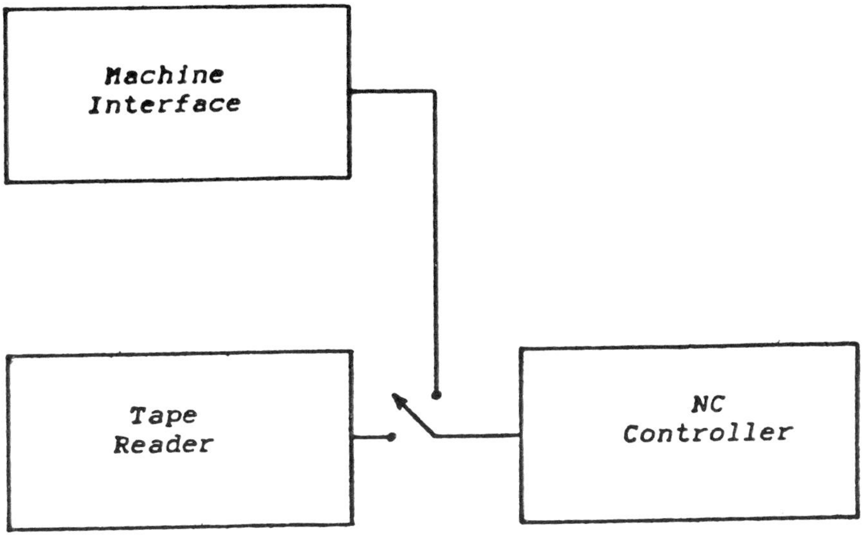

Figure 7 Typical Behind the Tape Reader Interface

One advantage of DNC is the elimination of the problems associated with the use of paper tape. Other advantages are time savings during proving, the upgrading of a conventional NC machine to a CNC-like machine, and the ability to gather statistics valuable to management.

These advantages and others make a DNC system desirable if it can be implemented economically. Up to the present time the major disadvantage of DNC has been its high cost. Computers capable of handling all of the demands placed on them have been expensive and uneconomical for small and medium sized shops. The system described below is designed to overcome these disadvantages.

3. DNC SYSTEM DESIGN AND IMPLEMENTATION

Our DNC coupler design is based on a modern 8 bit microcontroller, and the software has been written in assembly language. In its present form the system can perform the essential DNC functions listed below:

1) Download an NC program form the host to the DNC coupler.
2) Send the NC data to the NC machine as it requests data.
3) List the NC Program in the DNC local memory on a CRT for inspection by the operator.

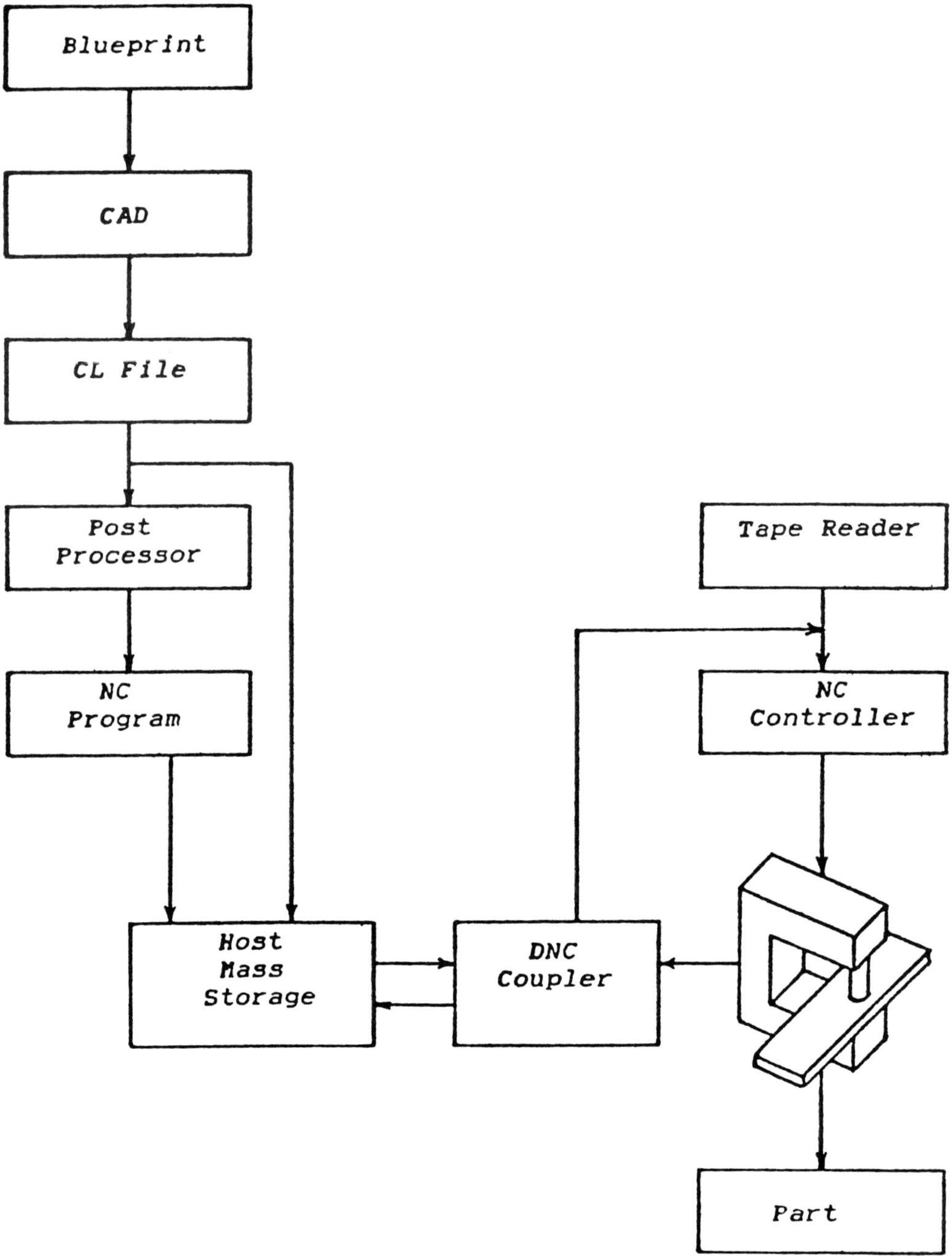

Figure 8 Part Production with a DNC System

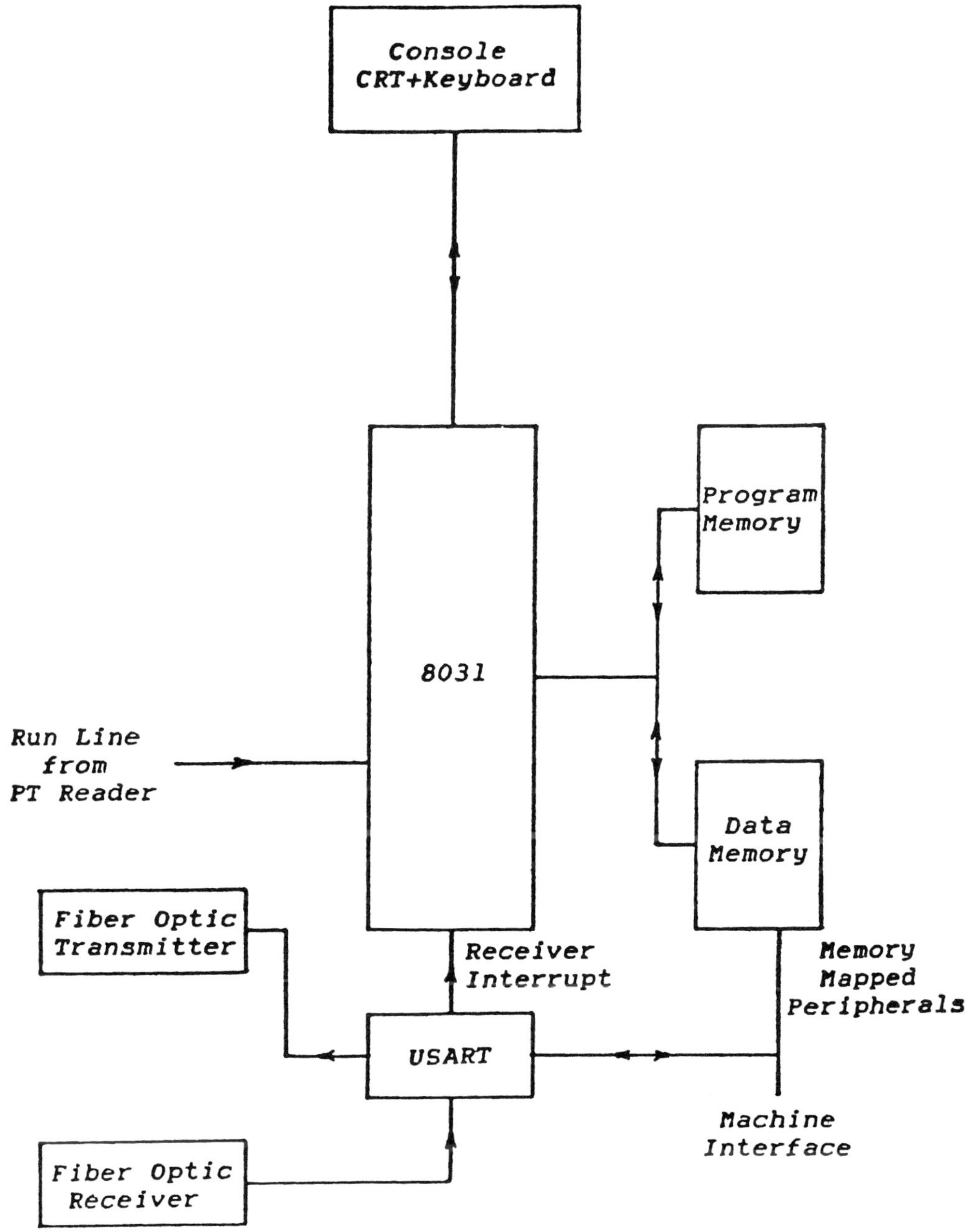

Figure 9 Block Diagram of a DNC Coupler

The coupler can also serve as a terminal for interactive communications with the host.

The hardware consists of the functional blocks shown in Fig. 9. The basis of the DNC coupler is an Intel 8051 family micro-controller. The host communication lines utilize fiber optic technology and the machine interface can be used on most NC/CNC machines with optical type tape readers.

The host for the DNC system is a VAX minicomputer which is connected to the CAD/CAM computer from which it receives NC part programs for mass storage on a hard disk.

The DNC coupler is connected to the host through one of its standard RS232 terminal ports. Thus to the host, the coupler appears to be a user at a terminal. Operations such as creating files, deleting files, and listing files can be performed by the coupler by sending standard operating system commands to the host through the communication lines.

Communication between host and coupler is carried out over an asynchronous full duplex fibre optic link that was developed specifically for this undertaking. The DNC coupler can use the optical lines directly through a fibre optic transmitter and receiver. The fibre optic system has an EMI immunity of 8000 V/m with a bit error rate of less than 10^{-9}. This means that data transmitted through the optical lines will very likely not be distorted, regardless of cable routing.

Another justification for the optical lines is the long transmission distances attainable: 1000 m without the use of repeaters. This is useful in a shop where the optical cables will likely be routed in a non direct fashion from host to coupler. The maximum distance for RS232 electrical cables is 15 m.

Other advantages of fibre optics are; (a) high data transmission speeds (5 MBd is easily attained), (b) the elimination of ground loops in the communication lines, because there is no electrical connnection between the source and the destination a ground level shift is of no consequence.

An Intel 8051 family microcontroller was chosen as the basis of the coupler for several reasons including its large number of I/O lines, large address space, speed, and multi source interrupt capability. The members of the 8051 microcontroller family are the 8051, 8751, and the 8031. The distinction between the members of the family is the configuration of the on-board program ROM. The 8051 has 4 kBytes of factory programmed ROM, the 8751 has 4 kBytes of UV eraseable EPROM, and the 8031 has no on-board ROM. In all other respects the microcontrollers are identical.

The 8051 is a non Von Neuman machine capable of addressing 64 kBytes of data RAM and 64 kBytes of program ROM. Included on-board the 8051 are 256 bytes of data RAM which includes four banks of eight registers, special function registers, and some directly addressable bit locations. Also on-board the 8051 are a UART for serial I/O, and two 16 bit timers/event counters. The 8051 recognizes five interrupt sources: two external interrupts, two interrupts from the on-board timers/event counters, and one from the

UART. Running on a 12 MHz. clock, fully 57% of the instructions execute in 1 us. and the two slowest instructions (multiply and divide) require 4 us.

The microcontroller, RAM, ROM, USART, and fibre optic terminals are accommodated on two small boards thereby avoiding unnecessary cluttering of the NC machine workspace.

4. CONCLUSIONS

The microcontroller-based system described above was installed and operated on a large milling machine. Performance conformed with design objectives, and system reliability was found not to be adversely affected. It is felt that the enhanced facilities that microcontroller based DNC affords will lead many future NC implementations to follow this general direction.

5. ACKNOWLEDGEMENTS

This undertaking was initiated with the advice and support of J.E. Crozier, Canadian Institute of Metalworking (CIM), Ancaster, Ontario. The system described was tested at CIM. Thanks are due to the personnel of CIM for assistance with the CAD system and NC machine control.

6. REFERENCES

1. Groover, M.P., Emory, W.Z., CAD/CAM Computer Aided Design and Manufacturing, Prentice-Hall, New Jersey, 1984, pp. 133-211.

2. Olesten, N.O., Numerical Control, Wiley-Interscience, New York, 1970, pp.366-368.

3. Patton, W.J., Numerical Control Practice and Applications, Reston Publishing, Virginia, 1972, pp. 15-16.

4. Hegland, D.E., "Numerical Control-Your Best Investment in Productivity", Production Engineering, March 1981, pp. 42-47.

5. Pressman, R.S., Williams, J.E., Numerical Control and Computer Aided Manufacturing, John Wiley and Sons, New York, 1977, pp. 292-293.

6. Optoelectronics Designers Catalog, 1984, Hewlett-Packard, pp. 6-2 - 6-70.

7. DMF-32 User's Guide, 1982, Digital Equipment Corp., p. 8.

8. MCS-51 Family of Single Chip Microcomputers User's Manual, 1981, Intel Corp.

9. Microcontroller Handbook, 1983, Intel Corp.

10. Microcontroller Applications Handbook, 1982, Intel Corp.

11. SDK-51 MCS-51 System Design Kit User's Guide, 1981, Intel Corp.

12. Private communication from R. Chang, CIM.

13. Microprocessor and Peripheral Handbook, 1983, Intel Corp., pp. 6-122 - 6-138.

14. MCS-51 Macro Assembler User's Guide, 1983, Intel Corp.

CHAPTER 14

MICROPROCESSOR-BASED REPETITIVE CONTROL

M. Nakano and S. Hara
Department of Control Engineering
Tokyo Institute of Technology
Oh-Okayama, Meguro-ku, Tokyo
Japan

ABSTRACT. Repetitive control is a new control scheme, in which the control variables follow the periodic reference commands with high accuracy. The asymptotic tracking property is achieved by locating the model 1/(1-exp(-Ls)) with appropriate digital filters in the closed-loop system. A design procedure of microprocessor-based repetitive control systems is proposed based on the derived stability condition and their properties. The method is applied to the control of a single axis electric servomechanism in which the controller is implemented on a 16-bit microprocessor. The experimental result shows that the tracking error can be reduced rapidly to very low level by repetitive operation.

1. INTRODUCTION

There exists a basic requirement in control systems such that the control systems should have ability to regulate the controlled variables to reference commands without steady-state error against unknown and unmeasurable disturbance inputs. Control systems with this property are called servomechanisms or servo systems. A lot of useful design methods have been developed for servo systems.

In servo system design, the Internal Model Principle proposed by Francis and Wonham(1975) plays an important role. The Internal Model Principle is a generalization of system type theory. It is well known that the output of a type L feedback system, i.e., the system has a $1/s^L$ element in the loop, may track a polynomial type reference input of degree L-1 without steady-state error, e.g., no steady-state error occurs for step type reference commands in type 1 feedback systems. The Internal Model Principle means that the controlled output tracks to a class of reference inputs without steady-state error if the generator for the references is included in the stable closed-loop system. For example, the generators for step inputs and sinusoidal inputs are 1/s and $\omega^2/(s^2+\omega^2)$, respectively.

In practical applications, we often encounter the case where the reference commands to be tracked or disturbance inputs excited are periodic signals, e.g., repetitive commands for mechanical systems such

N. K. Sinha (ed.), Microprocessor-Based Control Systems, 279–296.

as robots or disturbances depending on the frequency of the power supply. Any periodic signal can be generated by a linear free system including a time-lag element corresponding to the period with an appropriate initial function. The Internal Model Principle therefore reveals that the high accuracy asymptotic tracking properties for periodic exogenous inputs can be achieved by locating the model which generates a set of periodic signals. It has been first shown by Inoue et.al.(1981) that this conjecture is true and it has been verified that this scheme called repetitive control is useful for some applications. Furthermore, the extensions of this idea have been investigated by Omata et.al.(1984, 1985) and Hara et.al.(1985). This scheme is considered as a simple learning control. The analogous scheme has been developed by Uchiyama(1978), Arimoto et.al.(1984) and Kawamura et.al.(1984) and high accurarcy control was achieved by iteration of the control action with the same initial condition. In their methods, the reference command was corrected using the error signal obtained in the former trial.

First, we introduce the repetitive control system and state the principle and the stability condition which is closely related to the frequency domain optimal condition for Kalman filter or optimal regulator. The modified repetitive control system with appropriate compensators or filters is also introduced and the properties such as stability and characteristics of steady-state error and disturbance rejection are investigated in Chapter 3. Microprocessor are useful for realizing this scheme, since pure delay elements are required to construct the Internal Model or the generator for periodic signals. In Chapter 4, a design procedure and its implementation for microprocessor -based repetitive control are proposed based on the propeties obtained in Chapter 3. The controller implemented on a microprocessor contains nonrecursive digtal filters for compenstating the characteristics. The scheme is applied to the hybrid control of a single axis electric servomechanism, in which a 64-watt DC servomotor drives an inertial load through a gear train. The experimental result are discussed in Chapter 5.

2. PRINCIPLE OF REPETITIVE CONTROL

Any periodic signal with period L can be generated by a free system shown in Fig.1 with an appropriate initial function. It is therefore expected from the Internal Model Principle (Francis and Wonham 1975) that the asymptotic tracking property for exogenous periodic inputs may be achieved by locating the model $\exp(-Ls)/(1-\exp(-Ls))$ or $1/(1-\exp(-Ls))$ in the closed-loop system. This has been pointed out by Inoue et.al. (1981). These models are said to be repetitive controllers and the system with these models is named repetitive control system.

We now consider the simplest repetitive control system with the model $1/(1-\exp(-Ls))=1+\exp(-Ls)/(1-\exp(-Ls))$ depicted in Fig.2, where U, Y, R and E denote the control input, controlled output, reference command and error signal, respectively and $P(s)$ and $C(s)$ are the transfer functions of the plant and the controller for the basic servomechanism, respectively.

The following relations hold in the system:

$$E(s) = R(s) - Y(s), \tag{1}$$

$$Y(s) = P(s)C(s)V(s) + Y_o(s), \tag{2}$$

$$V(s) = E(s) + W(s), \tag{3}$$

$$W(s) = e^{-Ls}(W(s)+E(s)) + W_o(s), \tag{4}$$

where $Y_o(s)$ and $W_o(s)$ are the Laplace transforms of the initial responses of $P(s)C(s)$ and $\exp(-Ls)$, respectively. From these equations we have

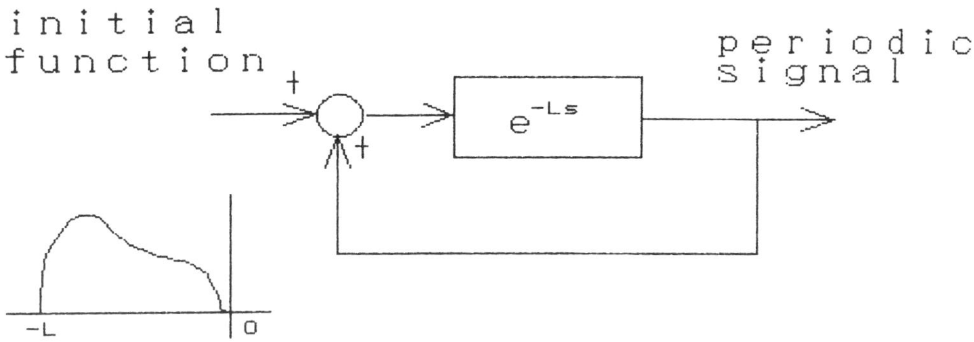

Figure 1. Generator of periodic signal

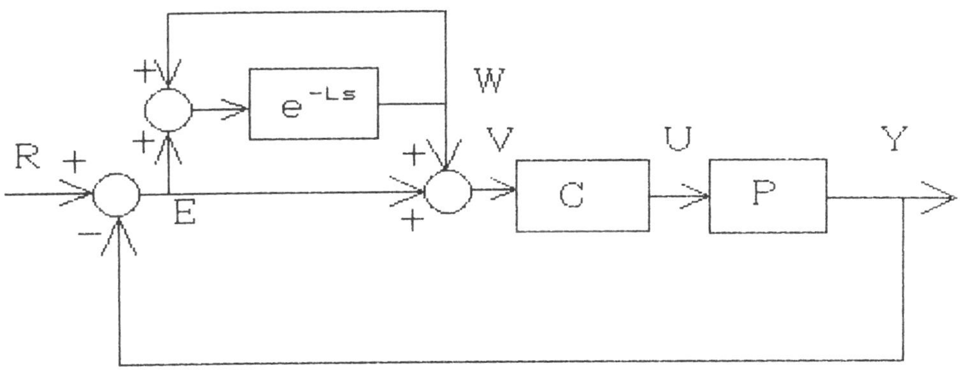

Figure 2. Block diagram of repetitive control system

$$E = \{ e^{-Ls}/(1+G_o) \} E + D_e$$
$$= \{ e^{-Ls}(1-G_c) \} E + D_e, \quad (5)$$

where G_o and G_c are the open-loop and closed-loop transfer functions of the basic servo system respectively and D_e is the equivalent exogenous input (see Fig.3). These are defined as follows:

$$G_o = PC, \quad (6)$$

$$G_c = G_o/(1+G_o) = PC/(1+PC), \quad (7)$$

$$D_e = \{ (1 - e^{-Ls})(R-Y_o) - PCW_o \} /(1+PC). \quad (8)$$

Suppose that the reference command r(t) is a continuous periodic signal with period L. In this case, if r(t) is bounded for $0 \leq t \leq L$, then we get

$$r_o(t) \triangleq \mathcal{L}^{-1}|(1-\exp(-Ls))R(s)| \in L_2, \quad (9)$$

since $r_o(t)=0$ for $t \geq L$, where $\mathcal{L}^{-1}$ denotes the inverse Laplace trasform. This and (8) imply that $\mathcal{L}^{-1}|D_e(s)| \in L_2$ under the assumption of the stability of $G_c(s)$. Therefore, the error convergence or stability condition for the repetitive control systm can be derived by applying the small gain theorem (Desoer and Vidyasagar 1975) to the system described by (5).

In the repetitive control system shown in Fig.1, if

1) $G_c(s)$ is asymptotically stable, (10)

2) $|1/(1+G_o(j\omega))| < 1$; for all ω , (11)

or $|1-G_c(j\omega)| < 1$; for all ω , (12)

hold, then

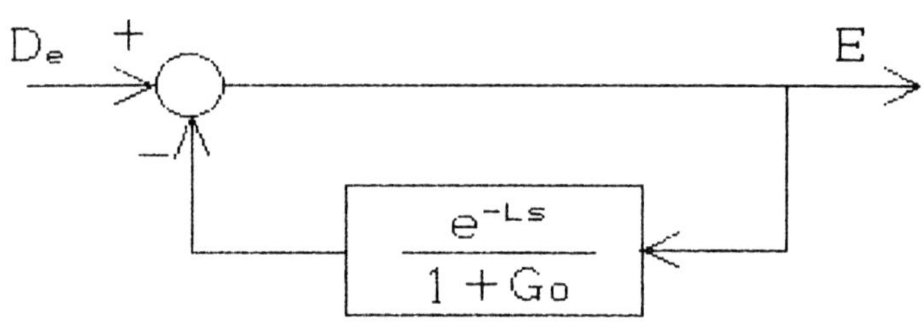

Figure 3. Equivalent block diagram of repetitive control system

$$e(t) \triangleq \mathcal{L}^{-1}|E(s)| \in L_2. \tag{13}$$

The above result is obvious, since the induced norm of L_2 operator $G(s)$ is defined as $\sup|G(j\omega)|$ and $|\exp(-j\omega L)| = 1$ for all ω. Note that (13) guarantees that the steady-state error converges to zero if e(t) is continuous. We also note that this sufficient stability condition is very close to the necessary one, since the phase of the time lag part $\exp(-j\omega L)$ in the equivalent system may have arbitrary value in the high frequency domain.

The conditions (11) and (12) require that the Nyquist plots of $G_o(j\omega)$ and $G_c(j\omega)$ remain the corresponding shaped portions in Fig.4, i.e., the loci $G_o(j\omega)$ and $G_c(j\omega)$ lie outside the unit circle centered at $-1+j0$ and inside the unit circle centered at $1+j0$, respectively. It is interesting that the circle condition (11) is same as the frequency domain optimality condition of Kalman Filter or optimal regulator (Kalman 1964). This fact suggests that the repetitive control system can be realized by using the technique of Kalman Filter with perfect regulation or optimal regulator with perfect observation for minimal phase plants, i.e., systems without unstable zero (Hara et.al. 1985).

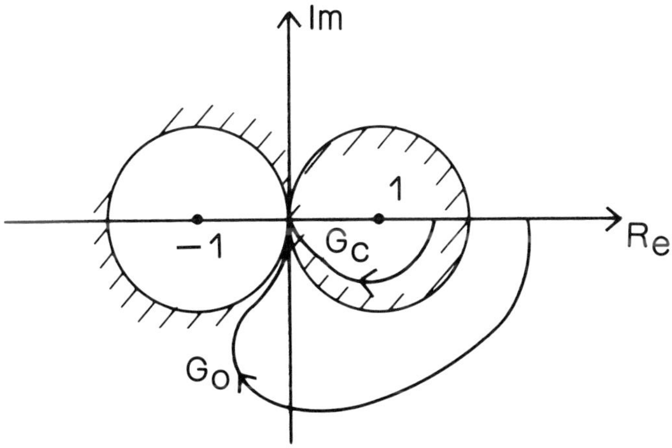

Figure 4. Stability condition for repetitive control system

3. CONFIGURATION AND PROPERTIES OF MODIFIED REPETITIVE CONTROL SYSTEM

The repetitive control system discussed in the previous chapter has the following two defects, since the circles of the stability conditions (11) and (12) are touched with the origin of the complex plane in the high frequency:

1) The stability condition (11) or (12) holds only for minimal phase plants with relative degree less than one.
2) The phase margin of the system is almost zero degree.

These are occured in the cause of the unrealistic demand of the perfect tracking for any periodic reference command including the higher order harmonic signals. We now consider a modified repetitive control system shown in Fig.5 to overcome this problem, where D(s) denotes the disturbance input. It is expected that the stability condition is relaxed by diminishing the loop-gain of the repetitive compensator, i.e., replacing exp(-Ls) with Q(s)exp(-Ls), where

$$|Q(j\omega)| \leq 1 \quad ; \text{ for all } \omega \quad . \tag{14}$$

Noting that the error may not converge to zero in the case where $|Q(j\omega)| = 1$. This implies that Q(s) determines the frequency band of the reference command to be tracked. Another compensator M(s) is introduced to specify the characteristics of the random disturbance rejection or the error convergence rate.

Assuming that all the initial conditions are zero for simplicity, the following relations hold in the modified repetitive control system depicted in Fig.5:

$$E(s) = R(s) - Y(s), \tag{15}$$

$$Y(s) = G(s)V(s) + D(s), \tag{16}$$

$$V(s) = E(s) + M(s)W(s), \tag{17}$$

$$W(s) = Q(s)e^{-Ls}(W(s) + E(s)). \tag{18}$$

These equation leads to

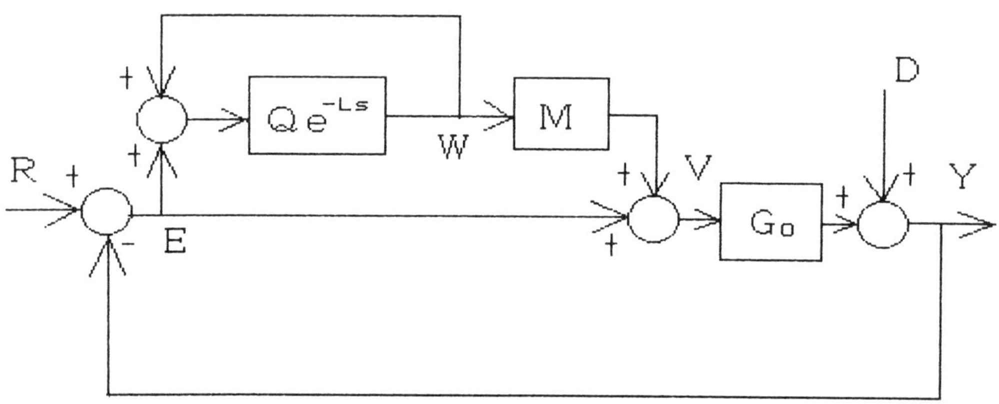

Figure 5. Block diagram of modified repetitive control system

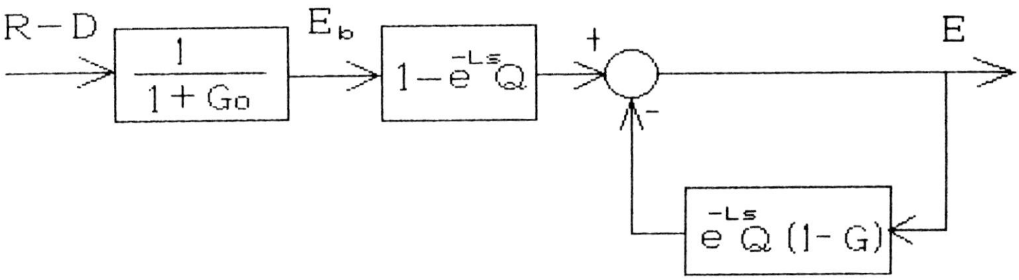

Figure 6. Equivalent block diagram of modified repetitive control system

$$E = e^{-Ls}Q(1+G_o)^{-1}\left[1+G_o(1-M)\right]E + D_m$$
$$= e^{-Ls}Q(1-G)E + D_m, \tag{19}$$

where

$$G = G_o M/(1+G_o) = MG_c, \tag{20}$$

$$D_m = (1-Qe^{-Ls})(R-D) \;/\; (1+G_o). \tag{21}$$

Eq.(19) represents the equivalent system shown in Fig.6, where $E_b(s)$ denotes the error for the basic servo system. In the following, we investigate some properties of the modified repetitive control such as stability and characteristics of steady-state error and disturbance rejection using this equivalent system.

3.1. Stability Condition

The stability condition, which is independent of the period L, is derived from (19) or Fig.6 (Inoue, Nakano and Iwai 1981, Hara et.al. 1985).

Assum that Q(s) and M(s) are asymptotically stable. If the following two conditions:

1) $G_c(s)$ is asymptotically stable, (22)

2) $|Q(j\omega)|\,|1-G(j\omega)| < 1$; for all ω (23)

hold, then the modified repetitive control system shown in Fig.5 is asymptotically stable, and then the error e(t) is bounded for continuous periodic signals r(t) and d(t) with period L.

It is seen from (23) that if the Nyquist locus of G(s) lies inside the circle in complex plane of radius $1/|Q(j\omega)|$ centered at 1+j0 then the system is stable (see Fig.7). Therefore, the stability condition (23) is satisfied for sufficiently small $|F(j\omega)|$ or $|M(j\omega)|$. In this case, however, the error convergence rate may be small and the steady-

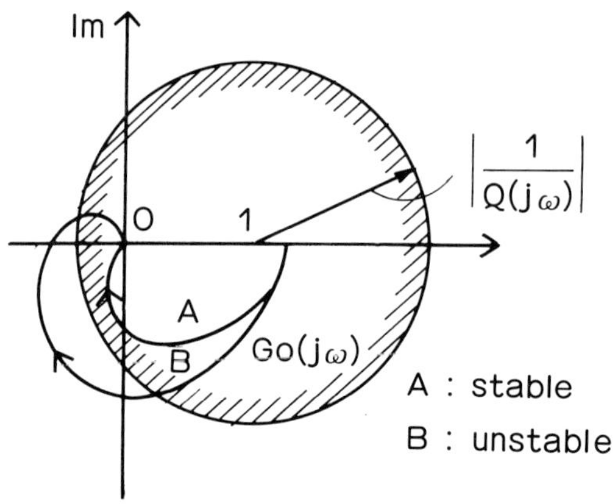

Figure 7. Stability condition for modified repetitve control systm

state tracking error may be large. This is a tradeoff in the repetitive control system design.

3.2. Characteristics of Steady-State Error

Once the compensator or the filter for band limitation Q(s) is included in the loop of the repetitive compensator, the error may not converge to zero as t infinity, since the feedback loop does not contain any model or generator for periodic signals. We, therfore investigate the steady-state tracking error for sinusoidal inputs $\sin(\omega_k t)$ with

$$\omega_k = 2\pi k/L \quad ; k = 0, 1, \cdots, \tag{24}$$

where $s=j\omega_k$ are the poles of the generator for the periodic signals with period L, $1/(1-\exp(-Ls))$. Using the relation described in Fig.6 and $\exp(-j\omega_k L)=1$, the gain of the steady-state error is written by

$$A_s(\omega_k) = \frac{1}{|1+G_o(j\omega_k)|} \cdot |A_b(\omega_k)| \quad , \tag{25}$$

where

$$A_b(s) = \{1-Q(s)\} / \{1-Q(s)(1-G(s))\} \quad . \tag{26}$$

Eq.(25) reveals that the tracking error can still be improved to a great extent in the frequency range over which $Q(j\omega_k)$ is sufficiently near to 1 or $|1+G_o(j\omega_k)|$ is sufficiently large. Note that the second condition is the same one required for the basic servo system.

3.3. Characteristics of Random Disturbance Rejection

It is obvious from Fig.6 that the disturbance inputs with the same period to the reference commands can be reduced to the sufficiently low level in the (modified) repetitive control system. However, it is not so for non-periodic disturbance inputs such as random signals with high frequency hands, which appear in many practical applications. The characteristics of random noise rejection are discussed in the following.

Assume that the reference r(t) is identically zero in Figures 5 and 6 and that the disturbance d(t) is a sample process from a stationary time series. It is easily seen from Fig.6 that the power spectra $S_o(\omega)$ for $e_o(t)$ and $S_e(\omega)$ are given by

$$S_b(\omega) = \left| \frac{1}{1+G_o(j\omega)} \right|^2 \cdot S_d(\omega) \tag{27}$$

and

$$S_e(\omega) = |H(j\omega)|^2 \cdot S_b(\omega) \ , \tag{28}$$

respectively, where $S_d(\omega)$ denotes the power spectrum of d(t) and

$$H(s) = \frac{1-Q(s)\exp(-Ls)}{1-Q(s)(1-G(s))\exp(-Ls)} \tag{29}$$

represents the transfer function from $E_b(s)$ to E(s). We now consider the average of $S_e(\omega)$ by taking the interval of angular frequency with its midpoint at an arbitrary frequency ω :

$$\omega - \pi/L \leq L \leq \omega + \pi/L \tag{30}$$

where the width of the interval corresponds to the period L. In this case, the average is represented by

$$\bar{S}_e(\omega) = \frac{L}{2\pi} \int_{\omega_L}^{\omega_u} S_e(\sigma) d\sigma \ , \tag{31}$$

where

$$\omega_L = \omega - \pi/L \ , \tag{32.a}$$

$$\omega_u = \omega + \pi/L \ . \tag{32.b}$$

Under the assumption of L be very large, we may suppose that the spectrum $S_b(\omega)$ and the frequency transfer functions $Q(j\sigma)$ and $G(j\sigma)$ are approximately constant over the interval, i.e., for $\omega_L \leq \sigma \leq \omega_u$,

$$S_b(\sigma) \simeq S_s(\sigma) \ , \tag{33.a}$$

$$Q(j\sigma) \simeq Q(j\sigma) \ , \tag{33.b}$$

$$G(j\sigma) \simeq G(j\sigma) \ . \tag{33.c}$$

From (28), (31) and (33.a), we have

$$\overline{S_e}(\omega) \simeq a_d^{\,2}(\omega)\cdot S_b(\omega) \ , \tag{34}$$

where

$$a_d^{\,2}(\omega) = \frac{L}{2\pi}\int_{\omega_L}^{\omega_u} |H(j\sigma)|^2 \, d\sigma \ . \tag{35}$$

Upon integrating $|H(j\sigma)|^2$ using (33.b) and (33.c), we get

$$a_d^{\,2}(\omega) \simeq A_d^{\,2}(\omega) = 1 + \frac{|Q(j\omega)|^2 |G(j\omega)|^2}{1-|Q(j\omega)|^2 |1-G(j\omega)|^2} \ . \tag{36}$$

Remark that $A_d^{\,2}(\omega)$ is greater than one whenever the stability condition (23) holds.

The above discussion yields the following properties with respect to the random disturbance rejection:

1) The power of the error signal to the disturbance input at angular frequencies near ω is increased by a factor of $A_d^{\,2}(\omega)$ by inserting the controller for the repetitive operation to the basic servo system.

2) The greater value of $|1+G_o(j\omega)|$, the lower value of $|Q(j\omega)\cdot G(j\omega)|$, or the smaller value of the phase of $G(j\omega)$, the superior disturbance elimination characteristics are generally achieved. The first and the second conditions are obvious and the last one follows from the discussion stated below.

The relationship between the gain of $A_d^{\,2}(\omega)$ and the phase of $G(j\omega)$ is considered. The second part of the right hand side of (35) can be rewritten by

$$\frac{|Q|^2|G|^2}{1-|Q|^2|1-|G|\exp(j\phi)|^2} \ , \tag{37}$$

where ϕ is the phase of G (the argument $j\omega$ is omitted for simplicity). It is seen from (37) that the minimum value of $A_d^{\,2}$ is attained when $\phi=0$ under the same gain of G, since the second part of the denominator of

(37) is less than one, i.e., the stability condition (23) holds. In this case, A_d^2 is expressed as

$$A_d^2 = \frac{|G|^2|G|^2}{1-|Q|^2|1-|G||^2} \quad . \tag{38}$$

4. DESIGN PRINCIPLE AND IMPLEMENTATION OF REPETITIVE CONTROL SYSTEM

The design of the repetitive control system shown in Fig.5 is to determine the controller for the basic servo system C(s), the dynamical compensator M(s) and the filter for the band limitation Q(s) so that the resultant control system has desirable characteristics.

M(s) and Q(s) should be implemented on a microprocessor combined with the dead time exp(-Ls), since a microprocessor is required to realize the dead time, which can be easily constructed by some delay elements. Fig.8 shows this type of controller, where the coefficients W_i(i=0, 1, ••• ,m) and a_i(i=0, 1, ••• ,n) are the parameters to be designed for M(s) and Q(s) respectively and T is the sampling period of the digital controller. When the controller C(s) is also implemented on a microprocessor, we have a repetitive control system with complete digital controllers. On the other hand, a hybrid control system is constructed if C(s) is realized by analog elements.

The propeties of the modified repetitive control derived in the preceding chapter lead to the design principle. The outline of the design procedure is as follows:

Step 1) Design C(s) so that the basic servo system has desirable characteristics such as responses to reference command, disturbance elimination and sensitivity and roboustness for parameter perturbations, i.e., $|G_c(j\omega)|$ is near to one over the sufficiently wide frequency band and the gain is diminished in the high frequency domain.

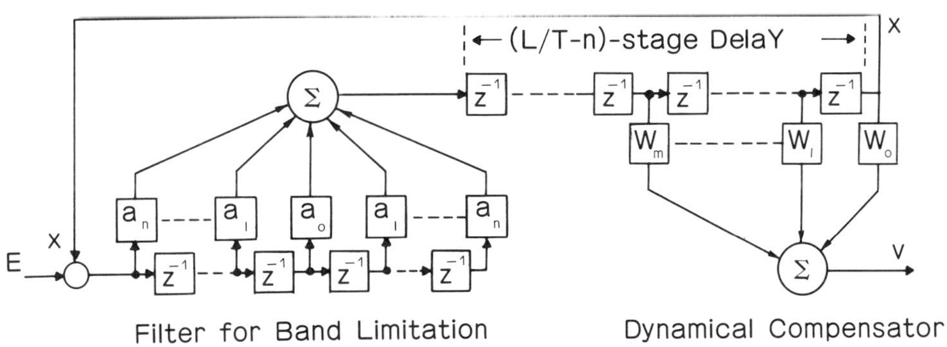

Figure 8. Block diagram of digital controller for repetitive control

Step 2) Determine M(s) so that G(jω) is real, i.e., the phase lag in $G_c(j\omega)$ is canceled out by the phase lead in M(jω) and has sufficiently wide low-pass gain characteristics. The former and the latter conditions are required for achieving the superior characteriscs with respect to disturbance elimination and error convergence rate respectively (Inoue, Nakano and Iwai 1981). Such a compensator M(s) can be constructed by a non-recursive digital filter as shown in fig.8, since the future values for its nominal input x(t) are available for the synthesis of the compensating signal.

Step 3) Construct the non-recursive filter Q(s), whose weighting sequence is symmetric, so that it has a possible wide frequency range within satisfying the requirements for stability and disturbance rejection. The symmetry of the filter guarantees Q(jω) is real or the phase is equal to zero, whcih yields the superior characteristics with respect to steady-state error.

5. APPLICATION

The proposed method was applied to a simple single axis electric servo-mechanism, in which a 64-watt DC servomotor with armature time constant 12ms drove an inertial load (about $4380\text{g}\cdot\text{cm}^2$) through a gerar train (the ratio 1:10). A basic servo system was constructed using the appropriate feedback from a tachogenerator on the motor shaft and from a potentio-meter on the load shaft. The step response to the basic servo system was measured to determine the transfer function. Neglecting the high-frequency damped oscillation appeared in the step response, we assumed that the dynamics of the basic servo system was represented by a second-order transfer function:

$$G_c(s) = 50^2 \;/\left\{ S^2+2\times 50s+50^2 \right\} \quad . \tag{39}$$

The controller for the experimental repetitive operation was implemented on a 16-bit micro-computer. 12-bit two D/C converters and one A/D converter were employed for the data transfer between the microprocessor and the basic servo system. A 1K-word memory area was shared for the implementation of the dead time elements corresponding to the period L. The interval of clock for data transfer and the sampl-ing period T were set at 0.01(sec).

The dynamical compensator M(s) and the filter Q(s) for band limitation were constructed by nonrecursive digital filters. M(s) was determined so that $G(j\omega)=M(j\omega)\cdot G_c(j\omega)$ has a low-pass gain with nearly-zero phase shift characteristics (see Fig.9). A non-recursive digital filter with a symmetric weighting sequence was used for band limitation. The filter gain depicted in Fig.10 was determined so that it has a possibly wide frequency range within satisfying the requirements for stability and disturbance elimination. The values of $|A_b(\omega)|$ and $A_d^2(\omega)$ are also shown in Fig.11. The steep slope for $|A_b|$ in the passband of the filter guarantees that the steady-state tracking error is reduced to a very low level in a wide frequency range. The plot of $A_d^2(\omega)$ shows

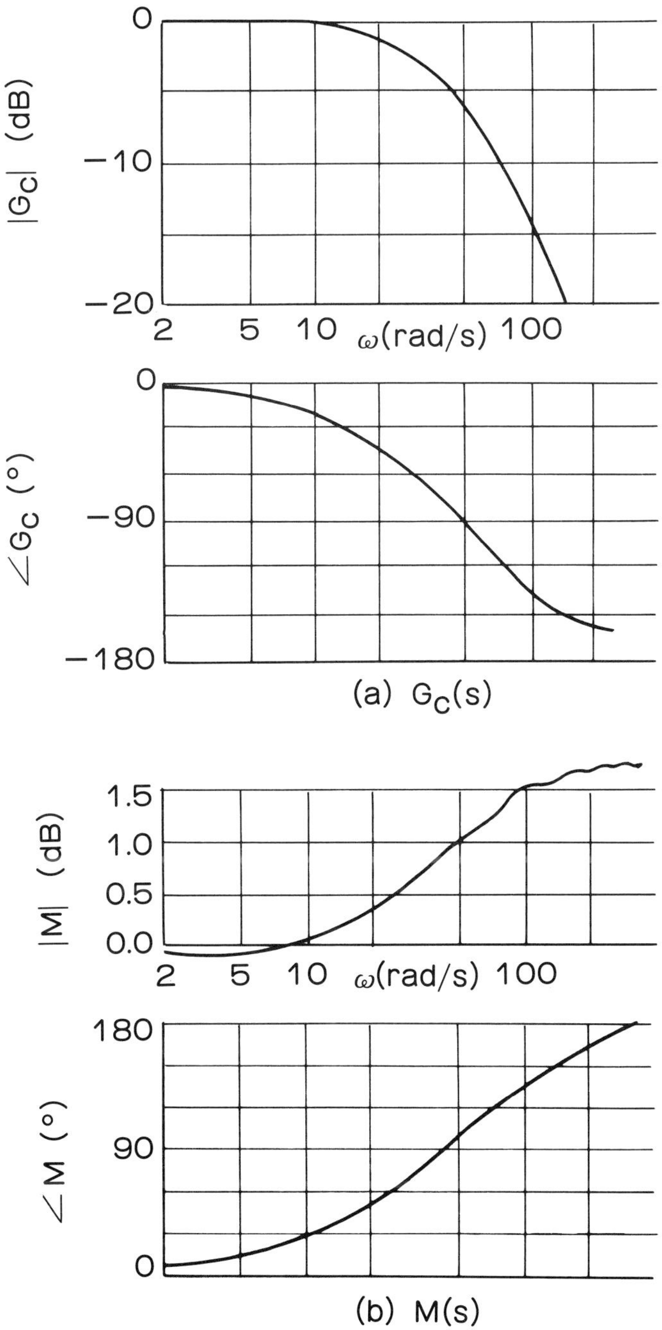

Figure 9. Frequency responses of closed-loop system of basic servo system $G_c(s)$ and dynamic compensator $M(s)$

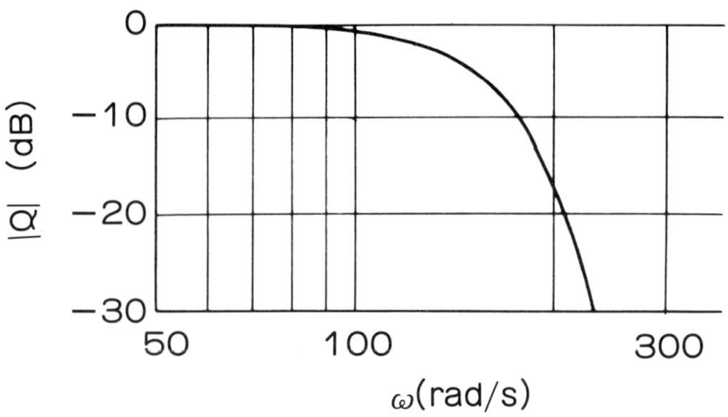

Figure 10. Gain characteristics of filter Q(s)

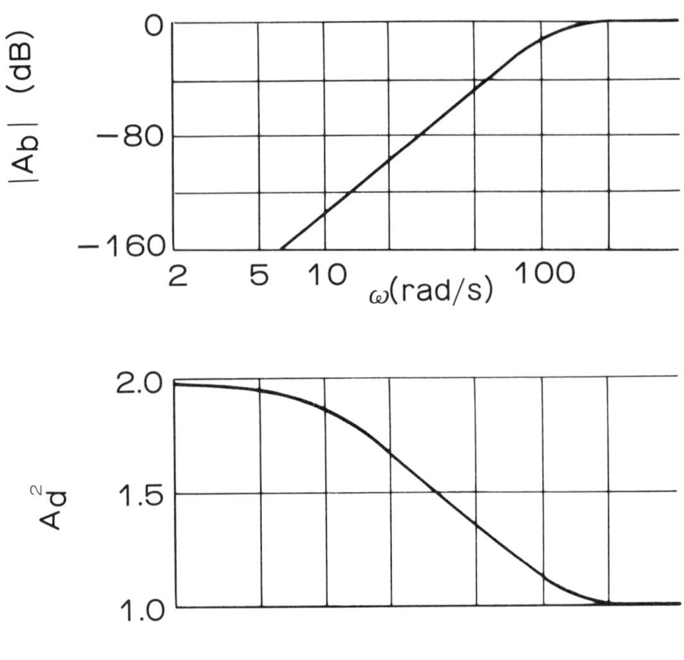

Figure 11. Values for indices of performance

that the deterioration of the performance of the disturbance rejection is diminishied in the high frequency domain, or dominant frequency range for random disturbances.

The experimental results for a reference command $V_r(t)$ with period 10.24(sec), shown in Fig.12(a), are as follows. Fig.12 also shows the waveforms of the error voltage $V_e(t)$ during the first, second and third periods. The waveform for the first period is the same one for the basic

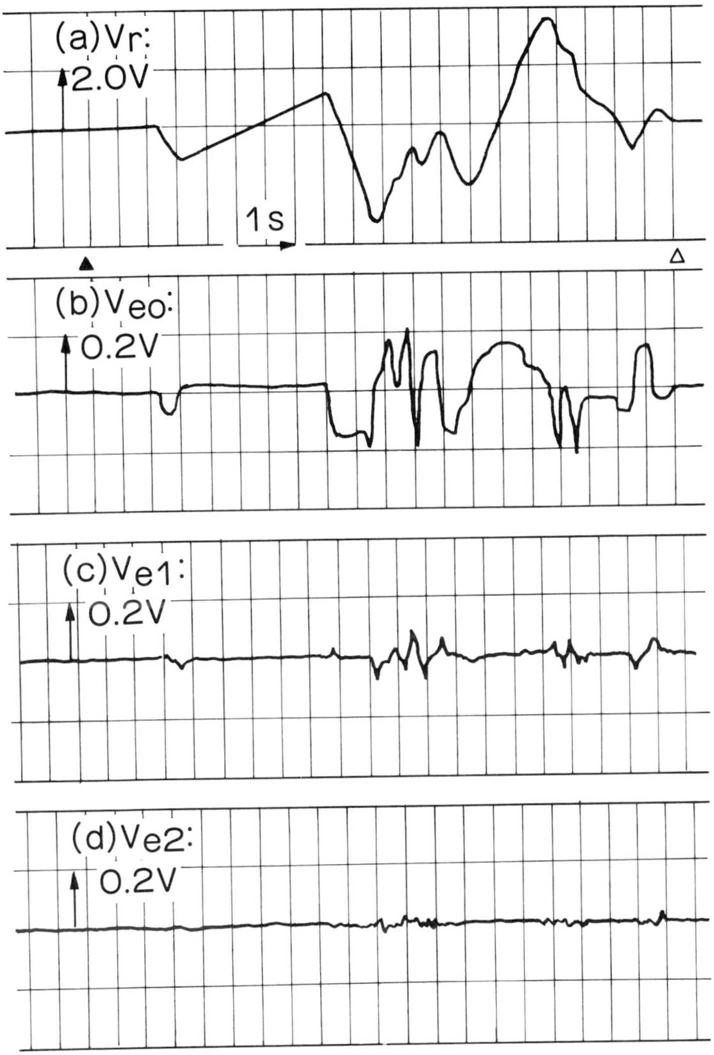

Figure 12. Reference command and error responses in the first three periods

servo system. The rapid error convergence, especially for the low frequency components, is achieved by repetitive operation. The error was monitored in the digital controller and rms (root mean square) value of the error voltage for each period was calculated. The relation of the rms error to the number of repetitions or periods is plotted in Fig.13, where k=0 implies the first period. This figure shows that the rms decreses to the level less than one percent of the rms for the basic servo system after tenth repetitive operations.

It should be noted that the proposed scheme can be also applied to practical systems whose dynamic models are precisely unknown, since the design procedure requires not the precise mathematical model but the frequency domain characteristics for the plant to be controlled. For example, Inoue et.al. (1981) have applied to the computer control of the 27-MVA thyristor power supply to the three main ring magnets of a proton synchrotron by assuming that the basic closed-loop model was represented by a simple third order disrete-time model. The 10^{-4} tracking accuracy required of the exciting current control was achieved after 16 cycles of a pulsed operation.

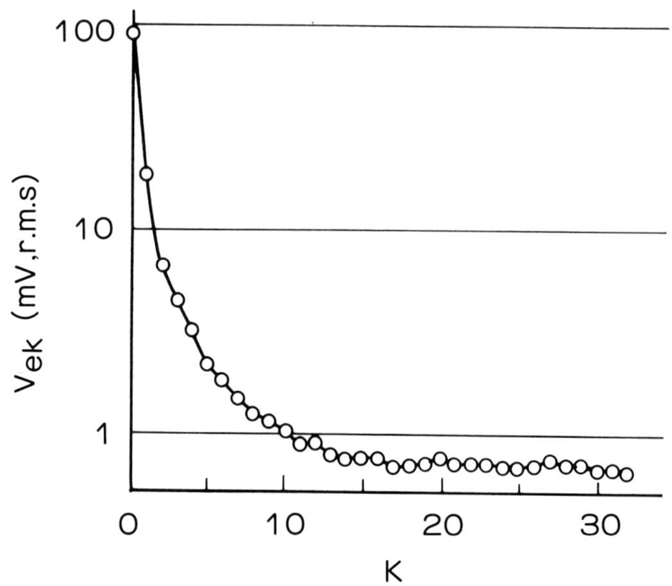

Figure 13. RMS error voltage for each period

6. CONCLUSION

A new control scheme named repetitive control has been introduced. The high accuracy tracking property for periodic reference commands is achieved by locating the generator for periodic signals in the closed-loop system. The characteristics of the (modified) repetitive control system such as stability, steady-state error and disturbance rejection have been investigated. A design principle of microprocessor-based repetitive control systems has been proposed based on the derived properties. The controller is constructed by the generator for the reference command, and the appropriate digital filters for specifying the characteristics and should be implemented on a microprocessor. The proposed method has been applied to the control of a single axis electric servomechanism. The controller was a hybrid ytpe, i.e., the analog controller was used for compensation of the basic servo system and the repetitive controller was implemented on a 16-bit microprocessor. The experimental result has showed that the tracking error can be reduced rapidly to very low level by repetitive operation.

The proposed idea is also useful for other dynamical systems such as both continuous and discrete-time linear multivariable systems (Hara et.al. 1985, Omata et.al. 1984), linear periodic systems (Omata et.al. 1985) and a class of non-linear systems (Hara et.al. 1985). The scheme is therefore applicable to controlled objects of many fields including repeated processes, which are commonly appeared in industrial production systems and mechanical control systems.

REFERENCES

S.Arimoto, S.Kawamura and F.Miyazaki : Bettering operation of dynamic systems by learning : A new control theory for servo mechanism or mechatronics systems, Proc. 23rd CDC, 1064/1069, 1984.

C.A.Desoer and M.Vidyasagar : Feedback systems : Input-output properties Academic Press, 1975.

B.A.Francis and W.M.Wonham : The internal model principle for linear multivariable regulators, Appl. Math. and Opt., 2-2, 170/194, 1975.

S.Hara, T.Omata and M.Nakano : Synthesis of repetitive control Systems and its application, to be presented at 24th CDC, Florida, 1985.

T.Inoue, M.Nakano and S.Iwai : High accuracy control of servomechanism for repeated contouring, Proc. 10th Annual Symp. Incremental Motion control Systems and Devices, 285/292, 1981.

T.Inoue, M.Nakano, T.Kubo, S.Matsumoto and H.Baba : High accuracy control of a proton synchrotron magnet power supply, Proc. 9th World Congress of IFAC, XX-216/221, 1981.

R.E.Kalman : When is a linear control system optimal, Trans. ASME. Ser. D : J. Basic Eng., 86, 1/10, 1964.

S.Kawamura, F.Miyazaki and S.Arimoto : Iterative learning control for robotic systems, Proc. IECON'84, 393/398, 1984.

T.Omata, M.Nakano and T.Inoue : Application of repetitive control method to multivariable systems, Trans. of SICE, 20-9, 795/800, 1984 (in Japanese).

T.Omata, S.Hara and M.Nakano : Repetitive control for linear periodic Systems,Trans. of IEEJ, 105-C, to be published, 1985(in Japanese).

M.Uchiyama : Formation of high-speed motion pattern of a mechanial arm by trial, Trans. of SICE, 14-6, 706/712, 1978 (in Japanese).

CHAPTER 15

REAL-TIME PROGRAMMING LANGUAGES FOR PROCESS CONTROL

S.I. Ahson
Department of Electrical Engineering
Indian Institute of Technology
New Delhi 110016
India

ABSTRACT. This chapter presents an overview of real-time programming languages that are finding increasing acceptance in process control applications. Major concepts and notations of high-level languages and concurrent programming are discussed and their implementations in Concurrent Pascal, Modula-2 and Ada are described. The programming language FORTH is also presented. Although a nonconcurrent language, FORTH contains certain features which facilitate the construction of software for process control. Examples from Concurrent Pascal, Ada and FORTH are given.

1. INTRODUCTION

For many years now process control software has been written almost exclusively in assembly language. By using assembly language, a control engineer can generate compact and efficient code. However, assembler programs are not portable and are expensive to write and maintain. In order to alleviate this situation, considerable attention has been given to the development of high-level real-time programming languages (1). With the advent of microcomputers, a great deal of interest currently exists in developing and using real-time languages for process control.

The combined use of blocks, procedures, functions and closed-form program control constructs provides structure to modern high-level programming languages. Real-time programming languages require additional constructs for describing concurrency (2). These constructs provide a means for representing several asynchronous tasks within a single program and provide efficient mechanisms for inter-task communication and synchronization (1,2). Real-time process control systems can be designed as a collection of co-operating tasks. Design of such systems requires great skills in software engineering (3).

The provision of run-time security is of paramount importance in process control. A run-time failure may be very expensive in terms of loss of production and may lead to accidents. Since an exhaustive testing of real-time software is not possible, the intrinsic security of the

N. K. Sinha (ed.), Microprocessor-Based Control Systems, 297–316.

language used plays a vital role. The intrinsic security partly comes through strong typing (1,2) and partly through recovery from error conditions using exception handling (3,4) mechanisms. The language must also include mechanisms to react to events as they occur in real-time.

Pascal (5) - the first programming language that was designed to support the concept of structured programming - is the base language for almost all real-time languages. Sequential language constructs in these languages have been derived from Pascal. We commence our study in section 2 by describing the Pascal language. In section 3, we introduce process and monitors (6) and show how these are used in Concurrent Pascal (7). This section is followed by a description of Modula-2 (8) in section 4. Ada (9) - the U.S. Department of Defense language for embedded computer applications - is presented in section 5. In section 6, we describe some experience gained in the use of FORTH (10) for process control. Finally, section 7 closes the chapter with a brief summary.

2. PASCAL

This section provides a brief introduction to Pascal. The language description that follows is based on reference 5.

Pascal is a structured programming language. Key elements that make Pascal a structured language are: declarations and procedures. A typical Pascal program has the overall structure shown in Figure 1. It consists of a heading and a block concluding with a period. The heading gives the name of the program and the file names which the program accesses as parameters. For example, program control (input, output); identifies the program name as 'control' which reads from the standard file 'input' and sends its results to the standard file 'output'. A program may access additional files by specifying them in the program heading. The program heading is followed by a list of declaration statements that assign meanings to the programmer - defined identifiers (symbolic names) appearing in the program.

```
program  programname (input, output);
     Declarations
          lables
          constants
          types
          variables
          procedure and function definitions
begin
     Executable statements
end.
```

Figure 1. General structure of a Pascal program

The word program is reserved by the Pascal syntax rules. It cannot be used in a program with meaning different than as prescribed by Pascal language. There are thirtyfive Pascal defined reserved words.

The list of statements of a block must be preceded by the reserved word begin and followed by the reserved word end. Semicolon is used as a separator between statements. A compound statement is formed by enclosing a sequence of simple or compound statements between delimiters begin and end. The executable statements of Pascal are grouped into four broad classes: assignment, data-processing, program control and input/output. Procedures provide the basic means for abstraction. A procedure is assigned a name and is executed whenever it is named in the program. The same general structure is used for procedures as for the main program. Each procedure defines its own local constants, variables, data structures and sub-procedures. The scope of an object defined within a procedure is limited to that procedure and all sub-procedures contained in it. A procedure can call itself. This characteristic is known as the recursive property. Pascal also provides functions. A function is destinguished from procedures by the fact that it computes or returns a single value and like an ordinary variable its type must be declared in the heading. Functions may also be recursive in nature. The general structure of procedures and functions is shown in Figure 2. A user-defined function or procedure can be defined employing the definition structures. Pascal has a number of pre-defined procedures and functions.

```
procedure procedurename (lfp);
    Declarations
        labels
        constants
        types
        variables
        other procedure and
        function definitions
    begin
        Executable statements
    end;
```

(a) Procedure definition

```
function functionname (lfp):type;
    Declarations
        labels
        constants
        types
        variables
        other procedure and
        function definitions
    begin
        Executable statements
    end;
```

(b) Function definition

Figure 2. General structure of Pascal procedures and functions

The constant definition part of declarations contains a list of 'identifier = constant' and begins with the reserved word const. The type of each identifier is implied by the type of the constant with which it is equated. The variables definition part contains a list of identifiers associated with their types. The four standard types of variables are: Integer, Real, Boolean and Character. Variable declarations are preceded by the reserved word var. Note that all the variables used in a program must be declared before use. Examples of const and var declarations are given below:

```
const                                   var
     maxswitchno = 8;                        count, clock:integer;
     epsilon = 0.01;                         velocity, current:real;
     samplingrate = 20;                      start, stop:Boolean;
     pi = 3.1416;                            firstchar:char;
     blank = '    ';
```

A variable's type determines the set of values it can take, and the set of operations that may be applied to it. While all integer quantities assume integer values in the range -MAXINT to +MAXINT, real data are assigned the values of floating-point numbers whose range and precision are compiler dependant. Variables of type CHAR use the ASCII character set. Finally, Boolean variables are assigned values denoted by the words true and false.

The arithmetic operators +, - and * yield integer results when applied to integer operands. The division operator / yields real result when applied to integer operands. Division operators DIV and MOD produce a truncated quotient and the remainder. Multiplication and division operations have precedence over addition and subtraction in expressions. Values may be compared using expressions with the relational operators = (equal), < (less than), > (greater than), < = (less than or equal), > =(greater than or equal), <> (not equal). The logical operators and, or and not are the standard operators of Boolean algebra and apply to Boolean data. The relational operators are used for comparing data items of the same type and yield Boolean results.

A type declaration permits the assignment of properties to the range of values that can be assumed by the variable. A subrange type has a value set defined as a subrange of the values of an existing ordered type. An enumerated type is defined by listing all the possible values allowed by the type.

Pascal provides the facility of structured data types - array, record, file and set. Structured data variables have more than one element where each element is either a simple or structured data type. An array is an indexed set of data items of the same type (integer, real, character, Boolean). The record type has a number of components associated with an object. The fields of a record are listed between the reserved words record and end. The composite data type file consists of a stream of data items of the same type that are stored sequentially on an I/O device. Files are accessed by the standard procedures read, readln, write, and writeln. The set type is provided in Pascal alongwith the set operators of union, intersection and difference. Some examples of data types are given below.

```
type
     switchno = 0..maxswitchno;
     measurement = (level, pressure, flow, temperature);
     vector = array [ 1..20 ] of integer;
     name = array [ 1..10 ] of char;
     matrix = array [ 1..50,1..100 ] of real;
     student = record
          name : array [ 1..10 ] of char;
```

```
          rank : integer;
          marks : real
              end;
     datafile = file of integer;
     measset = set of measurement;
```

Pascal permits packed array to be used instead of array in order to economize storage space. The record type allows for a variant part, allowing a record structure to be formed based on the value of the variant part. The order in which label, const, type, var declarations occur is important, in that each can only be used after they have been declared. The programmer can use any constant within a data type definition, and any data type or constant when defining a variable.

Files are initialized and opened for reading and writing by the built in procedures reset (filename) and rewrite (filename). Another standard procedure is close (filename) which should be used whenever a file is finished with. The communication between a filebuffer and the file is through procedures get (filename) and put (filename). For example, to fetch data from a file called 'temp' and to print the data on the output medium, the following sequence of statements will be used: reset (temp); get (temp); write (temp↑);. An end-of-file function eof (filename) will return a true when the end of file has been reached. Special facilities are provided in Pascal to deal with the text files which contain just characters.

In a process control application, a particular set of statements may have to be repeated many times for some sort of iterative solution. It is then necessary to program loops. There are three different ways of programming a control loop in Pascal - the while, repeat and for structures. If it is intended that the loop is to be executed so long as some condition still exists, the while statement is appropriate.

    While condition do statement;

If, however, the loop is to be executed until some condition is met then the repeat construction can be used

    repeat statement until condition;

The while-do statement will execute the statement (simple or compound) if the condition (Boolean expression) is true. The repeat-until statement will execute the statement (simple or compound) until the expression following until is true. If the number of times a certain group of statements must be executed is exactly known (e.g. when processing data of type array), then the for loop construction may be used. The general structure of this statement is given below.

    for index := initial-value to/downto final-value do statement;

The three control constructs of Pascal (while, repeat, for) are considered to be key constructs in the design of well-structured programs.

Branch control statements of Pascal are if-then-else (two-way-branching), case-of (multiway branching) and the goto statement. The use of goto's is discouraged since their presence complicates program structure. The general structure of the if statement is given below.

    if expression then statement 1 else statement 2;

The else part may be absent and multiway branches can be constructed by nesting several if statements. The _case_ statement implements multiway branches more efficiently. The case statement has the general structure shown below.

case expression _of_ C1:S1;C2:S2;....Cn:Sn _end_;

The selection of statements S1, S2,....Sn is based on integer or character values of the expression.

Dynamic variables are provided in Pascal through the pointer type. Standard procedures (new and dispose) are used for the dynamic allocation and deallocation of variables.

Pascal has had a significant impact on computing in the 1970's. It has received considerable use in the teaching of programming and has influenced new language designs. Versions of Pascal have been developed for microcomputers. These are subsets of the standard Pascal (5) and incorporate extensions which permit linkage to assembly language routines and allow direct control of peripheral devices. These features are very useful for control applications (20).

3. CONCURRENT PASCAL

When a microcomputer is used within an industrial control environment, the application software could be structured as a collection of co-operating sequential routines which are executed concurrently. For example, the tasks: 'measure plant variables', 'report alarms to operators', 'print report for management', can be executed concurrently. The programming technique used to control concurrent operation is called multiprogramming. In order to explicitly indicate that a routine will be executed concurrently it is declared as a process. A process runs asynchronously with respect to other processes.

In order to achieve successful co-operation, concurrently executing processes must communicate and synchronize(11,12)at certain points in time. A monitor is a mechanism that implements synchronization and communication. It consists of a data structure which is shared among processes and a set of operations (procedures) that processes can perform for accessing the data structure. The monitor guarantees that only one process may be executing a procedure in the monitor at a particular time, and this procedure must be completed before another process is allowed to enter into the monitor. When a monitor procedure is executing it has exclusive access to the permanent shared variables.

A process is conceptually similar to a procedure. It has a name, a set of declarations and a set of sequentially executable statements delimited between _begin_ and _end_. A process can be specified with parameters, priority and access rights to data. Some form of abort mechanism is required in a real-time environment. Process creation is explicitly denoted by a statement. A number of identical processes may be created through type declaration as an array or dynamic data object.

The general structure of a monitor is shown in Figure 3. It is written as a set of permanent (global) variable declarations followed by a set of procedure declarations. The monitor has a body (_begin...._ _end_) which is a sequence of statements that is executed immediately when the program is initiated. The body of a monitor is used to give initial values to the monitor variables. A procedure of the monitor is invoked by executing the statement "monitorname.procedurename (parameters)" in a process. Communication is only through the parameters of the procedure. The monitor guarantees that only one process may be executing a procedure in the monitor at a particular time. When a procedure of the monitor is executing, it has exclusive-access to the permanent shared variables. Other calls to the monitor will be delayed in a queue for later access.

```
monitorname monitor;
var declarations of permanent variables
procedure A (parameters)
    var declarations of variables local to A
        begin
            code for A
        end
procedure B (parameters)
    var declarations of variables local to B
        begin
            code for B
        end
other procedure definitions
begin
    code for initializing permanent variables
end;
```

Figure 3. General structure of monitor

Monitors first appeared in a practical programming language Concurrent (7) Pascal in the year 1975. This language has received widespread use in real-time systems and has inspired other language designs, such as Pascal Plus (13), Mesa (14) and Modula (15). Concurrent Pascal is a multiprogramming extension of Pascal. Sequential programming constructs in Concurrent Pascal are identical to those of Pascal. Additional features are provided for supporting concurrent processes. Monitors are used for the control of shared resources and for system-structuring. The language has been mainly used for writing operating systems and process control systems.

Processes and monitors in Concurrent Pascal are declared by 'type' definitions. A process consists of three parts: a declaration of its access rights to shared data and resources, a declaration of its private data structures and a sequential program which defines the actions of the process when it is executed. A monitor consists of four parts: a declaration of its access rights, a declaration of data structures to be shared between processes, a set of synchronizing operations and an initialization operation.

Defining processes and monitors as data types makes it possible to use several instances of the same component type in a system. It is possible to define constants, data types and local procedures within monitors and processes. The local procedures can also be called within the system type. Procedure definitions cannot be nested and cannot be recursive.

A monitor performs the short-term scheduling on processes for exclusive access to permanent shared variables. To perform long-term scheduling calls on the monitor are delayed in a queue for later access. A process which is within a monitor procedure may suspend itself by executing the statement delay (x). This places the calling process in the queue x, which may only hold one process at a time. Other processes may then enter the monitor and reactivate the queued process by executing a continue (x), whose effect is to force the calling process to leave the monitor immediately and resume the process waiting in the queue. If x is empty the call has no effect.

Our example in Concurrent Pascal is a program for transferring data from one process to another by means of a buffer stored on a disk. This small system has two processes: a producer process and a consumer process, and a buffer monitor which ensures mutual exclusion.

```
program  producerconsumer;
    const  size of buffer = ...;
    type   page = array [0..sizeofbuffer] of integer;
    type   producer = process (b1:pagebuffer);
           var  block:page;
           cycle  read (block); b1.send (block) end;
    type   consumer = process (b2:pagebuffer);
           var  card:page;
           cycle  b2.receive (card); write (card) end;
    type   pagebuffer = monitor;
           var buf:array [1..n] of page;
    head, tail, length:integer;
           p,q : queue;
procedure entry  send (content:page);
begin if  length = n then delay (p);
           buf (tail): = content;
           tail: = (tail mod n) + 1;
           length: = length + 1;
```

```
                continue (q)
        end;
        procedure entry  receive (var content:page);
        begin  if  length = Ø  then delay (q);
                content: = buf (head);
                head: = (head mod n) + 1;
                length: = length - 1;
                continue (p)
        end;
        begin head: = 1; tail: = 1; length: = Ø end;
            var x:producer, y:consumer, z:pagebuffer;
        begin  init  x, y, z  end;
```

Figure 4. Producer-Consumer in Concurrent Pascal

The producer wishes to append a page to the buffer. It calls the monitor procedure 'send'. Similarly, the consumer calls 'receive' procedure to obtain a page. The operations on the variable block or card (of type page) by both processes do not interfere with each other.

A monitor has a special procedure call mechanism to guarantee mutual exclusion. A 'class' in Concurrent Pascal is similar to monitor but can only be accessed by a single process. The concept of 'class' as abstraction mechanism was originally introduced in the programming language SIMULA (16).

Storage allocation in Concurrent Pascal is completely static. The lack of recursion and commands for dynamically creating new processes and monitors are some other drawbacks of Concurrent Pascal.

4. MODULA-2

Modula (15) was developed shortly after Concurrent Pascal for computer-based systems, including process control systems. The language is Pascal based and contains multiprogramming features: modules, monitors and processes. A module is a structuring tool similar to 'class'. If a module is shared by more than one process, then it is called 'interface module'. Interface modules are identical to monitors and guarantee mutual exclusion. Modules which contain machine specific mechanisms for referencing peripheral hardware are called 'device modules'. A process uses an I/O device by calling a procedure of a device module.

Programs in Modula are partitioned into several modules each module containing constants, types, variables, procedures and functions. Identifiers within a module can be 'exported' outside by declaring them in an 'export list' and similarly identifiers can be 'imported' if they have been listed in an 'import list'. The actual data structure within a monitor module is not accessible from outside the module.

Modula has inherited almost all the sequential features of Pascal. These include in particular the elementary data types and data structures. Program control statements include the familiar *if*, *case*, *repeat*, *while*, and *for* statements and the additional *loop* statement. Every control structure ends with an explicit termination symbol. The *loop* statement is necessary to express the continuous repetition of cyclic processes, where no termination is specified. An *exit* statement may be used to specify termination of the enclosing *loop* statement.

A module definition in Modula-2 (8) is subdivided into two parts: the *definition* part and the *implementation* part. The definition part contains export list and the declarations of exported objects. An importer of a module needs to know the definition part only. The exported objects can be read-only variables and procedures. The definition part keeps procedure headings only, the details of the module (body) belong to the implementation part and are hidden and thus protected from misuse. Frequently used modules are kept in a library in pre-compiled form.

Processes in Modula are defined like procedures and not as system types. Synchronization of processes is achieved by using a signalling mechanism. There is a 'signal' data type and three standard procedures: Wait(s), Send(s) and Awaited(s). Modula-2 has the lower-level notion of coroutines that provides explicit transfer of control between processes. Transfer of control is realized by the *transfer* statement (Figure 5). Execution of a *transfer* B statement in coroutine A involves the following steps.

(a) The location of the statement following the *transfer* statement in coroutine A is stored in the resume point location for A.

(b) Control is transferred to coroutine B by accessing the resume point location for B at which execution of B should be resumed.

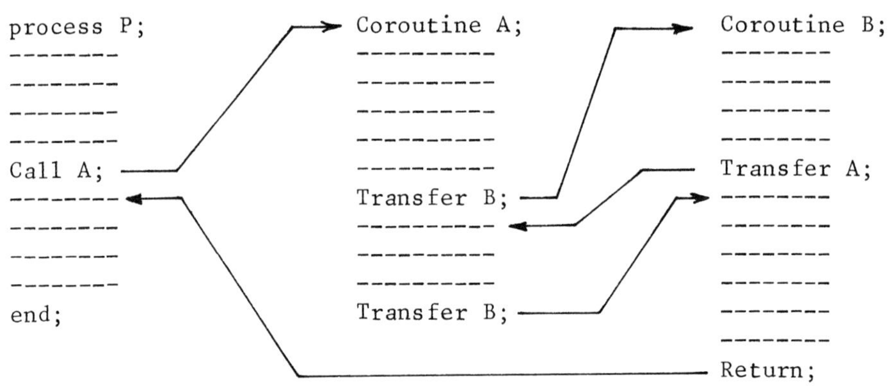

Figure 5. Coroutines

Note that the *transfer* statement serves as the only way to transfer control between coroutines, but a *call* is used to initiate the coroutine computation and *return* is used to transfer control to the calling process. In Modula-2, to create coroutines procedure NEWPROCESS is used and control is switched between processes by a call to the system procedure TRANSFER. Modula-2 is thus more suited to organize concurrent processes that run on a single processor and when transfers are completely specified.

5. ADA

The Ada (9) programming language was developed by the U.S. Department of Defense. Ada provides all the major features of a modern real-time language: structured constructs, strong typing, data and operation abstraction, concurrent processing, exception handling, low-level facilities and separate compilation.

The basic lexical style of an Ada program reflects the influence of Pascal. In addition to subprograms (procedures and functions), Ada offers program units 'tasks' and 'packages'. An Ada program consists of one or more of these compilation units. A task defines actions that are logically executed in parallel with other tasks and a package holds data types, objects, subprograms, tasks and other packages. While packages (Ada construct corresponding to modules of Modula-2) enforce abstraction, information hiding and modularization; tasks implement concurrent processing on a single-processor, a multiprocessor or a network of computers.

Tasks, packages and subprograms consist of two separate parts: the specification and the body. The specification identifies the information visible to the user of the program unit and the body contains the unit implementation details. The two parts of a program unit are separately compilable. Figure 6 shows the general structure of Ada program units.

Subprogram	Package	Task
procedure NAME (lfp) is	package NAME is	task NAME is
specification part	specification part	specification part
begin	end NAME;	end NAME;
body part	package body NAME is	task body NAME is
end;	body part	loop
	end NAME;	executable statements
		end loop
		end NAME;

Figure 6. Ada program units

Ada has adopted the ALGOL60 scoping rules. When nesting of subprograms is used, identifiers from an outer enclosing context are visible in the inner context. In addition to local and global variables a block may use the identifiers of a package x by declaring the statement 'use x' in the declarative part. A package may use other subprograms and packages by using a 'with y' statement where y is the name of the imported package/subprogram. The following data types are available in Ada:

(a) scalar : integer. real, enumeration, Boolean and character
(b) structured : array, record
(c) access type for dynamic data
(d) subtype and derived types
(e) private type

Ada includes the familiar program control structures *if*, *case*, basic *loop*, *loop* prefixed by an iteration clause (*for* or *while*). The basic loop structure is a sequence of statements that are bracketed by the reserved words *loop* and *end loop*. Loop identifiers are optionally used and must occur at both ends of the loop structure. *Case* and *if* structures are also closed constructs. The reserved word *others* may be used as the last alternative in a *case* statement. Also included in Ada is the *goto* statement for providing an explicit transfer of control. The assignment statement, the *null* statement, subprogram calls, and the *return* statement implement the structure for sequence. Block identifiers are optionally used. Blocks begin with the keyword *declare*. Labels, block identifiers and loop identifiers are exceptions to the explicit declaration rule of Ada.

Ada permits concurrent execution of a number of different activities through its tasks. The declaration of a task automatically initiates it. A task is generally enclosed between *loop* and *end loop* such that it executes indefinitely. For task termination statements *terminate* and *abort* may be used. A task can be declared as a type if several identical tasks are needed.

Ada tasking model is based on CSP (17). Inter-task communication is achieved by means of *entry* calls and *accept* statements. The task specification contains entry declarations for all points of communication with other tasks. For each entry declaration in a task specification, there must be a corresponding accept statement within the task body. A calling task issues an entry call to communicate with another task. The two tasks are first synchronized in time and space and then communicate. The explicit mechanism of synchronization, and message passing is referred to as 'rendezvous'. Semantically, an entry call is similar to a procedure call. A task entry may be called from any point at which subprogram call is permitted.

If one task is ready to enter or accept before the other is at the rendezvous point, the calling task can wait until the rendezvous is completed. Only one task is permitted to rendezvous at a time. All other tasks wait in a FIFO queue. A *select* statement is available for handling multiple accept statements within a task. It allows several alternatives separated by *or*. The alternatives are prefixed by *when* clauses which establish what condition should be true for an alternative to be in a condition for execution.

The generic facility in Ada provides for parameterized subprograms and packages. A generic definition defines a template for a common algorithm for different data types. A user must instantiate a generic unit and then invoke it as a program unit. The exception handling facility in Ada provides for error recovery during program execution. There are five predefined exceptions that cover a broad range of run-time errors. User defined exceptions can also be handled. Certain low-level facilities in Ada allow the programmer to specify features for length and size. Interrupts are treated as hardware-generated entry calls. Ada provides a mechanism for tying specific interrupts to particular entries.

Ada language stresses embedded computer application areas, and is most appropriate for process control. A small example including a package description is given in Figure 7. It illustrates how Ada can be used for continually monitoring temperatures from sixteen independent sensors and for controlling the mean temperature. The 'scan-data' task performs the data acquisition function. The main program - procedure 'controller' - implements an on-off controller using calculated mean-value of temperatures. The entry 'read' is called by the main program to get the measured temperatures. The main program 'controller' cycles at a fixed period determined by the constant 'cycle-time' declared in the package.

```
package  plant-monitor is
  maximum-sensors : constant : = 16;
  subtype sensor-number is integer range 0..maximum-sensors;
  subtype sensor-reading is integer range 0..255;
  cycle-time : constant duration : = 10.0;
    task  scan-data is
      entry read (n : in sensor-number; r = out sensor-reading);
      entry end-of-conversion;
      for end-of-conversion use at 16 # 0011 #;
    end  scan-data;
end plant-monitor;
package body plant-monitor is
  task body scan-data is
    type bit is (off, on);
    type status-register is
          record
          start-of-conversion : bit;
          channel : sensor-number;
          end record;
    sr : status-register;
    br : sensor-reading;
    for sr use
          record
            start-of-conversion at 0 range 0..0;
```

```
            channel         at    0 range 12..15;
          end record;
    for sr use at 16 # 00F0 #;
    for br use at 16 # 00FF #;
    begin
      loop
        accept read (n : in sensor-number; r : out sensor reading) do
          sr : = (on, n)
          accept end-of-conversion;
          sr : = (off, 0)
          r  : = br;
        end read;
      end loop;
  end scan-data;
end plant-monitor;
-- once the above package has been defined, it can be used in the main
program as follows:
with plant-monitor;
use plant-monitor;
procedure controller is
    set-point : constant : = ....;
    type control-signal is (off, on);
    for control-signal use (off = >16 # 0000 #, on =>16 # FFFF #);
    controller-output : control-signal : = off;
    for controller-output use at 16 # 0110 #;
    begin
      loop
        delay cycle-time;
        declare
          sum : integer : = 0;
         begin
          for i in sensor-number loop
            scan-data . read (i, r);
            sum : = sum + r;
```

```
        end loop;
          mean-temp : =                 ;
            if mean-temp < set-point then
              controller-output : = on;
            else
            if mean-temp >set-point then
              controller-output : = off;
            end if;
       end;
     end loop;
   end controller;
```

Figure 7. Ada program for data-acquisition and control

6. FORTH

The programming language FORTH (10) was invented by C.H. Moore for controlling a telescope. It is also suitable for process control applications. This section gives an overview of FORTH and illustrates its use through a multiloop control system (18,19) example. A software package consisting of routines for feedback, feedforward, and cascade control is described.

FORTH is a modular language containing predefined operations in a dictionary of 'words'. These words are executed by invoking their names and can be listed by their dictionary addresses. New words can be defined by the programmer using existing words and these words become an integral part of the language. Most words expect the data to be available on the top of a parameter stack and they leave results on the same stack. Access to any memory location is allowed, and constants and variables can be defined. FORTH's post-fix notation is efficient for parameter passing and for linking words together.

Structured programming constructs are available in FORTH. It has all the normal control structures: if---then, if---else---then, begin---while---repeat, begin---until, begin---again, do---loop. FORTH provides data types constants and variables using defining words constant and variable. The other defining words are:, code, user, < builds, does> . It is possible to create new data structures like arrays and records.

FORTH has assembly language capability built into it such that it can handle most time-critical operations directly. The programmer may directly address any memory location and alter its contents. Data can be represented in any number base and can be manipulated on the stack in a variety of ways.

FORTH is fast because it is implemented as a threaded code interpreter. The compiled program is just a string of addresses, each pointing towards at lower levels. The execution consists of passing control successively to each address in turn by the inner interpreter. FORTH is thus very convenient for process control. It is now available for almost all microcomputers with minor differences.

The AIM65 microcomputer has an 8K ROM-based FORTH compiler/interpreter. A software package for a multiloop control system (Figure 8) has been developed using the AIM65. The microcomputer is interfaced via R6522 Versatile Interface Adaptor (VIA) as shown in Figure 9. The VIA contains two 8-bit I/O ports (PA and PB) and four status/control lines (CA1, CA2, CB1, CB2) which are used to connect a 12-bit eight channel data-acquisition system (model HDAS-8 of Datel/Intersil) and an 8-bit digital-to-analog converter. A monoshot has been used for generating the start-of-conversion (SOC) pulses of proper duration.

The control words defined are CASCADE, FEEDBACK and FEEDFORWARD. The data-acquisition word DAS is used by all three control words. The word DAS expects the channel number to be available on the parameter stack. For details of DAS see reference 21. The following channel assignments are used.

channel 0 : output of actuator
channel 1 : load disturbance
channel 2 : controlled output variable

To achieve a foreground/background mode of operation, a counter S-CNT is used. The following sequence runs indefinitely after the multiloop controller is started by typing the word MULTILOOP. Figure 10 shows the output response for a step change in the set-point.

```
: MULTILOOP   o o   begin   FEEDBACK   S-CNT
  @   o   do   FEEDFORWARD   CASCADE   end?
  loop   again ;
```

(Newly defined words are in upper case and AIM65 FORTH dictionary words are in lower case.)

The details of word FEEDBACK which implements a difference equation of the following form

$$m_n = K_o e_n + K_1 e_{n-1} + K_2 e_{n-2} + P_1 m_{n-1} + P_2 m_{n-2} + P_3 m_{n-3} + P_4 m_{n-4}$$

is given below.

```
: FEEDBACK  E2 @ K2 @ * E1 @ dup E2 ! K1 @ * + SP @ 2 DAS - dup E1 ! KO
  @ * + SF1 @ /mod swap SF1 @ 2 /1 - ROUND
  M4 @ P4 @ *
  M3 @ dup M4 ! P3 @ * +
  M2 @ dup M3 ! P2 @ * +
  M1 @ dup M2 ! P1 @ * +
  64 /mod swap 31 ROUND +
  dup M1 ! dup . FBOP ! ;
```

(Output of feedback controller is FBOP, SP is set-point and ROUND is used for rounding-off.)

Details of other words can be seen in reference 21.

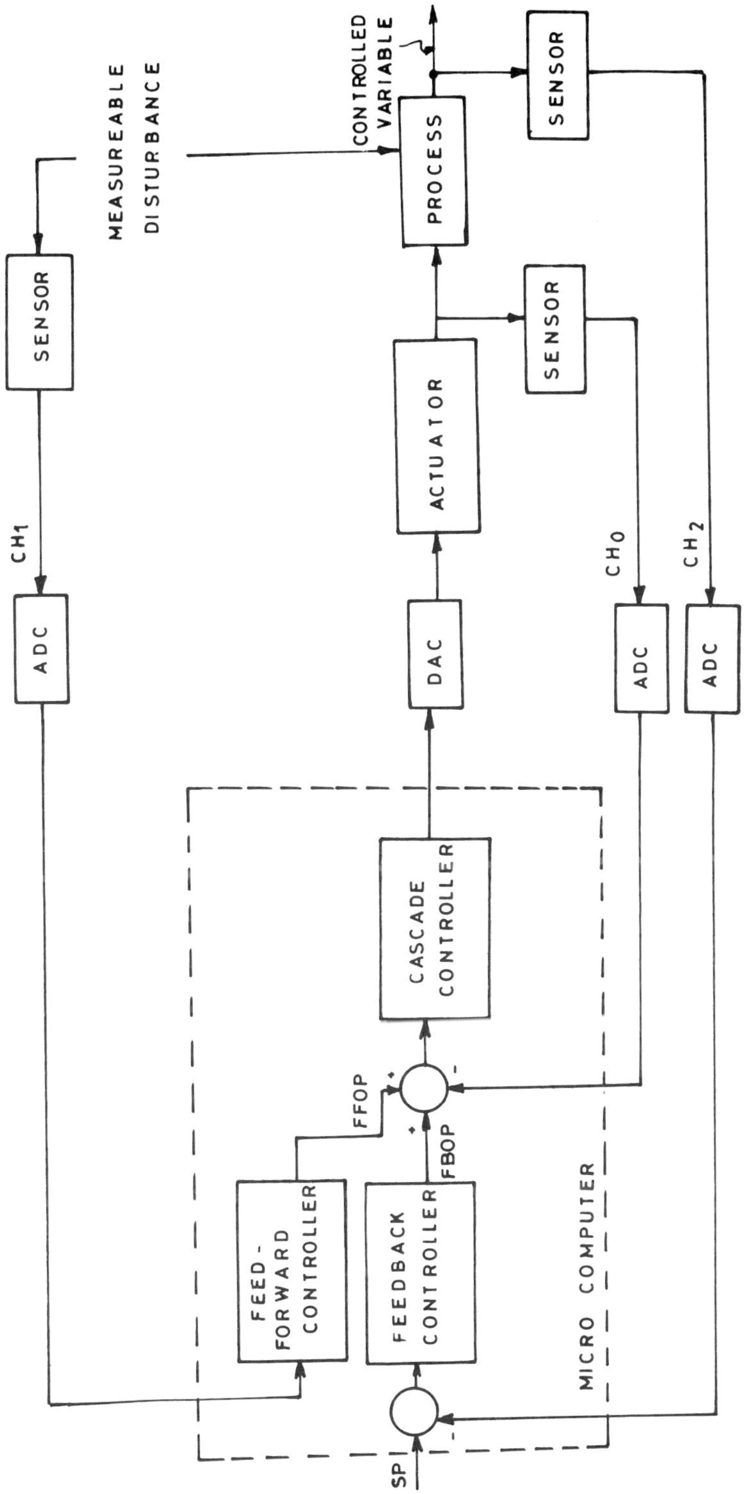

Fig. 8 Microprocessor - based multiloop process controller

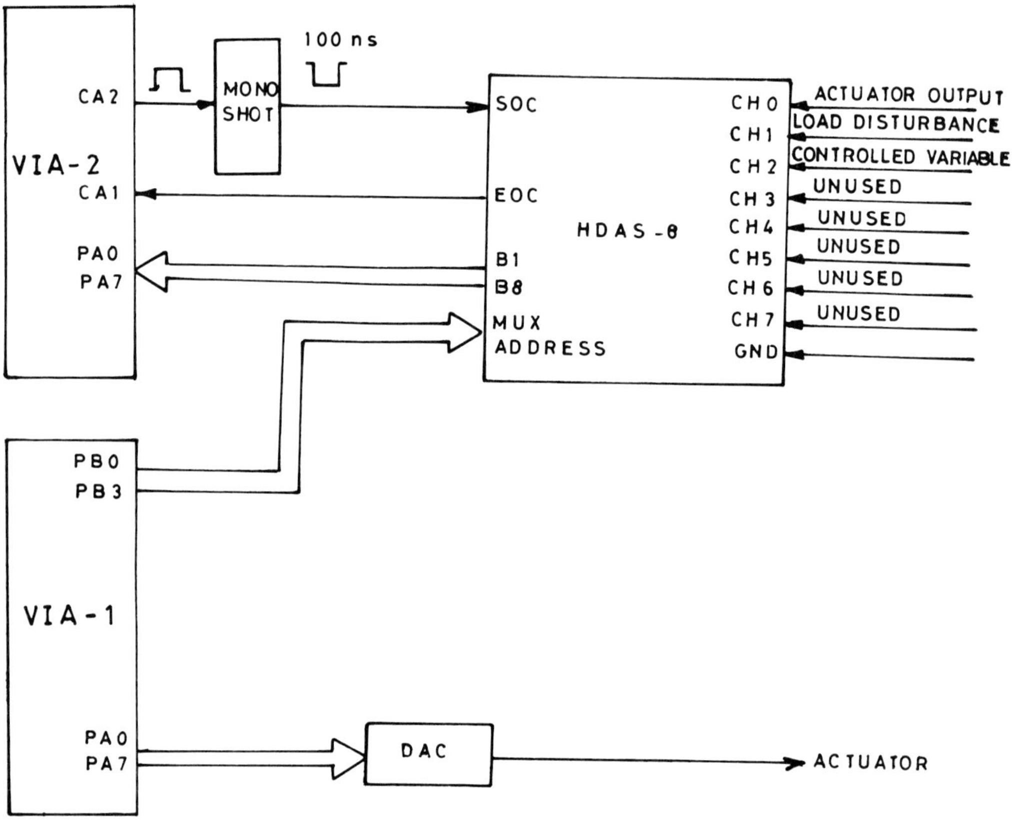

Fig. 9 Interface circuit schematic

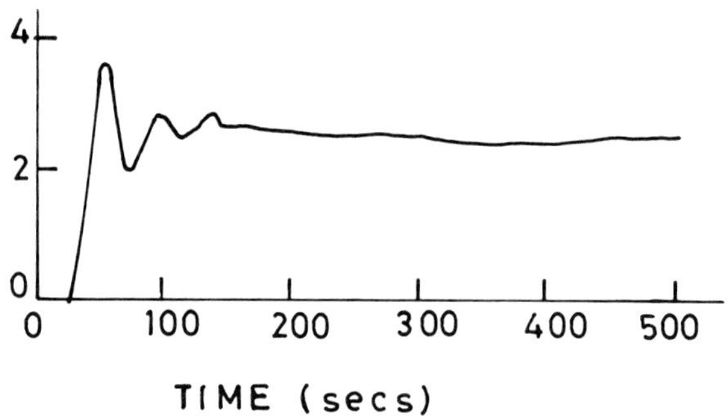

Fig. 10 Output response

7. CONCLUSION

Real-time programming languages are still evolving. The major problem facing the industry is to train control engineers such that they are able to implement microprocessor-based control systems using one of these high-level languages. The language selected should include the following features.

(a) Strong-typing, structured data types and efficient control structures.

(b) Abstraction mechanisms.

(c) Concurrent processing.

(d) Synchronization and communication mechanisms.

(e) Low-level facilities, representation specification.

(f) Exception handling.

(g) Separate compilation.

Ada provides all the above facilities. However, the complete language is too large and complex and its microcomputer-based versions are not yet widely available. Modula-2 is comparatively small and can be used on microcomputers. However, Modula-2 does not provide mechanisms for exception handling. Concurrent Pascal provides no mechanism for specifying hardware representation of data and for handling exceptions. Also, because of the lack of dynamic resources, concurrent Pascal implementations are inefficient for process control. FORTH does not support concurrency, neither does it have other features, but it certainly seems suited to process control applications. It has compactness and speed. Its only drawback is that it is difficult to learn. However, there are attempts being made to improve the syntax of FORTH.

REFERENCES

1. S.J. Young, Real Time Languages - Design and Development, Ellis Horwood Limited, 1982.
2. M. Ben-Ari, Principles of Concurrent Programming, Prentice Hall International, Inc., 1982.
3. S.I. Ahson, 'Software design methods for distributed computer control', Journal of Microcomputer Application, **6**, 287-293, 1983.
4. J.B. Goodenough, 'Exception handling: Issues and a proposed notation', Comm ACM, **18**, pp. 683-696, 1975.
5. K. Jensen and N. Wirth, Pascal User Manual and Report, Springer-Verlag, 1979.
6. C.A.R. Hoare, 'Monitors: An operating system structuring concept', Comm. ACM, **17**, pp. 549-557, 1974.
7. P.B. Hansen, 'The programming language concurrent PASCAL', IEEE Trans. Software Engineering, **1**, pp. 199-207, 1975.
8. N. Wirth, Programming in Modula-2, 2d ed., Springer-Verlag, 1983.
9. United States Department of Defense, Reference Manual for the Ada Programming Language, ANSI/MIL-STD1815A, February 1983.
10. L. Brodie, Starting FORTH, Prentice-Hall, 1981.
11. G.R. Andrews and F.B. Schneider, 'Concepts and notations for concurrent programming', ACM Computing Surveys, **15**, pp. 3-43, 1983.

12. P. Wegner and S.A. Smolka, 'Processes, tasks, and monitors: A comparative study of concurrent programming primitives', IEEE Trans. Software Engineering, 9, pp. 446-462, 1983.
13. J. Welsh and D.W. Bustard, 'Pascal Plus: Another language for modular multiprogramming' Software Practice and Experience, 9, pp.947-958, 1979.
14. C.M. Geschke, J.H. Morris and E.H. Satterthwaite, 'Early experience with Mesa', Comm. ACM, 20, pp. 540-552, 1977.
15. N. Wirth, 'Modula: A language for modular programming', Software Practice and Experience, 7, pp. 3-35, 1977.
16. O.J. Dahl, B. Myhrhang, and K. Nygarrd, The Simula 67 Common Base Language, Norewegian Computer Centre, Oslo, 1968.
17. C.A.R. Hoare, 'Communicating sequential processes', Comm. ACM, 21, pp. 666-677, 1978.
18. S.I. Ahson, S.S. Lamba, M.C. Chaube and N. Chandra, 'A microprocessor-based multiloop process controller', IEEE Trans. Industrial Electronics, 30, pp. 34-39, 1983.
19. S.I. Ahson, Microprocessors: With Applications in Process Control, Tata McGraw Hill, 1984.
20. S.I. Ahson and U.S. Triar, 'Distributed arithmetic for microcomputer implementation of control algorithms', IEEE Control System Magazine, 5, May 1985.
21. P. Waychal, FORTH with Personal Computers: A Tool for Process Control, M.Tech. Dissertation, Department of Electrical Engineering, IIT Delhi, 1984.

CHAPTER 16

INDUSTRIAL LAN USING DISTRIBUTED MICROPROCESSORS

O. L. Storoshchuk
Plant Engineering 077-905
General Motors of Canada
Oshawa Ontario
Canada L1G 1K7

B. Szabados
Power Research Laboratory
McMaster University
Hamilton Ontario
Canada L8S 4L7

1. INTRODUCTION

1.1. Objective

The objective of this work was to design and implement a local area computer communication network (LAN) to support computerized monitoring of production equipment in an industrial plant. The network was estimated to consist of 200 microprocessor data acquisition units (nodes) monitoring a total of 10,000 sensors (an average of 50 sensors per node) and a central host minicomputer. The maximum distance between nodes would not exceed one kilometer. If the transmission media required repeaters or regenerators, the minimum distance before they would be required would be one kilometer. Communication would primarily occur between the host minicomputer and the nodes. There would be minimal communication between the nodes themselves. The network was to be fault tolerant and was to continue operating in the event of a single point failure such as a transmission line breaking or node failure. The network was to consist of interconnected segments where the failure of any segment would not affect other segments. The typical message size was estimated to be 10 bytes consisting of an identification of a sensor, the time of change in condition of the sensor (occurrence or clearing of a fault) and the time of occurrence. Longer messages of approximatly 100 bytes consisting of parameter tables describing sensors and data base information, would be infrequently transmitted (whenever a node was powered up). The message traffic would typically be short bursts of an average of 10 messages per minute but with peak bursts of 10,000 (in event of power failure of entire plant). The transmission media must operate reliably under extreme electromagnetic noise conditions with a maximum bit error rate of 10^{-6}. The bandwidth was to be minimally 2 MHz because the nodes could transmit at a maximum rate of 2 MHz. The network was to require a relatively low level of technical expertise for maintenance.

N. K. Sinha (ed.), Microprocessor-Based Control Systems, 317–341.

1.2. Background

This task was undertaken in late 1979 when the General Motors Oshawa Car Assembly plant was replacing its paint facility with a new building and required a monitoring system of the process equipment. A market search for suitable computer hardware and LAN was conducted but did not locate any which met the desired criteria. Thus a decision was made to develop custom hardware and LAN. At that time a task force within General Motors named "Manufacturing Automation Protocol" (MAP) [1] was reviewing existing and emerging LAN standards as well as actively working with LAN standards committees and LAN vendors to develop a standard. Participation in this task force provided current literature as well as an insight into the LAN strategies being pursued by others. Based on this research and information, a LAN was designed and hardware to support it was developed. Software for a minimal implementation of the LAN was written and a laboratory test and demonstration system was built. This LAN was sufficient to support the system for monitoring of production equipment.

1.3. Open Systems Interconnection Reference Model

The OSI model [2] specifies a modularization of the network software based on functionality. Each of these modules is referred to as a "layer" which provides its specific network services to the layer above it while utilizing the services provided by the layer below it. Theoretically any layer can be replaced by a new layer without affecting the other layers or the user's perception of network operation. This allows upgrading and changing of the various layers individually without requiring a massive revision effort of the other layers. There are a total of seven layers. At the highest layer are the application programs and the lowest layer is the physical media over which the data is transmitted. The seven layers and their relationship to each other is shown in Figure 1. It should be noted that the model allows the use of null layers where the functions of those layers are not required.

1.4. Topology

The objective of the topology was to provide a flexible network to which stations could be readily added or removed. The reason for this lay in the dynamic nature of the plant equipment which is relocated each model year. It was desirable to do this without disrupting the rest of the network. Furthermore, the network should still function in the event of a single point failure such as cable breakage or node failure. The maximum distance between nodes was specified at one kilometer.

The topology selected was a hierarchical interconnection of networks shown in Figure 2. Each network consists of a three channel physical ring structure which is logically configured as a multipoint bus. On each network a primary station called a subhost performs the controlling functions while the other stations called nodes operate in a secondary capacity transmitting only when given access by the subhost. The transmission media is physically connected in a point to point manner. The signals enter each node where they are repeated and regenerated by "steering card" circuitry. The signals are passed on and also given to

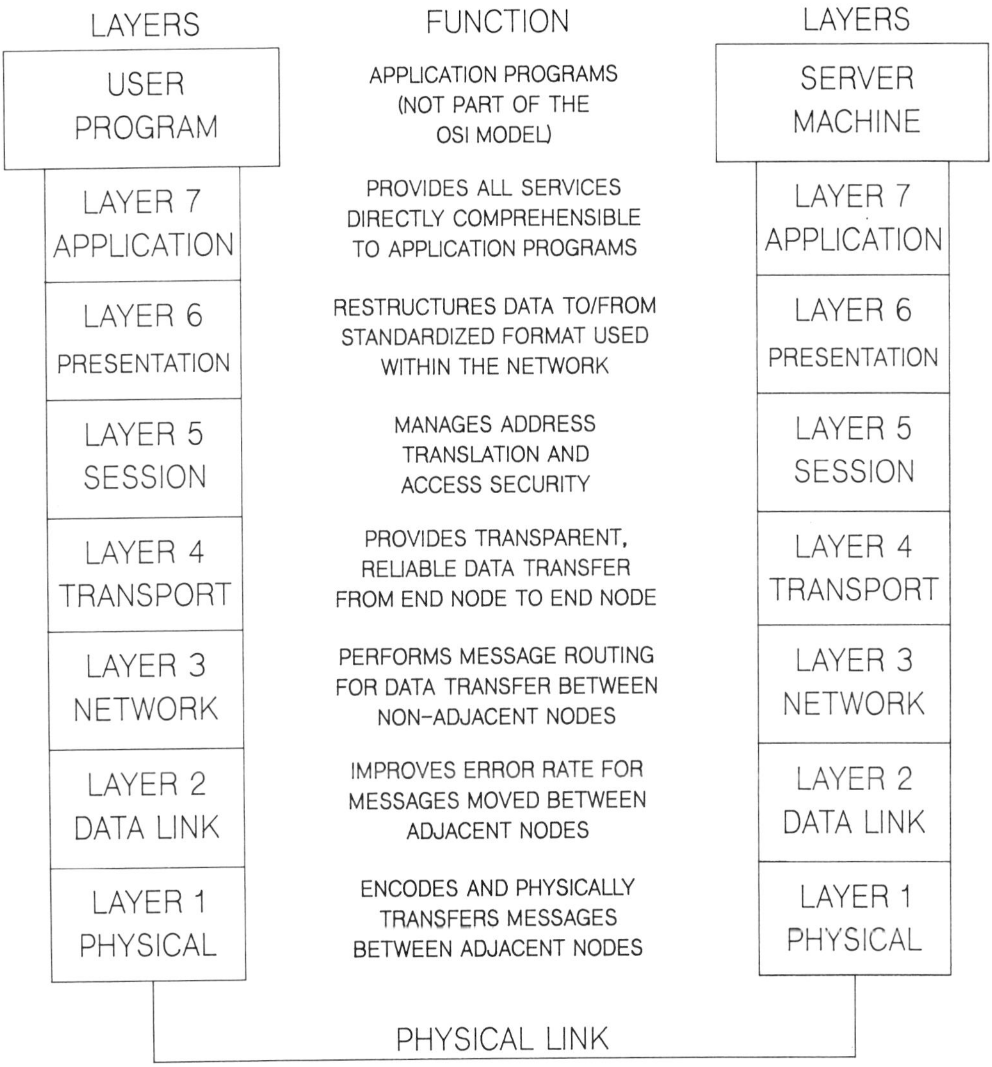

Figure 1 : OSI Reference Model

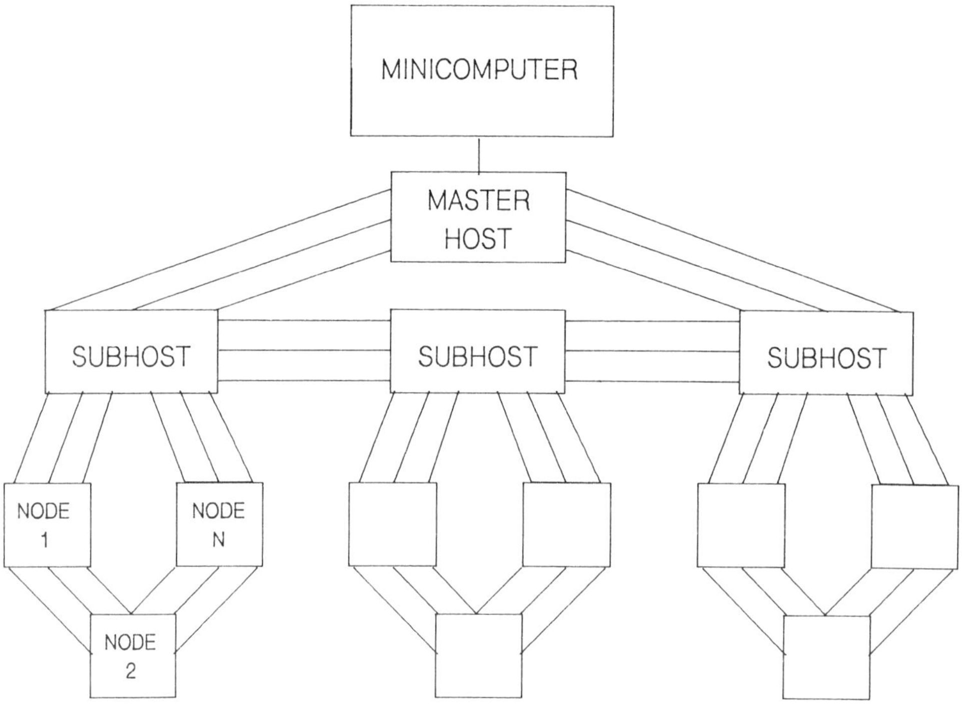

Figure 2 : Network Topology

the node, so in effect an electrical "T" is formed providing a bus structure. The signals are not stored and forwarded.

Active repeaters present a problem, if they are locally powered, of potentially bringing down the network when local power is lost. To overcome this problem, a separate 48V AC power cable was run with the communication cables. It provided a source of backup power to the repeater sections of the nodes when they lost local power. Using repeaters at every node resulted in an effectively unlimited network length since signal strength loss was not a problem (provided the one kilometer length between nodes was not exceeded) and the delay caused by a repeater was only two bit cells wide (two microseconds). Addition of nodes or cable length did not require calculation of line losses or impedence matching which would have been necessary with active taps.

To achieve single point fault tolerance, the three communication channels were configured for redundancy. Two of the channels were optical fiber while the third was RS422 [3] twisted pair. Since transmission can only be supported in one direction on optical fiber, the two channels carried data in opposite directions. Each node's steering card monitors the channels for the occurrence of a fault which can be recognized by the loss of light. When a fault was detected, the nodes on either side of it would redirect the data which would be transmitted in the direction of the fault into the opposite direction achieving a "fold-back". Thus, all the functional nodes on the network still had a communication path both to and from the subhost shown in Figure 3. The fold-back capability also allowed one to add or delete nodes without disrupting the rest of the network. The RS422 was used to provide a quickly repairable channel. Should new terminations be required on the optical fiber, which requires 24 hours, the subhost could utilize the RS422 in the interim. The RS422 needed to run at a reduced rate of 100 kilobits per second. When the subhost turns on the RS422, the node steering cards will sense the signal and automatically switch over and slow down from using the optical fiber.

An analysis of the installed network showed that the use of different steering cards for the primary and secondary stations could have been avoided. This distinction was designed primarily due to the primary/secondary link level implementation where difference does exist at a logical level, however, there was no need to bring it down to the physical level. In normal receive operation, the node card operates as a pass through repeater transmitting what it receives independently on each optical channel . In the transmit mode, it in effect "splits" the ring, sending out the node's data on both channels and ignoring the signals it is receiving. A subhost card always splits the ring in both the transmit and the receive modes which results in two splits whenever a node is transmitting. This implies that when a fault occurs on the network, the nodes on one side of the fault would not receive from the nodes on the other side. In the monitoring application this was acceptable since all communication always moved between the subhost and the nodes. Peer to peer communication between the nodes was not needed and could be accomplished by a store and forward through the subhost. In retrospect, there was no advantage in having the subhost split the ring when it received. It actually complicated the receiving process when a fold-back occurred because the subhost had to switch the channel which it was receiving on

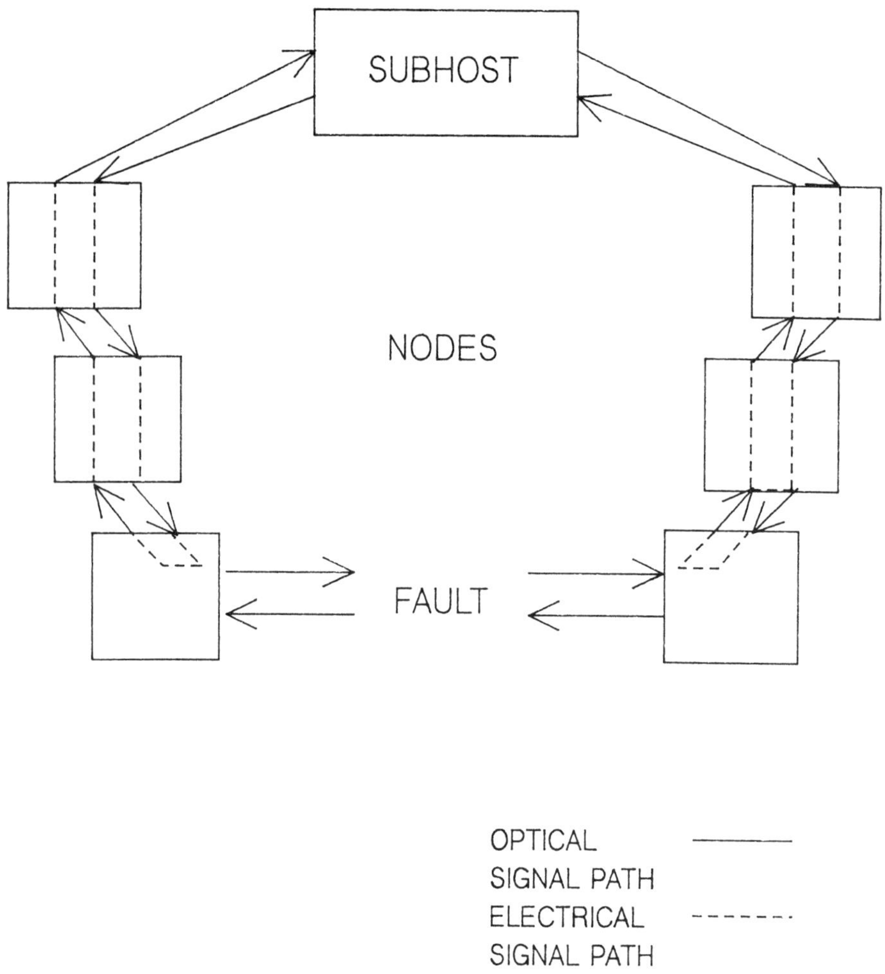

Figure 3 : Communication Fold-back

for the nodes on the opposite sides of the fault. This would not have been necessary if the split was omitted since the signal would always occur on both channels due to the foldback which would occur on both sides of the fault as shown in Figure 4.

A node steering card has been tested in place of a subhost and found to function without any problems. However, the control of the RS422 channel would have to be changed since only a subhost card was designed with the ability to activate it. The logic used to determine when to activate it was simply the exceeding of a number of consecutive unsuccessful receive retries on both optical channels. The hardware for activating the RS422 and the logic for determining when could be easily moved to the nodes. This would allow peer to peer communication in the presence of a fault plus it would eliminate the dependency on a single subhost since its functions could be assumed by any node. The gateway function of a subhost could also be duplicated if one of the nodes was connected to both its own network as well as the central interconnecting network. In future networks, it is recommended to use node steering cards in the subhosts and to connect a node to the central network for backup.

1.5. Hardware

When the hardware was designed three years ago, only one manufacturer could supply light receivers and transmitters as well as packaged receiver and transmitter electronics which would operate at the required speeds. The microprocessor hardware was capable of transmitting synchronous data at one megabit per second. The steering cards encoded the clock with the data using a modified manchester code which resulted in a rate of two megabits per second.

2. MICROPROCESSOR HARDWARE

2.1. Overview

Analyzing the link level protocols which were popular at the time the harware was designed (BYSINC, DDCMP, HDLC) and the available protocol support chips, it became obvious that the microprocessor would carry a high overhead. The support chips only performed the framing, bit or byte stuffing, address detection and error detection. Even though these functions did relieve a time consuming burden, they still left the major portion of the logic needed to implement the link layer with the microprocessor. Furthermore, it was envisioned that the network and transport layers would also be implemented on the same microprocessor. To prevent the network software overhead from impacting the application software, a dual processor architecture was chosen where a separate processor would be dedicated to each. To allow some degree of media independence, the signal encoding/decoding, receivers/transmitters and related electronics, were placed on separate cards. Should the media or fault recovery need to be changed, the microprocessor card servicing the link layer, would not have to be altered.

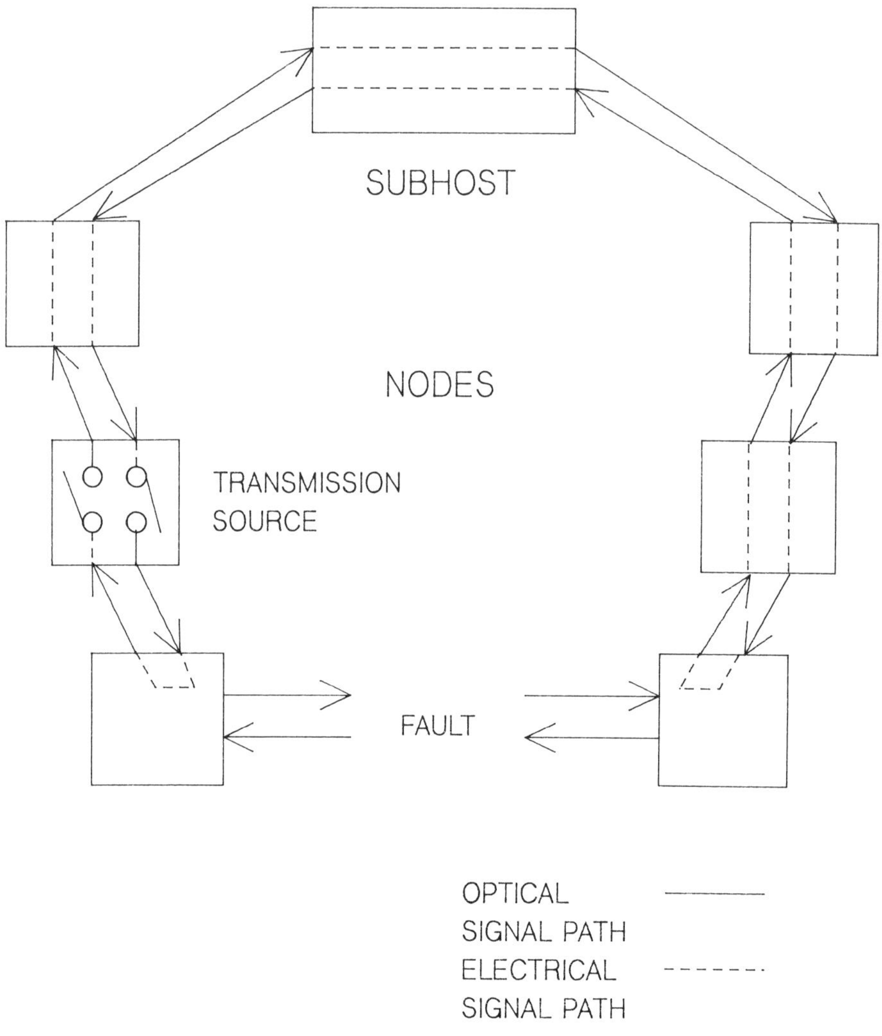

Figure 4 : Single Split Ring

The corporation had standardized on the Motorola 680X family of microprocessors in their products, resulting in availability of software and hardware expertise which could be drawn upon for assistance. Thus, the most advanced Motorola microprocessor available at the time, the 6809 was selected. The 6809 system clock cycle consists of two basic phases. During the first phase, internal operations and the outputting of address and control signals takes place, while actual bus data transfers occur during the second phase. A tightly coupled dual processor architecture was implemented by running two 6809 processors sharing a common system bus on opposite phases of a common clock as shown in Figure 5. To prevent bus contention, each processor was kept from accessing it during its respective first phase. Thus the processors could run at full speed and communicate with each other through shared memory without needing bus arbitration. The 6809E is best suited for this since its clock is externally generated. The 6809 is more difficult to use because it generates its own clock from an externally supplied "four times" clock and the timing relationships between them are not specified by the manufacturer. Unfortunately, the 6809E was not available in time. The microprocessor hardware was installed and operated in the plant for approximatly two years and experienced considerable problems. The major source of these problems lay in the synchronization timing of the two processors. Subsequently, the hardware was redesigned by the author, replacing the microprocessors with 6809E parts and utilizing as many of the components as possible from the old cards (all integrated circuits were mounted in sockets)

To reduce the overhead in servicing the communication protocol, several support chips were added. To allow maximum data transfer rates, the Motorola 6844 direct memory access controller was used. It served a secondary function of supplying an end-of-message character to the communication protocol controller immediately after transferring a block of data which the microprocessor could not supply within the maximum allowed time. The Signetics 2652 communication protocol controller was selected since it could support a wide variety of protocols plus it was the only one, at that time, which could operate at two Mbit per second. To reduce the overhead of keeping track of elapsed time for timeout error recovery, a Motorola 6840 programmable timer was added. The maximum data rate which could be achieved with this hardware was one Mbit per second in full duplex or two Mbit per second in half duplex. After the link level software was implemented, it was interesting to note that the line turn-around (750 usec) was the limiting factor to actual data throughput and the transmission rate attributed very little. In addition, the monitoring application used very short messages of an average 20 byte length which made the transmission rate even less important since little data was sent per message exchange. Increasing the data transmission rate would not substantially improve the throughput unless more of the protocol overhead could be moved from the microprocessor to the protocol controller.

2.2. Communication Processor Block Diagram Description

2.2.1. General Architecture. The Optical Communication Interface card (OCI) block diagram is shown in Figure 6. The OCI can be divided into two

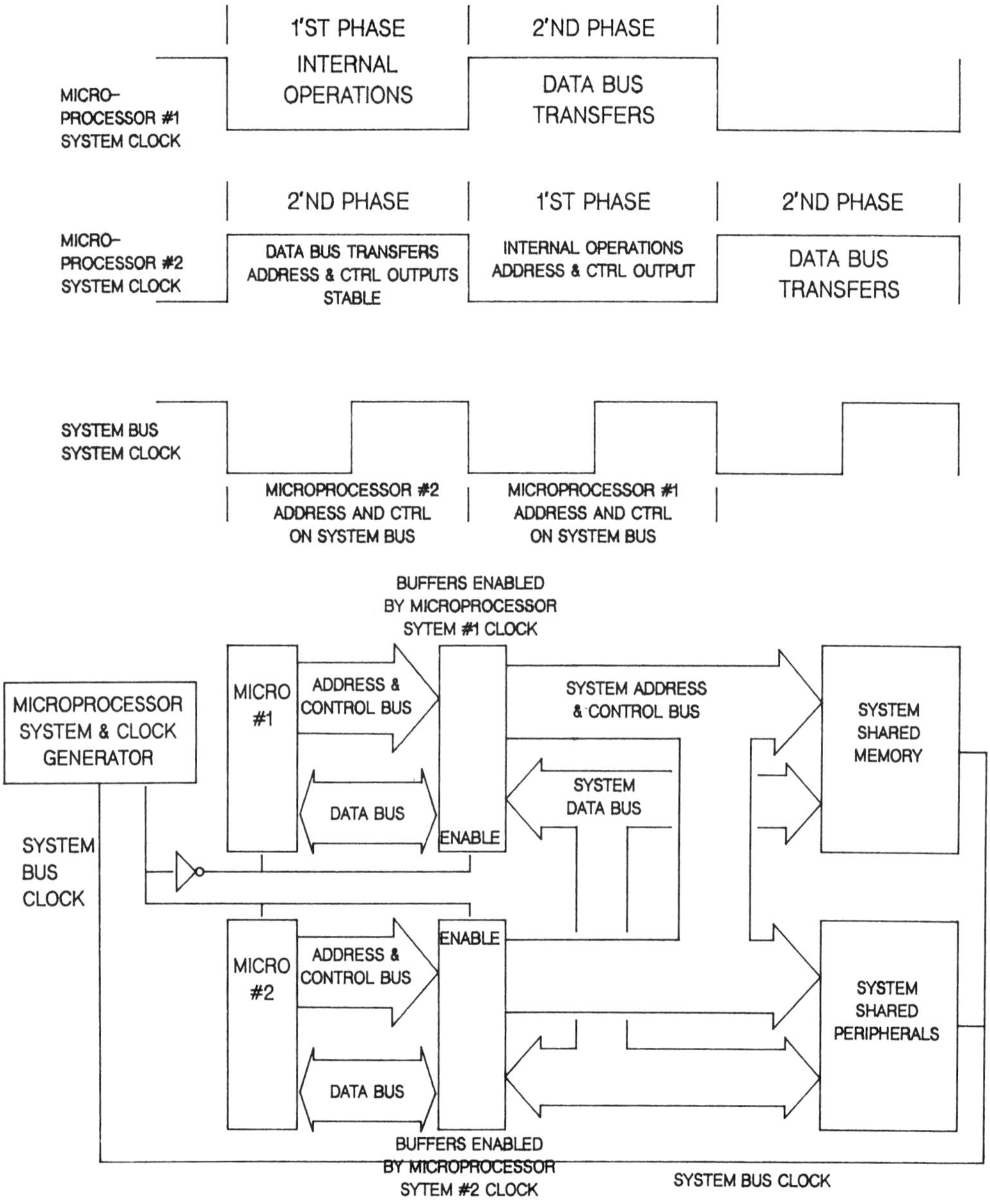

Figure 5 : Dual Processor Timing and Control Flow

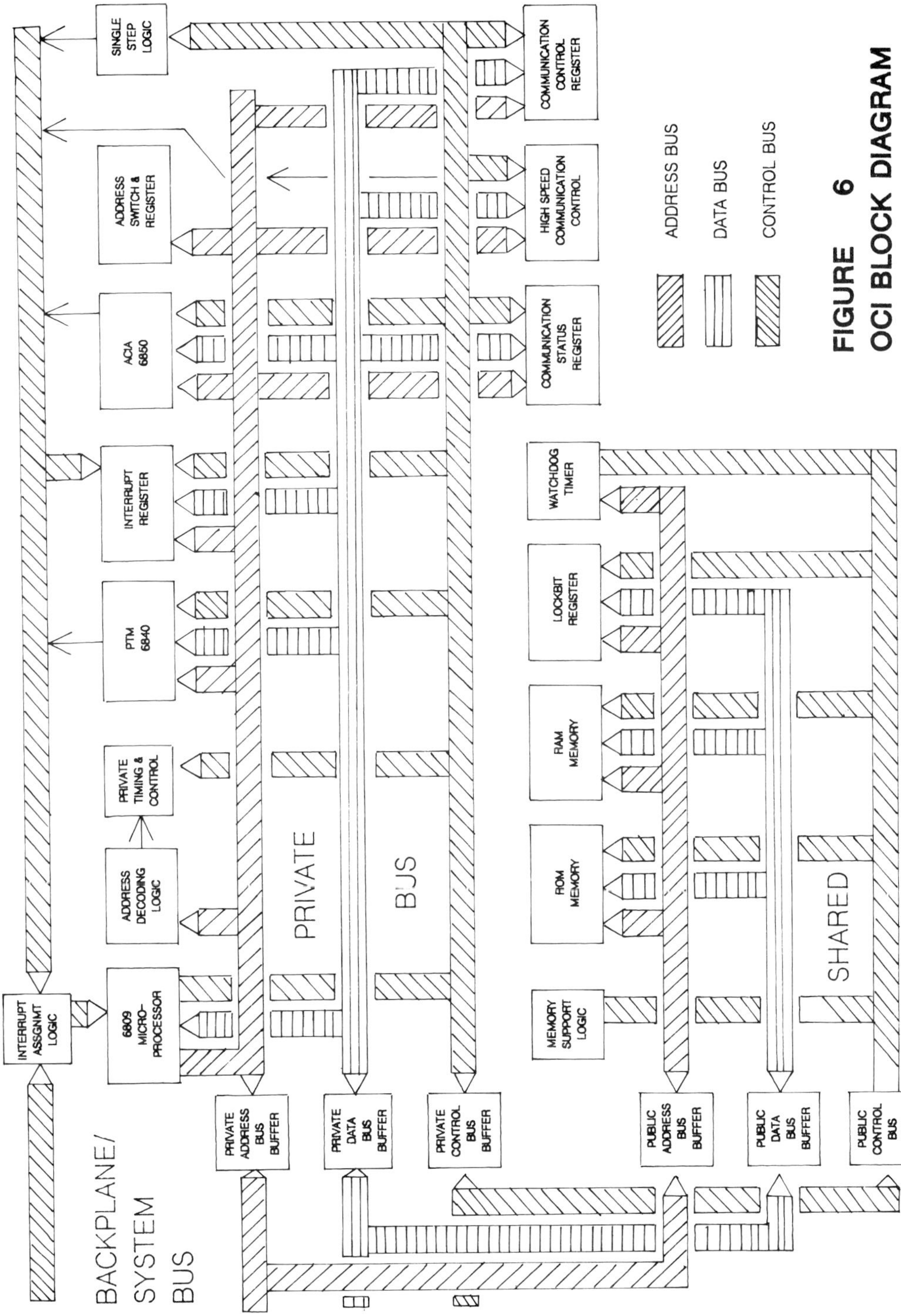

FIGURE 6
OCI BLOCK DIAGRAM

sections. One section contains the 6809E microprocessor, communication support chips, timing and control logic referred to as the "private" section, while the other referred to as the "shared" section contains memory, lock bits, watch-dog, timing and control logic. The private section bus activity is controlled solely by the onboard 6809E microprocessor and access to devices on this bus is limited only to it. Other devices on the system bus (backplane - motherboard which other cards are plugged into) are blocked from accessing the private bus by the private bus buffers. The private bus operates at a one MHz. rate permitting slower devices to be used than can be used on the system bus which operates at two MHz. The shared section bus activity is controlled by its timing and control logic and access to devices on this bus is available to any devices which can access the system bus. The memory on the shared bus can be configured to respond to both processors to allow communication between them through RAM or to save program code space for common programs or library code in ROM. Portions of the memory can also be configured to respond to only one of either of the processors to act as a private resource for that processor. This allows both processors to have their own private code or data space at the same memory locations in the memory map to yield more total memory for both than the address space permits as well as providing separate reset and interrupt vectors. A high speed microprocessor timing bus consisting of a ribbon cable connecting cards which need to synchronize their timing was implemented rather than routing these signals over the backplane which does not have suitable bandwidth.

The implemented dual processor architecture allowed both processors to run at full speed and communicate with each other through shared memory without needing bus arbitration, however, a problem did occur using this structure when implementing semaphores to control access to shared resources. In a single processor multiprocessing system, a process can guarantee that it has sole access to a sempahore by disabling the interrupt hardware. It can then check the semaphore and set it to lock access to a resource if the semaphore was not already set. By having the interrupts disabled, it is assured that in the time between checking of the semaphore and setting it, due to an interrupt, the semaphore will not be read by some other process, found clear and subsequently set. However, in the dual processor architecture, such a simple scheme will not work since there is no way of controlling the other processor's access to memory without resorting to more complex multi-flag handshaking and timing loops. To avoid this, a hardware read followed by a modify within the same half clock cycle was implemented. A read of a register referred to as a lock bit, would return a zero if it was not set and set it to a one during the same second phase of the microprocessor. The other processor might read the same lock bit in the following phase of the clock and would read a logical one indicating that the resource was already locked. A write to a lock bit would clear it to a logical zero.

2.2.2. Private Timing and Control. The private timing and control circuitry can be configured for either dual processor with or without CRT controller card. The timing is generated by two interacting state machines, shown in Figure 7, implemented with Programmable Array Logic (PAL) to allow flexibility in reconfiguration. The clock is provided by

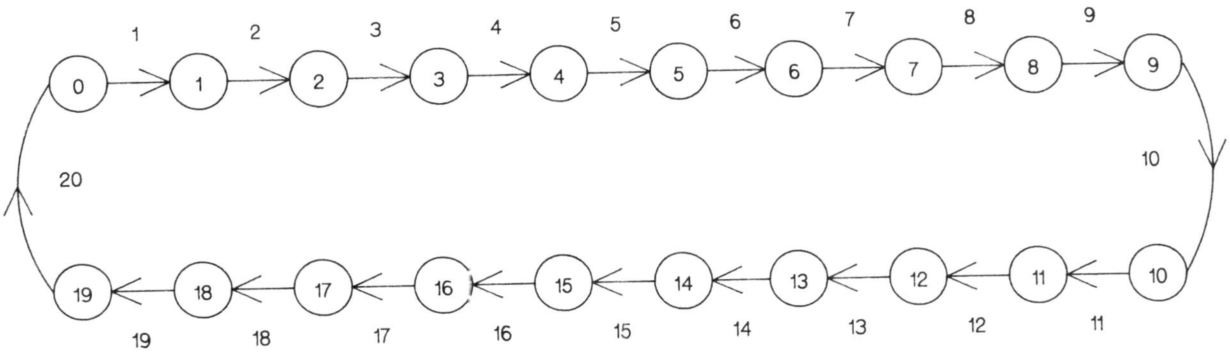

Non DMA State Machine

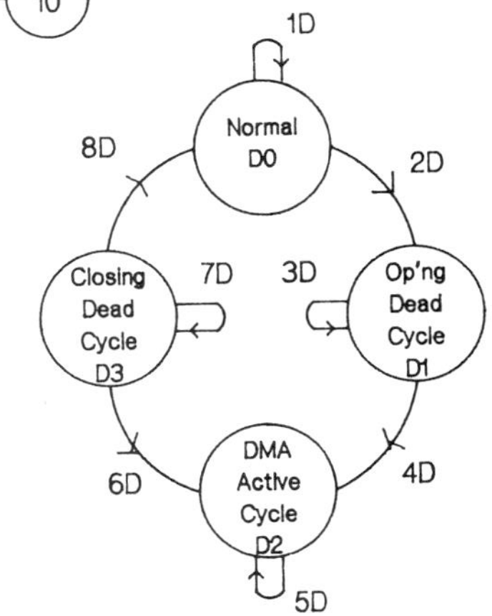

DMA STATE MACHINE

DMA State Machine Transitions

Transition	Inputs
1D	NON DMA STATE 0–18
2D	NON DMA STATE 19 and DRQST
3D	NON DMA STATE 0–18
4D	NON DMA STATE 19
5D	NON DMA STATE 0–18 or NON DMA STATE 19 and DRQST
6D	NON DMA STATE 19 and $\overline{\text{DRQST}}$
7D	NON DMA STATE 0–18
8D	NON DMA STATE 19

Figure 7 : 20 MHz OCI Timing Generator State Machines

one crystal oscillator for all the cards in the system and can be located on one of the OCI cards (in the case of dual OCI configurations) or on one of the CRT cards. A jumper allows the selection or omission of the crystal oscillator on the OCI card. For configurations without CRT cards a 20 MHz clock is used with a 20 state state machine for generating the majority of the timing signals. This is the fastest clock rate which can be used with readily available PALs and was chosen to allow the narrowest time adjustment of the timing signals (50 nanoseconds). For configurations with CRT cards the clock rate is reduced to 16 MHz. with a 16 state state machine shown in Figure 8 because the CRT cards are based on a 16 MHz. pixel clock rate and generate their timing signals based on this frequency. Timing signals associated with DMA are derived from a four state state machine shown in Figure 7.

2.3. Application Processor Block Diagram Description

2.3.1. General Architecture. The application processor, referred to as the Remote Input/Output (RIO) card, block diagram is shown in Figure 9. The RIO consists of an interface to the system bus and a "private" section which contains: a time of day clock calendar, medium speed RS-485 [4], or optionally the steering card can be used to provide fiber optics and regenerated RS-422 at a reduced speed (no DMA on RIO causing microprocessor to become the limiting factor based on how quickly in can supply the MPCC with transmit data and remove the received data), communication support, RS232 serial interface, programmable timer, parallel digital I/O, timing and control circuitry and the microprocessor. The private section is controlled solely by the RIO microprocessor and access to the devices on this section is limited only to it. The RIO can be configured in the dual processor mode with the OCI card, or it can be configured to operate in a single processor stand alone mode. In the dual processor mode, a 6809E microprocessor is used and the master timing is generated by the OCI card. In this mode, the private section operates at one MHz while the system bus operates at two MHz. In the single processor mode configuration without a CRT controller card, a 6809 microprocessor is used, which generates its own timing signals from a clock signal generated by an onboard crystal oscillator. In this mode, the RIO controls both the private section and the system bus which both operate one MHz or optionally (using a faster crystal oscillator and private section components) at two MHz. The single processor mode was implemented to provide an inexpensive medium speed remote I/O system. In the monitoring system, some machines in the plant did not have enough sensors to cost justify a dedicated dual processor node solely for them. In such cases one dual processor node's RIO would act as a primary (master) station on a network of up to 31 inexpensive single processor RIOs (1500 Canadian dollars for a RIO, power supply and enclosure) over an RS485 line. Each RIO would service a small machine which did not have enough I/O to justify a dual processor. For single processor with CRT card controller mode of operation, a 6809E is used. The CRT card supplies the timing signals for the microprocessor but the RIO controls the system bus. The system bus operation is limited to one MHz due to CRT controller card limitations (it operates as a dual processor with its own memory which is accessed on

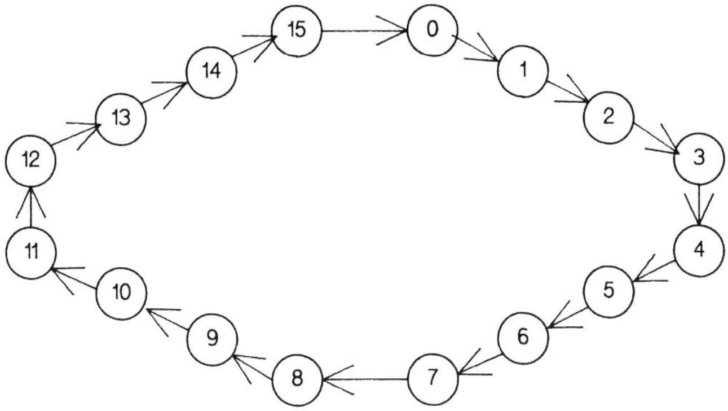

NON DMA STATE MACHINE

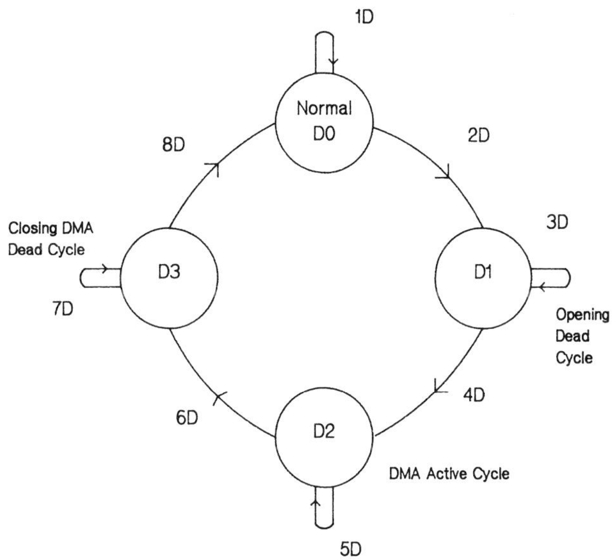

DMA STATE MACHINE

Transition	Inputs
1D	Non DMA State 0–14
2D	Non DMA State 15 and DRQST
3D	Non DMA State 0–14
4D	Non DMA State 15
5D	Non DMA State 0–14 or Non DMA State 15 and DRQST
6D	Non DMA State 15 and $\overline{\text{DRQST}}$
7D	Non DMA State 0–14
8D	Non DMA State 15

Figure 8 : 16 MHz OCI Timing Generator State Machines

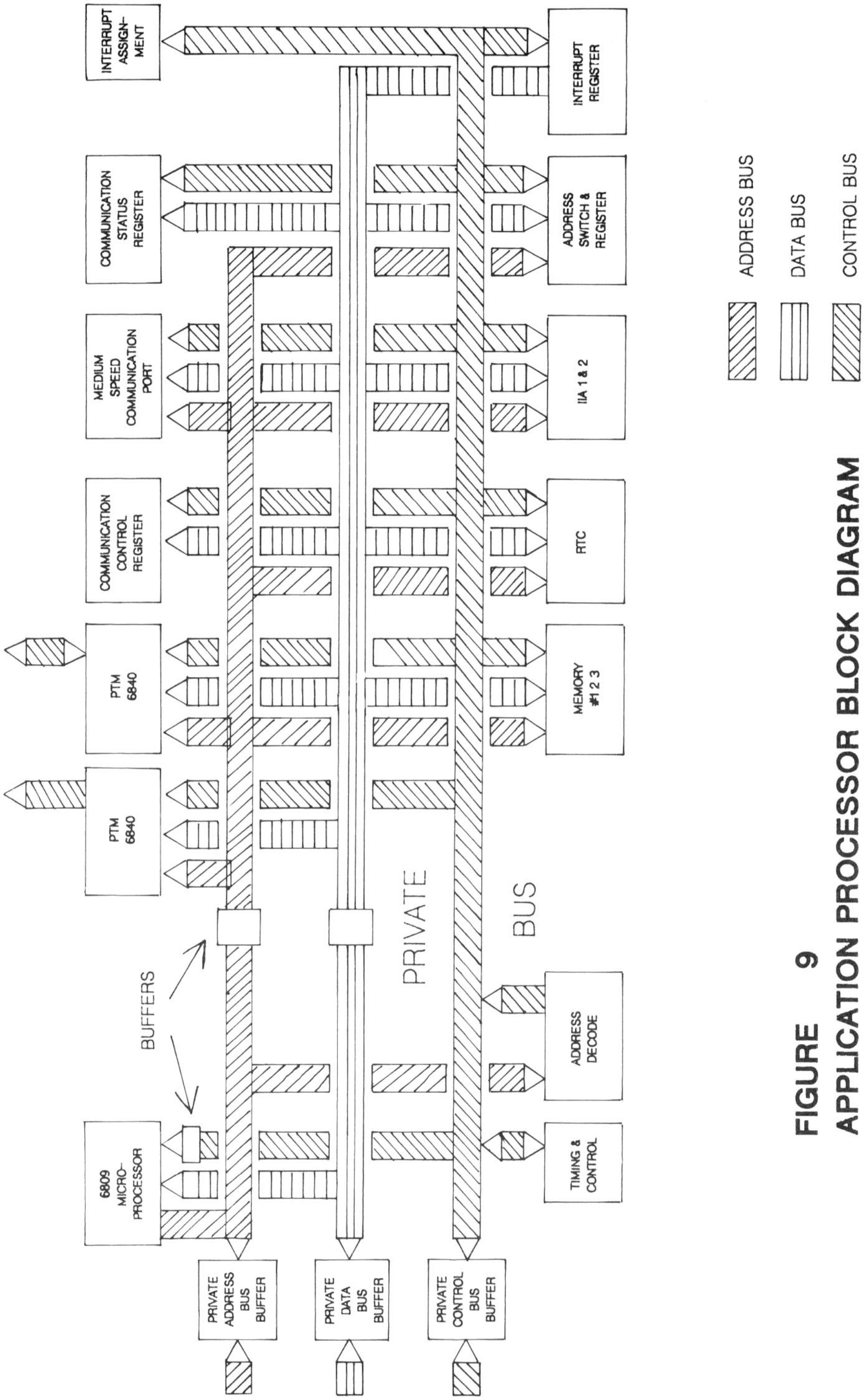

FIGURE 9
APPLICATION PROCESSOR BLOCK DIAGRAM

alternate half cycles by the RIO and CRT controller) which also limits the RIO private section to one MHz.

2.3.2. Timing and Control. The timing and control circuitry (TCC) can be configured for either dual or single processor modes. For single processor without CRT controller card configurations, an onboard crystal oscillator supplies the 4XE clock to the 6809. The 6809 internally generates the E and Q signals for the TCC from the 4XE clock. For dual processor applications, an OCI card supplies the E, Q, NCME and EDLY signals via the high speed timing bus. A 6809E microprocessor is used which uses the E and Q signals as timing inputs to synchronize its bus activities. For single processor with CRT card configurations, the CRT card supplies the E and Q signals. NCME and EDLY are not needed in this case because there is only one processor on the system bus (the CRT controller uses memory on the CRT card and does not have access to the system bus).

3. KERNEL EXECUTIVE

3.1. Introduction

Initially it was decided to implement the communication and application system without using a formal kernel executive other than an interrupt handling utility. However, as work progressed, this led to considerable complexity in program interrelation and extra code which was starting to approximate a kernel executive scattered among many programs. The various programs (processes) needed to execute concurrently, communicate with each other, synchronize with each other, handle interrupts and initialize in an orderly fashion upon power-up. It was felt that since a kernel was in effect being written in an informal, noncentralized fashion, it would be advantageous to separate the kernel functions from the processes and write a separate kernel executive. The kernel executive implemented was written in Motorola 6809 assembly language based on the Waterloo Port kernel.

3.2. Functional Design Specifications

3.2.1. General overview. The Kernel which is to be implemented shall have five basic functions. It will implement the multiprocess abstraction (contain the minimum amount of functions which must be implemented before coexisting processes can be implemented). It will provide synchronizing primitives and abstract interrupts as well as provide utility processes best managed by the kernel as well as power-up initialization. The implementation of the kernel shall not provide memory protection, allocation of I/O devices or allocation of memory upon request for process use.

3.2.2 Multiprocess abstraction. A process can be either ready (has all the required attributes and resources and can execute when its turn comes) or blocked (is waiting for some resource or attribute or synchronization mechanism and therefore can not execute). All processes which are ready shall be executed by the kernel as follows. Each process is assigned a priority. More than one process can have the same priority. A higher priority process shall be executed by the processor in preference to a process of a lower priority. In addition, if more than one process has the same priority, the process which has been waiting for execution the longest shall be executed first. In the case of multiple processes at the same priority (with the processor executing the one which has been waiting the longest) and a higher priority process becoming active, the higher priority process will be executed and the process which was previously executing shall be deemed, by the kernel, to become the one which has been waiting the shortest time among its peers (other ready processes at the same priority).

All processes which are not ready shall be in one of the following blocked states for the associated reasons:

1) Send Blocked - A message has been sent to which no reply has yet been received.
2) Receive From a Specific Process Blocked - A process is waiting to receive a message from a specified sender.
3) Receive From Any Process Blocked - A process is waiting to receive a message from any sender.
4) Event Blocked - A process is waiting for a specified external event to occur such as an interrupt (it sould be noted that software interrupts can be generated by processes to give the affect of an external event occurring).
5) Time Delay Blocked - A process is waiting for the elapse of a specified quantity of time (not for the occurrence of a particular absolute time).

3.2.3. Synchronization. To achieve process synchronization, the kernel shall use message passing rather than shared variables (flags, semaphors) [5]. The following message primitives shall be supported:

1) Send to a specific process
2) Receive from a specific process
3) Receive from any process
4) Reply to a process

Furthermore, to synchronize with an external event, a process will be allowed to request suspension of execution until the external event (interrupt) occurs. An extension of this concept will allow a process to suspend execution for a requested period of time to allow pseudo-time-scheduling (delay for a period of time rather than for the occurrence of an absolute time) as well as time delays which are useful for error recovery retry algorithms. A non-message passing synchronization technique is possible using a wait for external event where the event is the setting of a flag followed by the generation of a software interrupt by another process. This is handled in the same manner as hardware generated interrupts with the exception that memory rather than hardware is checked for the cause of interrupt.

The above send and receive synchronization primitives will result in the process becoming blocked. The reason for this blocking (rather than allowing a process to another send) is to avoid kernel controlled queues of buffers with their associated complexity. Where non-blocking message functions are required, special server processes shall be used to provide them [6]. The unblocking primitives shall be:

1) The reception of any message from any process when the receiving process is in the receive from any process blocked state.
2) The reception of a message from the specified process when the receiving process is in the receive from a specific process blocked state.
3) The reception of an acknowledgement from the destination process when the sending process is in the send blocked state (the acknowledgment is actually a reply message). This implies that whenever a process receives a message, it must send a reply to the originator to allow the originator to become unblocked.
4) Occurrence of a specified external event when the process is in a wait for event blocked state.
5) The elapse of the required quantity of time when the process is in a wait for requested period of time blocked state. (The alternate approach of waiting for the passing of an absolute time was avoided due to the complexities and necessary error recovery from time setting or changing).
6) The occurrence of an awaken a process command for a process which is waiting for the elapse of a quantity of time (the process had executed a delay for a specifed quantity of time but the quantity had not yet occurred when another process executed the awaken command to force the delaying process to run before the requested quantity of time had expired).

3.2.4. Interrupts. The kernel itself shall be treated as a separate entity from the user process (not executed as a subroutine). It shall be entered via a software interrupt and parameters will be passed primarily through the processor registers. The kernel will use its own stack to prevent forcing user processes to leave room for it on their own stacks.The kernel will save the entire user process volatile environment. These conditions allow the kernel to be written as an indivisable program (will run to completion uninterruptibly). Since it is indivisible, it does not need to be reentrent or recursive which will make its design simpler to code, understand, debug, change and maintain as well as keep its size smaller.

Interrupts will be serviced by separate special purpose (not integral to the kernel) interrupt handler processes, which will identify themselves to the kernel at any time (not only at system power-up or reset), and will be executed when the associated interrupt occurs, by the kernel. These interrupt handlers can cause an associated server process to move from the blocked awaiting external event state to the ready state. The interrupt handlers must clear the interrupt. Priority of interrupt handling shall be implemented using the priority assigned to the server process (with the exception that the actual notification, clearing of interrupt and possible data input/output queue managment which will be executed by the handlers in the case of simultaneous interrupts, will be executed by the handlers in the priority order assigned to the handler when it was made known to the kernel). The handler will request the kernel to unblock (put into the ready state) the server process awaiting this interrupt and the kernel will execute it based on its priority relative to all other ready processes.

3.2.5. Utility processes. The kernel will provide utility processes which it is best suited to handle. For this project only process creation shall be implemented. Implementation of process destruction is not required because it is felt that the requirements of the communication network software can be met by implementing an awaiting for time delay prematurely command. Undoubtedly, other important functions shall be identified and implemented in the future. One such function might be a process id (PID) to a process name directory lookup utility which would allow processes an easy method of determining another process's PID.

3.2.6. Power-on/reset initialization. Upon power-up or reset, the kernel shall initialize all hardware which it is aware of (timer, ports etc.). The kernel shall also create an idle process which shall run when no other processes are ready. The kernel will also create the first "creator" or "system initialization" process which will create the necessary processes which are to exist upon power-up. These in turn will create their own offspring processes as required.

3.3. Software Implementation Description

3.3.1. Process identifier. Each process was given a 16 bit process identification number (PID) upon creation of the process. This PID is passed back in a paramater to the creating "parent" process and in future enhancements will also be stored in the created process's process description block.

3.3.2. Process descriptor. Thirty two process descriptors (PD) were implemented, each of which contain the following information:

1) Process state - describes which ready or blocked state the process is currently in.
2) Process identification number (described previously).
3) The process's stack pointer which points to the location in the process's stack area where all of its other registers are saved to maintain the volatile portion of the process.
4) Queue linkage for the ready, send to me, free PD and time delay queues.
5) Head and tail pointers to the queue of processes which are send blocked on me.
6) The priority of the process.
7) The delay time before awakening the process after the previous process in the delay queue has awakened.
8) In the future it will also include a six ASCII character process name to implement a name to PID directory.

3.3.3. Send to me queue. When a process executes a send, it enters the send blocked state and is linked into the destination process's send to me blocked queue. The send processes are linked in the queue in chronological order. The first process to send will be placed at the head of the queue and subsequent processes will be linked at the tail. When the destination process executes a receive from any, it will receive from the process at the head of the queue. If it executes a receive from specific process, the queue will be searched from head to tail for the requested process. Processes are removed from this queue when the destination process executes a reply.

3.3.4. Active process. A pointer to the currently executing process's PD is maintained to keep track of it. This process is referred to as the "active" process.

3.3.5. Ready process queue. A queue of priority queue heads was implemented. All ready processes are linked into their respective priority queue. There are 16 queue heads, one for each implementation allowed priority level. In each head there is a pointer to the head PD of the priority level (the one which has been waiting the longest for execution at this priority) as well as a tail pointer (PD of the process at this priority level which has been waiting for execution the shortest time). These pointers are maintained to speed up the search for the next ready process to be executed as well as placing a newly ready process into the queue (at the head and tail respectively). Optionally, in future enhancements, a pointer to the next priority head which has ready processes in it will be implemented to speed up the search process. The ready queue was arranged in decreasing order of priority.

3.3.6. Process switching. Process switching shall load and store the volatile environment of the executing process and search for the highest priority ready process using the ready process queue.

3.3.7. Available process descriptor blocks linked list. All PDs which are not used at a given instant in time, are linked together in the available PD list. A pointer to the head as well as the tail was implemented. If process destruction is implemented in the future, it will be desirable to reuse a previously occupied PD block as late as possible (since it will have the same PD number but a new generation and it is desirable to not have them roll over back to zero).

3.3.8. Delay queue. All processes which are delay blocked for a quantity of time, are linked into the delay queue in increasing quantity of delay time order. When the first process is placed into the queue the hardware timer is started which interrupts when the delay time expires. The delaying process will be made ready by the hardware interrupt processing module and the timer will then be reinitialized to delay with the next process's delay time.

3.3.9. Interrupt table. When a process wishes to wait for an external event to occur, it will execute an "await interrupt number n" where n is associated with an entry in the interrupt table. The PD address is stored in the table entry or a zero is there when no process is awaiting the interrupt. When the interrupt occurs, an interrupt handler associated with this table entry will be executed by the kernel. The handler will inform the kernel whether or not the process should be made ready (ie. end of record received after a string of characters). Only one process can be associated with an interrupt at a time, however, this process can be an "event notifier" type which could notify many other processes of the interrupt occurrence by sending them messages.

3.2.10. Interrupt handler dispatch table. Each interrupt handler is stored in a table in the order which it was made known to the kernel. However, the handler entries are linked together in a list by their priority order. Upon the occurrence of an interrupt, they are processed in this linked list order. The information in each table entry consists of: the bit mask for isolating the "interrupt active" bits associated with this handler in an interrupt register or location of memory (for software interrupts), a flag to indicate if the interrupt register or memory location is to be checked, the interrupt table entry number where the address of the PD of the process awaiting this interrupt is stored, the priority of the handler, the start of handler execution address and a link to the next handler entry in the dispatch table by priority order.

3.3.11. Interrupts. The kernel was to be executed nondivisably for the reasons mentioned above. However, the longest pass through it (send a message) required 931 microseconds plus an additional 20 microseconds per byte sent. This was too long a time to have interrupts disabled. To avoid this time period, the processing of interrupts was allowed whenever the kernel was not modifying the ready or delay queues, modifying the interrupt table or processing interrupts or requests for awaiting occurrence of interrupts. To maintain an equivalence of nondivisability, a flag is set when the kernel is executing and interrupts are disabled only during the critical time of kernel execution. If an interrupt occurs during the kernel execution, the most which will occur, is the making of a process ready. If more than one process is made ready by interrupts during one pass through the kernel, the identity and priority of the highest priority one is saved and a flag is set. When the kernel finishes its pass, it checks for this occurrence and will perform a process switch if the process which it was treating as being the highest priority ready one is of a lower priority than the one awakened by an interrupt. Since an interrupt handler has been executed, it can save any data associated with the interrupt in a queue for consumption by the interrupt server process when it becomes active. When the kernel is not being executed on the behalf of the active process (an interrupt occurred while executing a process rather than the kernel), the interrupt will cause the process switch (if appropriate) to occur immediately.

4. CONCLUSIONS AND RECOMMENDATIONS

4.1. Physical Layer

At the physical layer dual optical fiber with a RS422 twisted pair was used in a topology consisting of a hierarchical interconnection of networks. A pilot installation which was in service for two years exhibited suitability to the plant environment. The optical fiber was inherently immune to electromagnetic noise, required only a low level of technical expertise for maintenance and had a higher bandwidth than required to not impact the throughput of the network. A unique design of an active termination and circuit switch powered by a separate 48 VAC source provided a redundant, fault tolerent bus structure, with a distance limitaion of one kilometer between nodes and an effectively unlimited total network length. Two types of circuits were used: nodes for secondary stations and subhosts for primary stations. As a result of the design and implementation the following recommendations are made: use only the node type of circuit for both secondary and primary stations since the subhost type is unnecessary, use metal instead of plastic connectors on the optical fiber, replace polishing of connectors with cleaving and use fast setting ultraviolet curing epoxies.

4.2. Microprocessor Hardware

To support the communication and application software and prevent one from impacting the other, a dual processor tightly coupled architecture was designed using Motorola 6809 microprocessors. To reduce the overhead in servicing the communication protocol a Signetics 2652 communication protocol controller was selected. A Motorola 6844 direct memory access controller was used to provide high speed data transfer between the communication controller and system memory. The hardware was installed and operated for approximatly two years during which it exhibited unacceptible synchronization timing problems between the two processors. The conclusion reached was to redesign the hardware replacing the microprocessors with 6809E parts which have well established timing relationships. The synchronization problems were successfully eliminated, the new hardware has been installed in the plant for half a year and has proven its reliability. The limiting factor to actual data throughput was found to be line turnaround delays caused by protocol software overhead. It is recommended to evaluate the new communication protocol chip sets, which have been announced to become available in the near future, which will implement up through the network and possibly the transport layers. These chips along with the newer 16 bit microprocessors could replace the current hardware to yield a much higher data throughput.

4.3. Kernel Executive

A kernel executive based on the Waterloo Port [6] kernel design was successfully written and tested. It is currently being applied to fulfill the industrial plant monitoring and communication network system needs consisting of: multiprocessing, process-to-process communication, synchronization, interrupt handling and power-up initialization. The functional design specifications have been met with the exception of interrupt handling. The design specified that the kernel was to be executed indivisibly (run to completion uninterruptibly). This would have resulted in interrupts being disabled for almost a full millisecond. This was judged to be excessively long in view of the high speed (one Megabit per second) fiber optics communication network used by the microprocessor and could have led to long communication line turnaround delays. To circumvent this problem and still maintain the characteristics of indivisibility, interrupts were allowed during kernel execution in all but critical table modification sections. During such an interrupt, interrupt handler dispatching would occur but process switching would not occur until normal (not caused by the interrupt) exit of the kernel. Thus for high speed processing, such as line turn around, a handler can take care of the time critical section and make the actual process which is waiting for the interrupt ready to run, but not execute it until the kernel is finished. In effect, the kernel does run to completion unaware and unaffected by the interrupt until just before exiting to execute what it considers to be the active process. At that time it only needs to check if a higher priority process was made ready by an interrupt in which case it would be made the active process and executed. In this manner reentrancy and recursion were avoided.

The application of the kernel has pointed out the following recommendations. A process name to process identification number (PID) directory should be implemented to avoid having the creating process (which is given the PID when it creates the process) distribute the PID to all other processes which need to communicate with the created one. A separate queue should be maintained in each PD block for messages against which a receive has been executed. Currently, the sending process remains in the send to me queue when the destination has executed a receive but has not yet replied. If the destination executes another receive any without first replying, it will receive the message again since the sender is still in the queue. If a separate previously received queue was maintained, this problem would be eliminated. In addition, if a receive from this previously received queue is also implemented, then the receiving process can receive and selectively process only the messages which it wants and leave the others buffered in the source process's buffers linked in the previously received queue. This would be done by executing a receive from any (which would cause a move from the send to me to the previously received queue) and checking if the information is to be processed immediately. If it is, then reply (which removes the source process from the previously received queue), otherwise execute another receive from any leaving the first message in the source's buffer. At a later time, the first message could be received again using the receive from previously received queue. Overall, the implemented kernel executive met the needs of the application and communication network processes.

REFERENCES

[1] General Motors Corporation. Manufacturing Automation Protocol.Warren, Mi., 1984.

[2] International Standards Organization. Reference Model of Open Systems Interconnection. ISO/TC97/SC16 N890, Gaithersburg, MA, August 1979.

[3] Electronic Industries Associated. Electrical Charac-teristics of Balanced Voltage Digital Interface Circuits RS-422. Washington, D.C., 1978.

[4] Electronic Industries Associated. Electrical Charac-teristics of Generators and Receivers for use in Balanced Digital Multipoint Systems RS-485. Washington, D.C., 1981.

[5] Cashin, P.M., "Inter Process Communication", Working Paper Reference 8005014, Bell-Northern Research, 1980.

[6] Cheriton, David R., The Thoth System, Stanford University, Elsivier Science Publishers, New York, NY, 1982.

CHAPTER 17

MICROCOMPUTER-EMBEDDED DISTRIBUTED CONTROL OF A SWITCHING AND COMMUNICATION SYSTEM*

Daniel Tabak
George Mason University
ECE Department
Fairfax, VA 22030

Joshua Etkin
Boston University
ECSE Department
110 Cummington St.
Boston, MA 02215

ABSTRACT: A new methodology for controlling and managing a complex distributed communication system is presented here. A model which describes the real-time behavior has been proposed, together with the recommended alternative for its implementation. The research is based on a rational design approach, supported by a simulation study, for evaluating the relative performance of different strategies. The main problem is related to the constraints of signal observability by all distributed units. It creates an uncertainty gap which ends up in a non-coherent view of the system state by all its units. The uncertainty gap can be minimized by the creation of a new control architecture. The communication coordinator section in every unit should carry all of the mechanistic management activities, leaving only strategic decisions to the control section. Another promising method for reducing the uncertainty gap, is the integration of the reconfiguration technique into the distributed system, which leads to the adaptation of the system to the varying work profile. The simulation study indicated a potential bottleneck in the centralized scheduling approach, and recommended the dynamic allocation method as the prefered one. The minimal-queue unit is selected for running the new task, provided that the source unit queue has reached a threshold level above its average queue length.

1. INTRODUCTION

1.1 Computer-Embedded Structure

Technology advances broadened the computer application spectrum in both directions. Upward, toward managing and controlling real-time complex systems, and downward, toward supporting cost performance and

This article is based on the Ph.D. thesis by J. Etkin, advised by Prof. D. Tabak, submitted as a partial fulfillment of the Ph.D. degree requirements at the Ben-Gurion University of the Negev, Beer Sheva, Israel. D.Tabak is an Abrahams-Curiel Professor (on leave).

N. K. Sinha (ed.), Microprocessor-Based Control Systems, 343–367.

modularity. These two directions can be accomplished by the same basic solution, namely, distribution of tasks among processors, for parallel execution and control [10,26].

There are three ways to achieve parallel execution and match the system's structure with the functional characteristics.

(a) Multiprocessor organization. Its goal is to attain a unified system, based on centralized, tightly coupled processing elements, communicating through a shared memory [13,19]. Multiprocessor architecture has a straightforward implementation, where each processor handles a simpler task mechanism than that of a mainframe CPU. Provided that it is correctly managed, the main drawback lies in the interprocessor communication mechanism. Connection via a shared memory has the severe disadvantage of potential bottlenecks [17].

(b) Computer networks. This configuration provides the solution for resource sharing demands. One class of networks connects geographically separated computers for data-base sharing. It is characterized by slow speed packet communication and sophisticated routing scheme [12,20]. Another class connects office equipment, residing in local environment. It is characterized by high speed, possibly broadcast transmission [1,11]. Network architecture is easily implemented by layering a network operating system (N.O.S.), on top of a host computer O.S. [11,17]. On the other hand, computer independence is a main drawback, because it restricts the real-time application space [9,17]

(c) Distributed computer system. This class is based on near proximity, physical distribution of loosely coupled computers. System control is decentralized among all autonomous computers, acting together to achieve a coherent goal [2,17]. It enables the overall throughput enhancement by preventing the bottleneck caused by the message flow into the centralized memory. It is done by performing the tasks at the physical spot where they were generated. Distributed systems involve various aspects, such as a naming scheme, resource allocation and protection, interprocess communication by messages and decentralized system-wide control [17,19].

The fundamental characteristic of physical distribution of hardware, software and data, implies a variable delay between the transmission and reception of messages. It ends up with a possibly different state representation as viewed by all autonomous computers, a major factor to be considered at the control strategy level [2,17,19].

The switching and communication field is a typical example of the described architectural solutions. From a non-computerized, voice only switching, two decades ago, it evolved via a centralized uniprocessor achitecture into todays integrated, multiprocessor implemented, voice/data machines [22]. This trend led to new and surprising solutions to problems, previously impossible to solve. Self diagnostics and system's survivability, up to the extreme of dynamic reconfiguration, are part of these subjects [18].

1.2 The Research Goals

Two research goals have been targeted:
- An in-depth study of the distributed computer architecture field, in order to overcome some of its current drawbacks.
- Selection of the architecture able to exhibit maximal throughput under varying constraints.

We chose to present the distributed computer concepts and the methods for it's efficient implementation, via a realistic model of a complex real-time machine - a large switching and communication system. This research is based on a rational design discipline, supported by a simulation study of the main alternative strategies.

The effort was geared toward a new system-management concept, based on a non-conventional relation among the system components.

This strategy is based on several software concepts:
- System management is based upon partial knowledge, by all computers, of the global state at any unit. This confronts correctness of operation against efficiency, an issue which is resolved by a set of task allocation policies.
- Narrowing the uncertainty gap is feasible due to a fast mechanism for task categorization.
- The tasks are running to completion on the computer to which they were originally allocated. These computers perform a "pool", queing for new tasks.

An architectural method that improves efficiency is the delegation of mechanistic management activities from the upper level control to the communication-coordinator unit. This implies a new protocol procedure for transferring information between these two units, which deviates from the standard approach [7]. An analogy can be drawn between the distributed routing methods in a network and a decentralized management of the distributed system. All these methods are based on a new hardware implementation of the switching system.

2. THE EVOLUTION OF SWITCHING AND COMMUNICATION SYSTEMS

2.1 The Twofold Evolution in the Switching Field:

a) <u>Architectural evolution.</u> The expected lifetime of the switching machines is around 20 years. This implies issues of reliability, modularity and system reconfiguration, in order to adapt the system to varying demands.
- The first generation of these systems has been based on a uniprocessor control and analog swithing (Pulse Amplitude Modulation or Space Division). An example is the TBS family, designed in Telrad, Israel, and based on VLSI space switches and LSI-11, a 16 bit processor [3].
- The second generation is using a centralized master control and peripheral slave controllers. They incorporate an internal digital switching (PCM), while outside signals remain analog. The well known examples for this generation are SL-1 of Northern Telecom and the Rolm family of switches [15,21].

- The third generation, today in it's infant stage, is based on a uniform management of voice, data and office processing functions. The integrated digital stream of data, voice and signalling is transmitted up to the user premises. This stream is routed via the digital switch to the outside networks (voice or data). The control is structured in a hierarchical form, with peripheral satellite controllers. The structured software is written in HLL and runs under multitasking O.S.. The best documented example is SX2000 of Mitel [22].

b) Functional evolution: From voice-only basic functions, into an integration of voice switching, data communication and office related processing issues.

- Modern voice processing functions are oriented toward cost saving and time saving (i.e. increase efficiency) activities. An example for a time saving function is a 'Camp on trunk', which queues the request until a busy trunk will become idle and then seizes the trunk. An example of a cost saving function is the 'Least cost routing', which selects the optimal long distance route, depending on system location, destination address, time of the day, etc.
- Data communication methods progressed immensely from it's initial phase of point-to-point connection, into Long distance and Local Area classes of networks [24]. Long Distance networks have been designed mainly for data-base sharing, while Local Area network's prime goal is resource sharing and transactional communication. Current research effort concentrates on network interconnection, which relates to gateways and protocol transformation.
- Office processing functions are geared toward increasing productivity, by interconnecting various stand alone machines. In addition there are several specific functions that support commercial applications (e.g. electronic mail), military environment (e.g. encription) or medical needs (e.g. real-time image processing).

2.2 The Importance of Distributed Architecture and it's Challenge

The framework for the fourth generation of switching and communication systems is proposed here. Inherent within the distributed architecture is the potential to overcome some of the drawbacks of other computer-embedded architectures [17,19]. Another important factor is the ability to reconfigure the system in order to adapt it to functional and economical needs [14,17,23]. The main research activity in this field progressed immensely in the last decade. It started up with an Inter-process communication mechanism, what is known today as the Local Area Networks [11]. On top of it one has to deal with naming, synchronization, authentication, and of course decentralized control. The model for distributed system has been established in the Munich Seminar [17]. The parameters are illustrated, in polar coordinates, in fig. 1. The radius coordinate spans the functional logic layers, from the bare machine, through the O.S. and up to the application layer. The polar angle coordinate defines subjects common to all layers, such as representation and error control. The third (top) coordinate deals with global attributes, such as task scheduling and overall consistency.

The limitation of signal observability by each unit of the distributed system, due to the varying delay among these units, increases the constraint for the coherent behaviour of the system (i.e. concurrency operation and equal responsibility of each unit). The importance in the specific research, is in setting down a solution skeleton which resolves some of the drawbacks for the switching area, and sets the ground rules for the entire real-time field.

3. DISTRIBUTED CONTROL OF A SWITCHING AND COMMUNICATION SYSTEM

We adopted, as a research vehicle, the concept of a large switching and communication system, which incorporates complex real-time data processing. Most of it's functions belong to a class of 'statistical real time' systems. It demands a reasonable response time, such as a dial tone response for telephone offhook condition or interactive terminal handling. Other functions demand fast real-time response or heavy throughput (e.g. electronic mail or encypherment). These different classes of activity imply different response strategies, where every task can be executed anywhere in the system. The architecture also supports the survivability concept, where the system should continue functioning in the case of a failure.

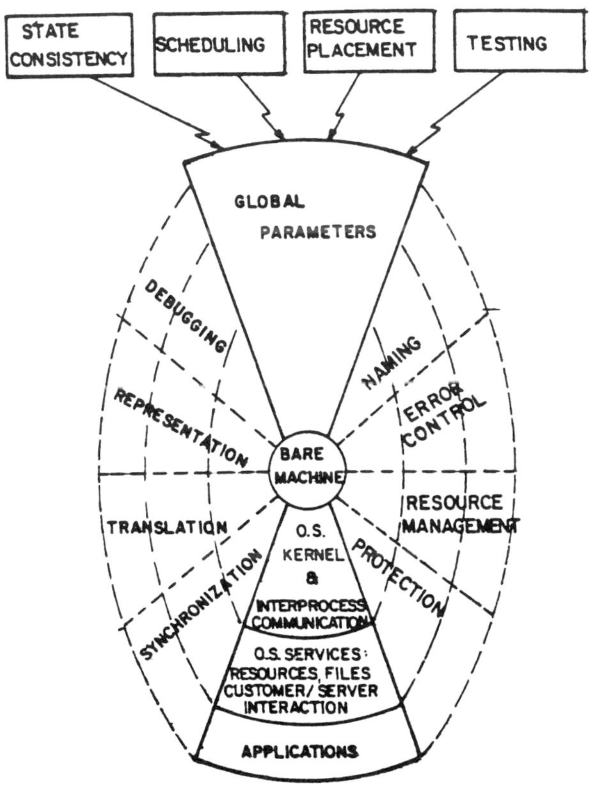

Figure 1. Distributed system model

Five different configurations can support this application:
(a) centralized architecture, based on a uniprocessor.
(b) centralized architecture, based on a multiprocessor organization
(c) distributed architecture, where every base unit is based on a uniprocessor.
(d) distributed architecture, where every base unit is built out of multiprocessors.
(e) distributed architecture with a central scheduler.

The research goals, as stated above, directed us toward the last three structures. The basic model contains a collection of Base-Units, uniprocessor type, connected via the communication sub-network (fig. 2). The research has been limited to the hierarchical model of a multiprocessor per each Base Unit (fig. 3).

A simulation study, based on the last two architectures, has been conducted. The basic goal of the simulation study was to examine the relative merits of several allocation strategies, in regard to performance and bottleneck behaviour.

Figure 2. The basic simulation model

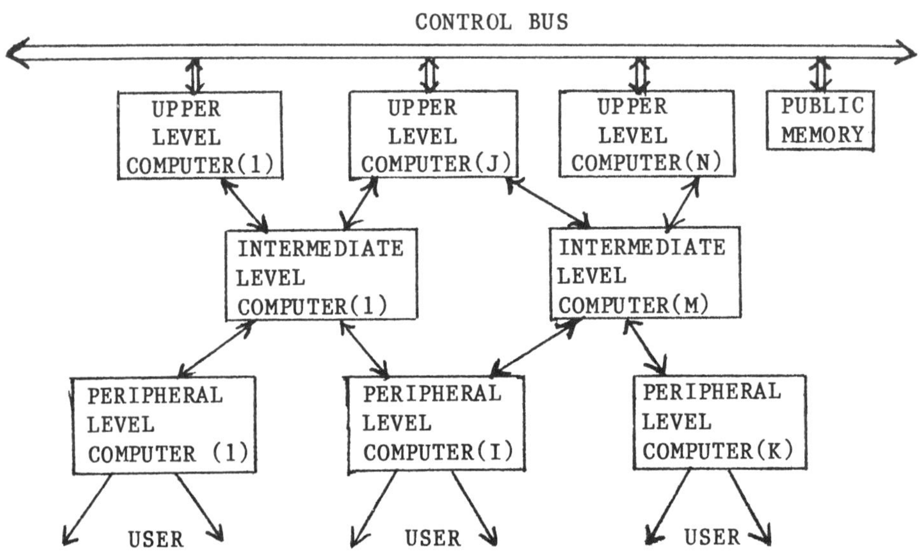

Figure 3. Base unit hierarchial model

3.1 - The Analytical Method

One can start with two extreme strategies:

- An autonomous model, where each task is executed at the Base Unit where it was generated.
- A central queue strategy, where all tasks queue and wait to be executed by any idle computer.

Analytically, one can view the authonomous model as m, single-server, different queues. The interarrival rate of this model is exponentially distributed.

The general expression for a single server queue (M/G/1) [16,19]:

$$L = \rho + \frac{\rho^2}{2(1-\rho)} \cdot \left[1 + \left(\frac{\sigma_S}{E(S)}\right)^2\right] \quad ; \quad \rho = \lambda E_{(S)}$$

where:

L = Mean number of tasks in the system

ρ = Computer utilization

E(S) = Mean service time

λ = Mean arrival rate

σ_S = Service time standard deviation

Note: S defines a system parameter

For an exponential service time (M):

$$L = \frac{\rho}{1-\rho} \quad ; \quad W = \frac{E(S)}{1-\rho}$$

W = Mean time of completing a task in the system.

For m such autonomous queues we have:

$$\lambda_1 = \frac{\lambda}{m} \quad ; \quad \rho = \lambda_1 E_{(S)} = \frac{\lambda E(S)}{m}$$

and then we get:

$$W = \frac{E(S)}{1-\rho} = \frac{E(S)}{1-\frac{\lambda E(S)}{m}} = \frac{mE(S)}{m-\lambda E(S)}$$

For the second model (centralized queue and exponential distribution) we have the Erlang's C formula, which states the probability of having an idle processor when a new task arrives [19].

$$C(m,\rho) = \frac{\frac{(m\rho)^m}{m!} \cdot \frac{1}{1-\rho}}{\left[\sum_{k=1}^{m-1} \frac{(m\rho)^k}{k!}\right] + \frac{(m\rho)^m}{m!} \cdot \frac{1}{1-\rho}}$$

where computer utilization is $\rho = \frac{\lambda E(S)}{m}$.

We get:

$$W = \frac{C(m,\ \rho)}{m} \cdot \frac{E(S)}{1-\rho} + E(S) \quad ; \quad L = C(m,\ \rho) \cdot \frac{\rho}{1-\rho} + m\rho$$

To evaluate analytically the performance of the simplest model, an equilibrium ought to be reached between new arrived tasks and serviced tasks. Moreover, a network of queues has to be built, setting some simplified assumptions. This simplification, and our goal to examine some complex architectures and allocation strategies, led us to select the simulation as the method to be pursued.

3.2 The Simulation Models and Their Evaluation

The principal simulation study compares four allocation strategies:
(1) Distributed system with centralized scheduler.
(2) Distributed system, with scheduler in every Base-Unit, which allocates every new task to the minimum-queue Base Unit.

These two models involve a price (or penalty), due to the need to communicate with other Base Units, in order to know the system-wide state. A graph of the performance as a function of this price (penalty), as well as performance versus the load, has been obtained as a result of the simulation study.

In order to avoid this penalty (which causes an overhead), and yet gain some level of efficiency (over the autonomous model), we presented a third allocation strategy.
(3) Distributed system which allocates a new task randomly to any Base Unit, as long as the immediate queue length in the Base Unit reaches a certain threshold above the average queue length in this Base Unit. This is a local decision in every Base Unit, and is based on the statistical assumption that the probability of such a long queue at that instance of time, in other Base Unit is very low.
(4) The fourth allocation strategy is a combination of the two last ones. The scheduler in every Base Unit selects the minimum-queue Base-Unit, only if it's own queue is above the threshold.

Some of the simulation results are illustrated in fig. 4. A graph of the throughput (of systems having these four allocation strategies), as a function of penalty is presented. The graph indicates that the centralized strategy is very sensitive and has a potential bottleneck: the queue length for the central scheduler. The other three strategies have a cross over point: the strategy that selects always the minimum-queue is the best under very low penalty, while the strategy based on threshold is the best when the penalty grows. The recommendation is to adopt a dynamic strategy, i.e. choosing the minimal-queue base unit only above a certain threshold, and changing the threshold level dynamically. The same pattern and recommendation holds for the simulation that measures througput versus load. The complete-system simulation process is illustrated in fig. 5.

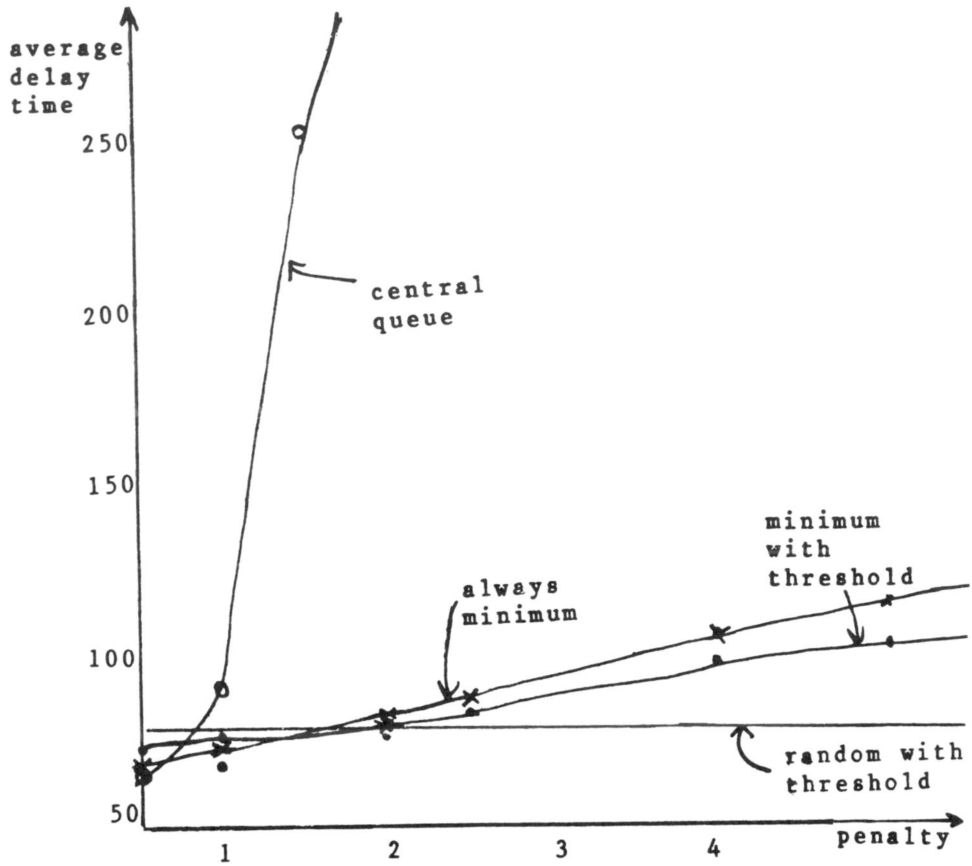

Figure 4. Allocation strategies: Performance as a function of penalty

4. CONTROL AND IMPLEMENTATION OF THE DISTRIBUTED COMMUNICATION SYSTEM

The specific conceptual model of the system, is a direct outcome from the rational design considerations and the simulation results. The model emphasizes the principles for a non-conventional relationship between the control layer and the communication coordinator unit. This structure helps to achieve the goals of the distributed system management. Both the conceptual model and the methods for its implementation will be illustrated. [4,6].

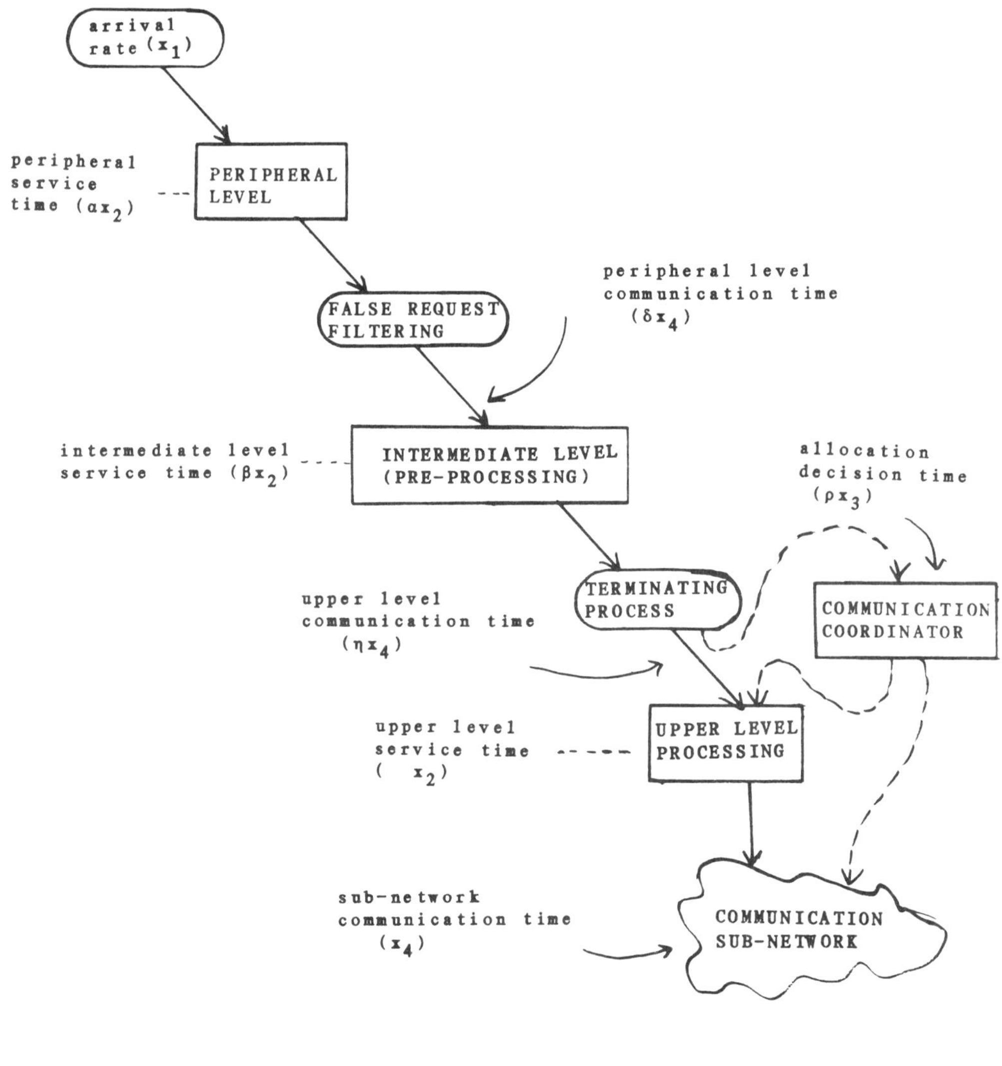

Figure 5. The complete simulation model

4.1 The Control Structure

The distributed switching and communication system is built out of Base Units (BU), separated physically and logically one from the other. Each BU is composed out of the following sub-systems (fig. 6):

a) The peripheral interface circuitry and controllers, sensing and supervising every activity of the functional modules (regular and diagnostics). It filters the false events from the true ones, and generates the tasks accordingly. The proposed implementation is based on Intel 8051 (an 8 bit controller) for every peripheral card. It serves as a slave processor, connected to a polling bus, and communicates upward toward the master controller, through an internal serial mechanism (187KB/S).

b) The intermediate control level which performs two functions: It serves as a communication cluster concentrator, where the 8051 master manages the serial bus polling to every peripheral card. The other function of the intermediate level is to act as a pre-processing unit. It manipulates every task and terminates it when possible, without the involvement of the upper level control. Only tasks that should be handled on a global basis are transferred upward. The intermediate level is implemented using Intel 80186 -a 16 bit processor. Several of them are connected to the upper level control by the Multibus, where each 80186 is controlling several 8051 master controllers, communicating through a common memory. The reason for choosing the 80186 was its very high level of integration: the inclusion of several I/O functions on the same chip. The 8051 was chosen for similar reasons, particularly its 4 kbyte ROM on the chip and its serial communication port.

c) The upper level control, based on a centralized multiprocessor configuration, serves as a control supervisor (intra BU and system-wide), and as a functional processing element. It also interfaces the communication coordinator module, which supports inter and intra Base Unit messaging. All BU communication is carried out via a "mail box" memory and an interconnection sub-system, which connects all modules mentioned above, and supports reconfiguration of the system architecture.

In addition to its control flow architecture, each BU contains all functional modules, such as analog and digital workstation interface, outward voice and data interface, sensor interface and diagnostic hardware. All these modules are connected by the switching unit, non blocking circuit switch-PCM type, which is favored for data communication embedded in voice traffic. The composite digital signal (voice-data-messages) is transmitted up to the user premises. The schematics of this information stream is illustrated in fig. 7.

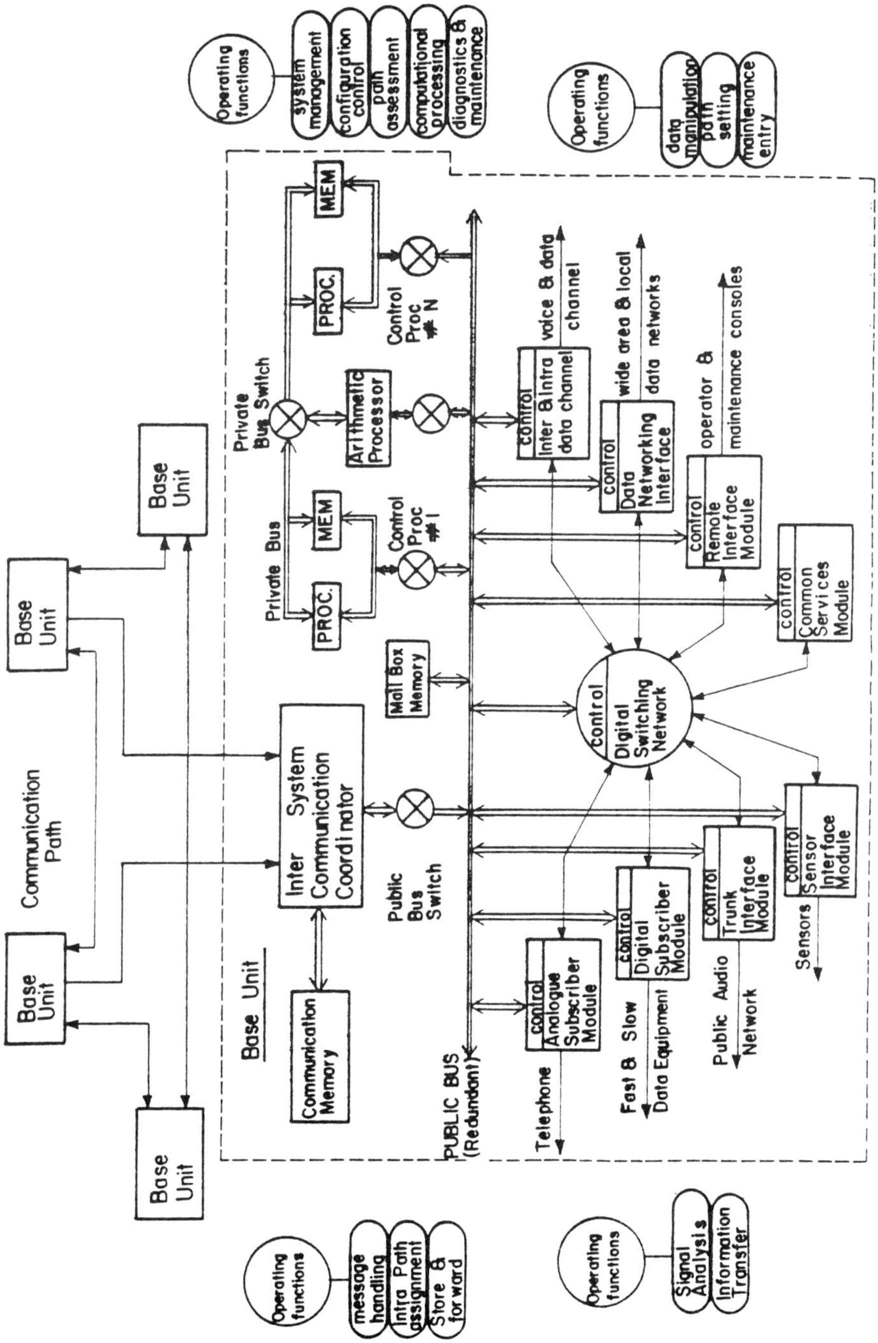

Figure 6. Schematic model of the distributed communication system

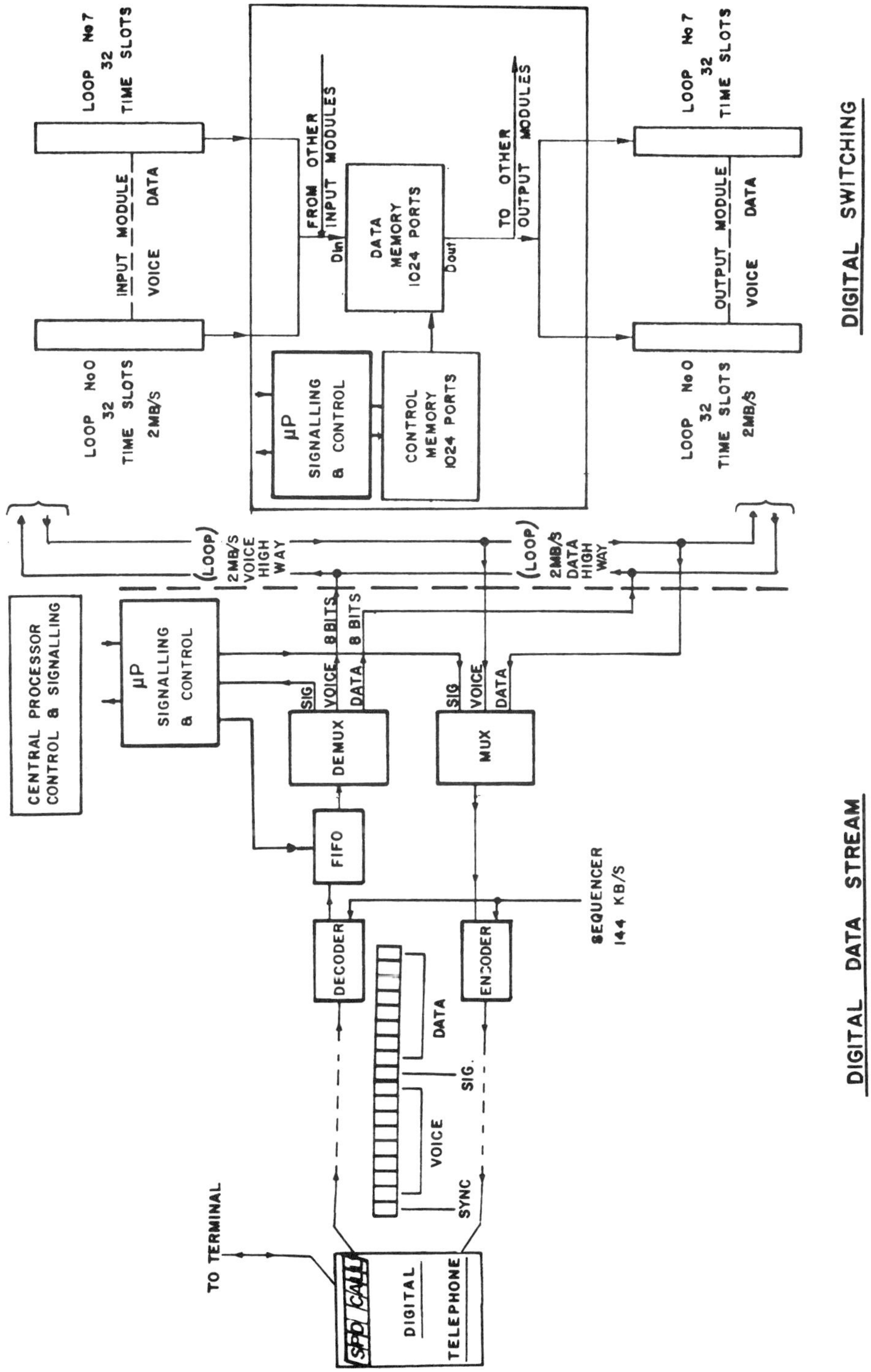

Figure 7. Information stream transmission and switching

4.2 Upper Level Control Considerations

There are two alternatives for implementing the upper level control, taking into account the concepts of the global distributed system [6]:

- One alternative is based on a conservative attitude, using the Intel 80286, a 16 bit processor, as the Base Unit high level control. The 80286 computers establish a master/slave relationship with the intermediate control level, communicating through the Multibus. The entire control structure is an interrupt driven mechanism, i.e. even the high level control responds immediately when a task is activated (fig. 8).
- The second alternative arose due to some disadvantages of the first proposal. In short, it does not reflect a transparent multiprocessor behavior. The interrupt-driven mechanism involves the upper control level with too many real-time details and causes an unneglectable overhead. Moreover, this alternative lends itself to a disjoint approach (between the communication-coordinator and the control) which does not enable the desired integration between the two units. The outcome of these drawbacks is a proposal for another realization of the

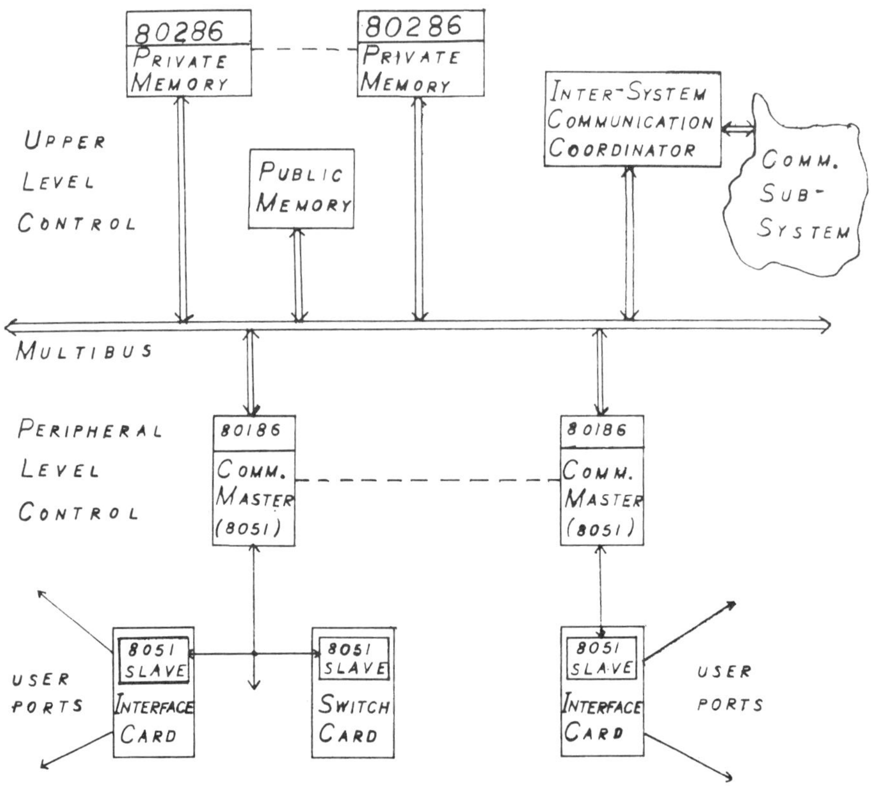

Figure 8. Implementation using iAPX286 family

upper level control, based on the Intel iAPX432 family -a 32 bit processor [6,25]. The General Data Processors (GDP - 43201/02), are interconnected in a multiprocessor configuration. Another family member, the 43203 Interface Processor (IP), serves as a buffer between the intermediate level (interrupt driven) and upper level control (message driven) (fig. 9). The object oriented architecture of the iAPX 432 supports a high degree of protection, due to the capability based

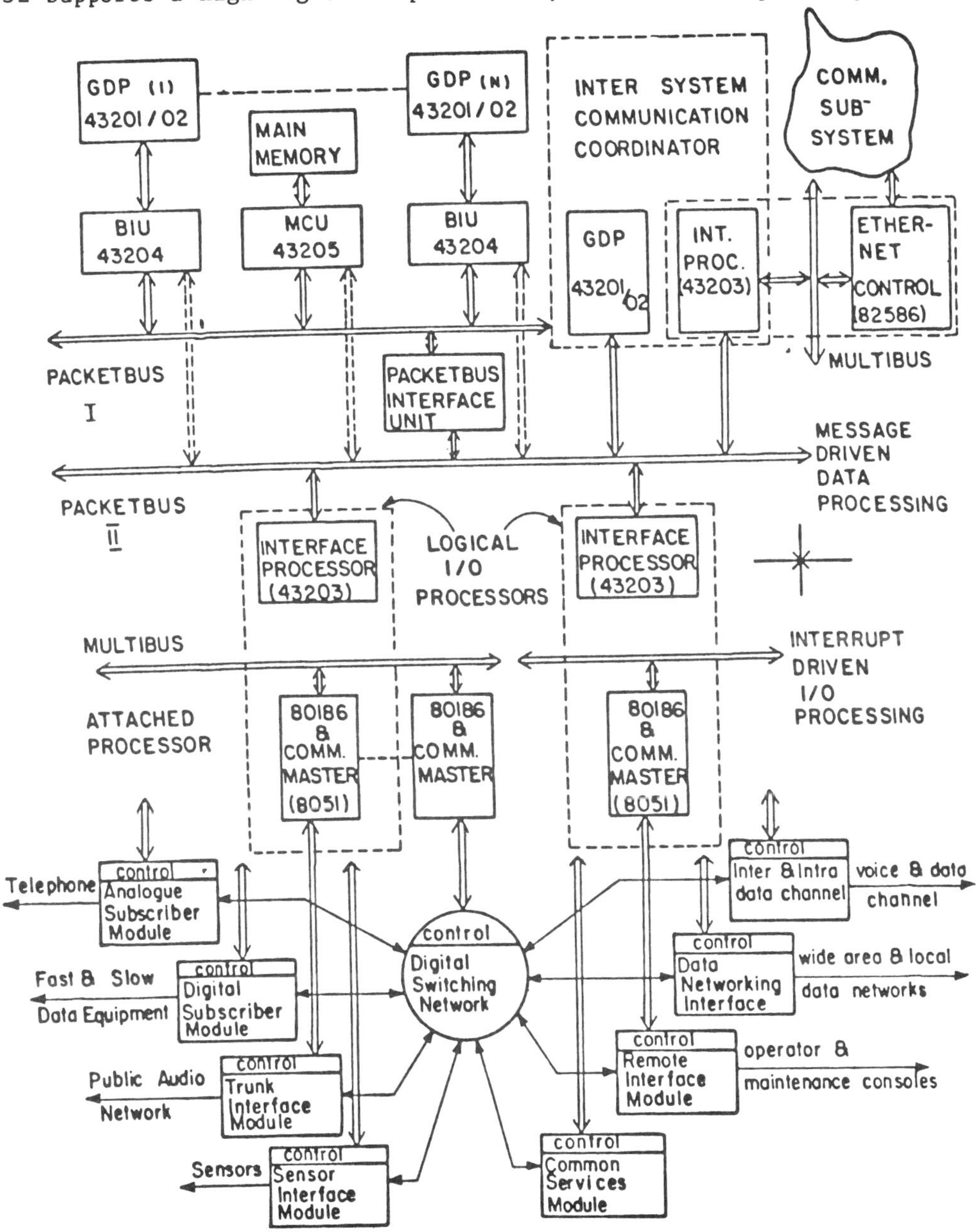

Figure 9. Improved implementation using iAPX432 family

addressing mechanism. It's multiprocessor operation is transparent in the sense that one can add as many processors without changing the software code. It supports very efficiently data transfer among the upper level control, intermediate level and the communication-coordinator, due to the unified treatment of software and hardware, based on the abstraction of object presentation. It is known that Intel has withdrawn its support from the 432 system and it does not market it any more. The realization option, shown in Fig. 9, should be understood as follows: The authors propose to utilize a system similar in concept to the iAPX432, whenever such a new system is announced, after having overcome the shortcomings of the 432 system.

4.3 The Communication/Control Relationship

A well designed distributed system is based on it's good inter-process/interprocessor communication. The implementation model utilizes the communication-coordinator module, which monitors internal and external messages and incorporates itself when needed [5].
There are three alternatives for its incorporation:
(1) Message separation: Intra-Base-Unit messages will flow directly to the destination processor, leaving the communication-coordinator to handle only outcoming (and incoming) messages. This minimizes the response time for internal messages, on behalf of overloading the control section.
(2) Transparent communication: Initially, every message is transmitted onto the communication sub-network, and then handled according to its destination. This method enforces simplicity and cost saving, on behalf of flooding the network.
(3) Composite communication: The control handles all messages (intra and inter-base unit) under a unified procedure. The communication-coordinator unit prevents network flooding, by blocking internal messages from going onto the network. The conceptual operation of this recommended procedure is illustrated in fig. 10.

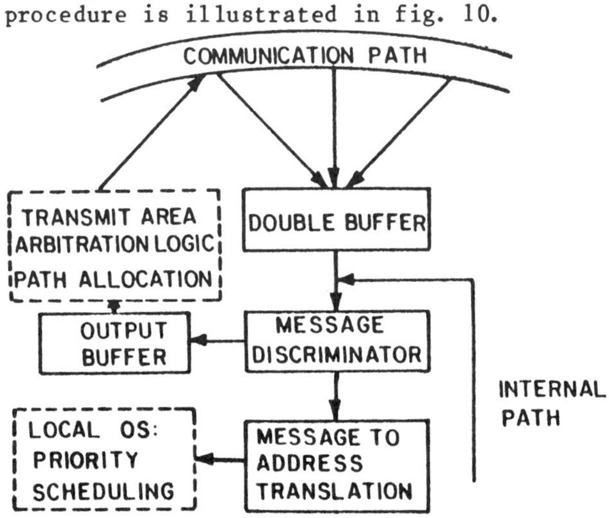

Figure 10. Conceptual operation of the communication coordinator

Two methods for the control-communication relationship can be considered [7]:

(a) The disjoint approach: Each sub-system performs it's own functions and reflects other modules needs, by a well defined protocol. It prevents a real-time feedback, i.e. one module does not influence the other's module dynamics. Fig. 11 illustrates schematically this structure. According to this approach, the communication-coordinator module performs only "classical" functions (transsmission, synchronization, error control). The internal protocol between the communication-coordinator and the control unit adopts the OSI seven layer model. The outcome of the disjoint approach is a poor signal observability, due to the extra delay and control overload. It ends up by increasing the constraints, i.e. minimizing the distributed management potential. In some applications, such as military systems where security is of high importance, a hostile object can easily flood the system with false messages. This is so due to the fact that authentication is treated at a high layer of the OSI model.

(b) The integral approach: The main principle is the sharing of management responsibility. The communication-coordinator carries out the mechanistic functions (e.g. allocation), leaving the upper level control with strategic decisions, and enabling it to perform efficiently the user functions. All messages are routed initially to the communication-coordinator module, which distributes them, based on its internal partial knowledge. In order to perform all these activities, the module has to incorporate a powerful computer. It is integrated from one side with the upper level control. Its other side is connected to the ETHERNET, an implementation of the communication sub-network. The protocol used to access the ETHERNET is the CSMA/CD with an acknowledgement mechanism. Moreover, we propose to add another unit, the packetbus interface unit (fig. 9), to allow very efficient message transfer among

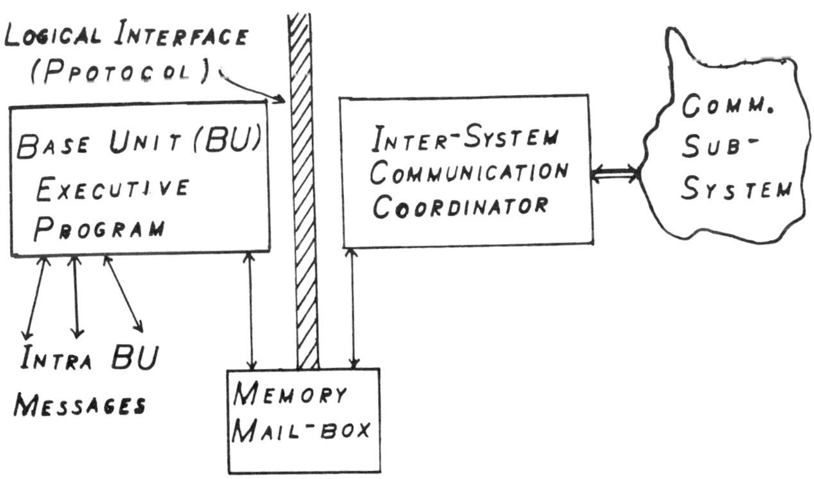

Figure 11. Contral/communication relationship: the disjoint approach

all mentioned blocks of each base unit, by direct transfer of information between the busses. The protocol between the control and the communication-coordinator is based on a minimal standard kernel, while all other functions are 'tailor made' to the specific application. For example, the inter-process communication can be considered standard, but error control and protection is an option.

5. THE DYNAMIC ASPECT OF THE SYSTEM

Thus far, the static skeleton of the distributed communication system has been emphasized. In order to improve the system adaptability together with upgrading it's performance and fault tolerance behaviour, we will focus the discussion on two aspects of dynamic nature: system reconfiguration and the method for operating work stations in a user friendly approach.

5.1 Reconfiguration of the Communication System

Reconfiguration is the process of adapting the system architecture to varying constraints. It is done by changing the relations between the system's functional units. The system can even reconfigure itself into a different processing organization [14,23]. Reconfiguration can be generated due to circumstances that cause one part of the system to be heavily loaded (e.g. military attack). Another reason is the adaptation of the system to peak type activities. The conceptual solution is to change the relationship between the peripheral units, and set-up a different priority scheme (fig. 12).

Another aspect of reconfiguration refers to circumstances where the system faces an irregular peak activity, that demands fast response. Such an activity is needed for a major control decision, invoked by an on-line statistics which indicates an abnormal behaviour. Fast response is also demanded for special type tasks (e.g. image processing). The system reconfigures its resources into a special control or processing block, for the duration of the specific task.

The third class of problems refers to fault conditions. This invokes reconfiguration for a fast control decision, which defines the new system's structure. Reconfiguration to the new structure should consider not only the bypass of the faulty unit, but also the attainment of the maximum performance out of the remaining resources.

The conceptual architecture is based on the hierarchical implementation model, where system's reconfiguration implies that every control layer should be dynamically managed. A particular processor ought to be bypassed when a fault occurs, and the system should be able to adapt itself to the varying processing needs. In order to support the dynamic flexibility, each peripheral card should contain three communication ports. One port interfaces to the original link and the other two ports should interface the two adjacent links. The software-selected transmission path is chosen to bypass the faulty 8051 (master or slave) and to enable load balancing. Only one communication manager is needed for the three bidirectional switches (fig. 13).

The same rationale holds for the 8051 master at the intermediate level. When an erroneous 80186 at this level is detected, the data is routed from the 8051S via the 8051M to an adjacent 80186. The information is routed via a different Multibus, through the Interface processor (43203) to the upper level control.

The architecture of the iAPX432 family, both the software and interconnection mechanism, supports reconfiguration at the upper level control. It can handle transparent multiprocessing and bypass faulty elements. The upper level processor area can be changed from multiprocessor architecture into an array processor, by incorporating the communication-coordinator module as a master controller and by adding a new switching and masking element between each two packetbuses.

The notion of reconfiguration can even be extended to the global system, by allowing the change of each Base Unit function into relay or monitor ones.

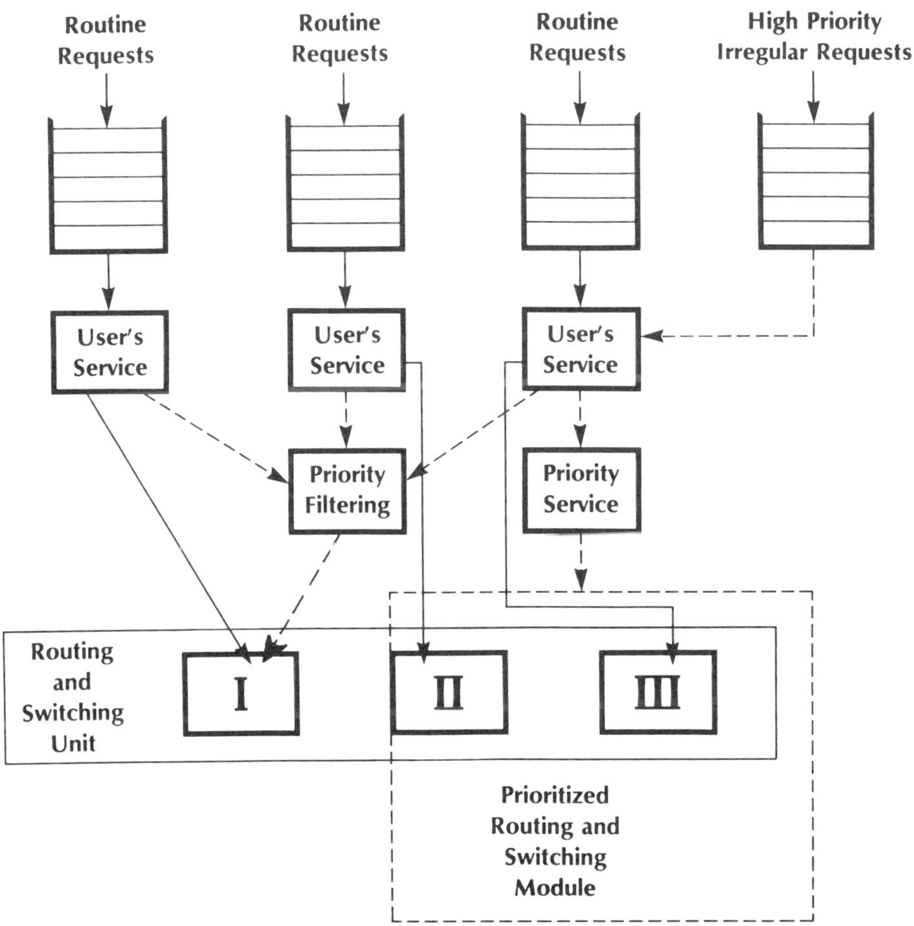

Figure 12. Prioritized reconfiguration of system wide routing

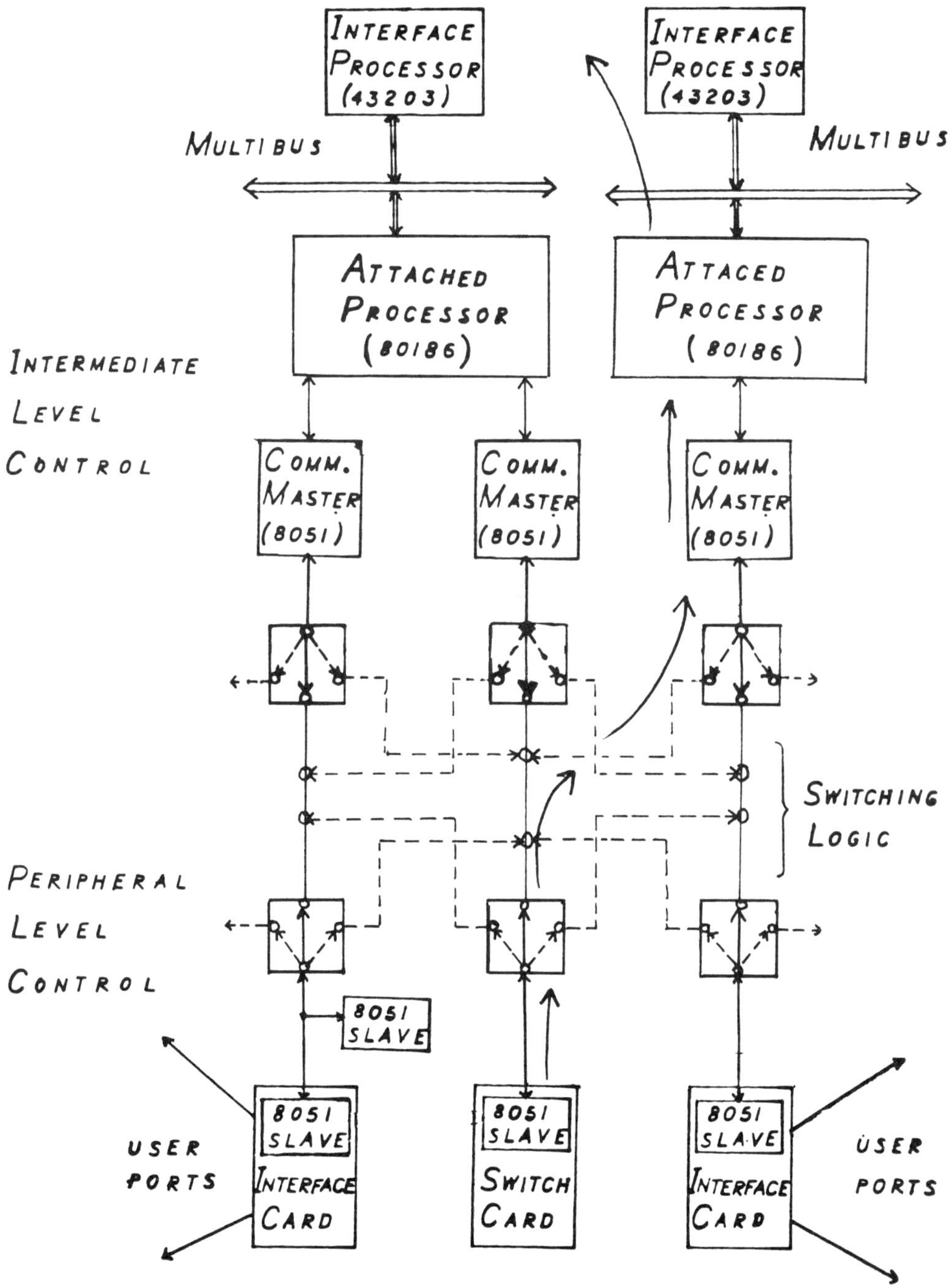

Figure 13. Control layer´s bypass mechanism

5.2 The Terminal Equipment - a User Friendly Approach

The system is attached to various types of work stations, supporting its functionality in voice, data and message processing. They address the man-machine interface problem, by utilizing the concepts of visual prompting.

Selection of the next step is done by depressing the appropriate "soft key" beneath the display. Each key performs the function that currently appears on the display. This function selection results in an easy walk-through the steps of complex functions. The function-space contains hundreds of different functions, but every step is limited to an average of three to four alternatives.

The display unit in the work stations serves various purposes:
- Displaying prompting messages
- Interactive connection to the internal data-base.
- Interogating internal and external message sources.

The basic display station contains six fields of display, organized in three rows. When the station is in the prompting mode, the upper display row indicates the current state, and the four remaining fields are displaying the next step's alternatives. The display serves other functions when interrogating the internal data-base, such as searching through an electronic telephone directory. Fig. 14 indicates the initial prompting display, and the fast selection of any character.

When the character is selected, the user can scroll the names forward and backward, and then call the selected subscriber by pressing the button beneath his displayed name.

The intelligent work station is based on a full screen CRT and is capable of interrogating data-bases. The prompting mechanism of the basic station is emulated in its original form, in order to provide a unified operation. Man-machine interface is based on a touch screen concept (i.e. all alternatives are displayed and the user selects by touching the appropriate displayed name). This method, as well as the "soft key" one, overcomes the psychological barrier of operating a sophisticated work station by using a conventional keyboard.

6. SOFTWARE STRUCTURE OF THE DISTRIBUTED COMMUNICATION SYSTEM

The main drawback of a distributed system is the incompleteness of signal observability, due to the unpredictable delay among all Base Units. It can end up in having non-complete, non real-time and noncoherent view of the system, by the different Base Units.

There are two ways to overcome this problem:
- One way is to reduce this uncertainty, without increasing the complexity of other subjects. The method proposed in this research, the new control communication relationship, is a promising one.
- Another way to overcome the signal observability constraint, is to have a balanced software strategy - correctness versus effectiveness. One main issue in this strategy is the method for task allocation.

6.1. Task Allocation Strategy

An analogy between packet routing in a computer network and task allocation in a distributed system can be drawn. Both are based on local knowledge, with adaptation to the system-wide traffic. Each mechanism can be built in a centralized or distributed manner. One can combine the centralized and distributed decisions, based on the Delta routing mechanism [24].

The method of adapting the allocation strategy to the global traffic profile can totally be based upon local parameters, such as the threshold method (above the average). This method is similar to the "hot potato" routing protocol [24], which transfers the packet to the link that owns the shortest queue.

Another way for adaptation to global traffic, is to use a special field in each message. It transfers the status of the source and the intermediate nodes to the destination.

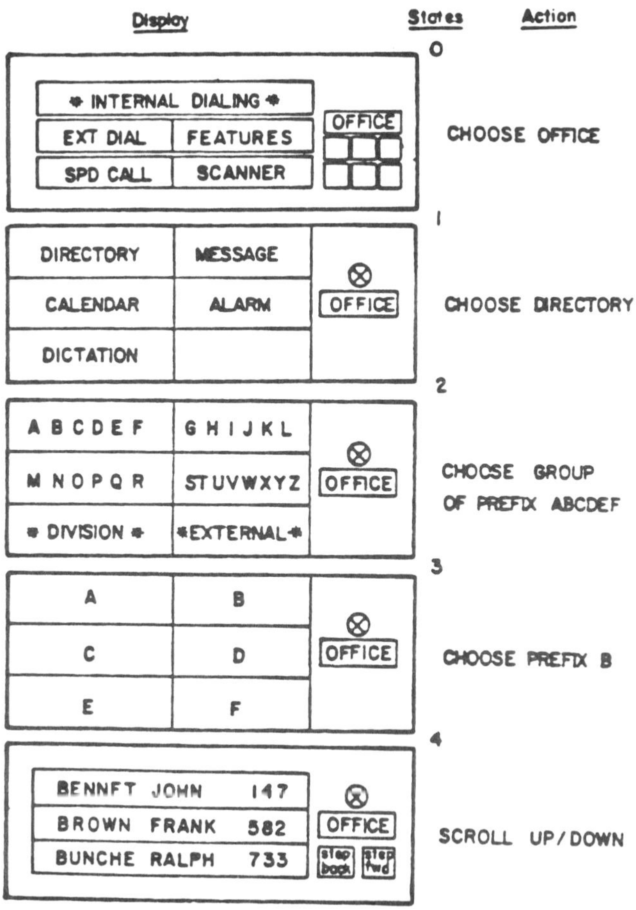

Figure 14. Search algorithm by the electronic telephone directory

The recommended method is to allocate the message to the Base Unit, rather than directly to the specific processor. The communication-coordinator at the desination BU should verify which is the best processor to be used.

As a guiding method we distinguish three allocation classes:
(1) Non time-critical tasks (that can be regenerated). They will be routed immediately to the destination.
(2) Irreversible (damaged) tasks. This implies resource allocation at the destination prior to sending the task itself.
(3) Time-critical tasks. They are executed at the same Base Unit where they were generated, on behalf of queued tasks.

A prerequisite for an efficient execution of this allocation method, is good knowledge of the global system's behavior. Since this is not possible due to constraints of signal observability, a 'synthetic' view of the system should be presented. This view is based on a fast categorization of newly arrived tasks. This method, which groups the tasks into predefined boxes, diminishes the uncertainty gap and gains a relatively good hit ratio for the correct global state presentation. It is based on the fact that the system is not sensitive to a single mistake, i.e. if a task is non-optimally allocated, it will not influence overall performance. The fast 'synthetic' categorization serves also the statistics programs. There can be a prediction of behavior and even decision concerning system reconfiguration, due to a sudden change in the profile of the tasks.

7. CONCLUSION AND FUTURE DIRECTIONS

Distributed systems are hard to manage and control, due to constraints in signal observability. A method that overcomes some of the drawbacks has been proposed. Both the new relationship between the control and the communication-coordinator, and the software method that supports it, are crucial in controlling the distributed communication system. We choose to present a realistic model, rather than an abstract one, in order to reach a concrete solution, based on the specific profile of tasks. It seems that the proposed method is applicable for a wide range of real-time systems.

Another software method that has a potential to shrink the uncertainty gap is based on eliminating the time sharing mechanism. This is based on the fact that the probability of having a free processor in the distributed system increases (as compared to uniprocessor), while eliminating the complexity and overhead of time sharing.

A future research direction is in the area of global O.S., with practical considerations toward Network O.S. . Another direction lies in the potential applications of distributed systems, such as distributed data based and real-time process control.

References

1. D.D. Clark, K.T. Pogran, D.P. Reed, 'An Introduction to Local Computer Networks', Proc. IEEE, Vol.**66**, No. 11, pp. 1497-1517, November 1978.
2. P.H. Enslow, Jr., 'What is a "Distributed" Data Processing System', Computer, Vol. **11**, No. 1, pp. 13-21, January 1978.
3. J. Etkin, M. Kuzman, B. Levy, A. Galil, 'Design Principles and Implementation of Business Communication System (TBS-64)', 11th IEEE Conference in Israel, pp. E3. 6/1-5, 1979.
4. J. Etkin, D. Tabak, 'Distributed Computer-Based Communication Systems', 13th IEEE Conference in Israel, pp. 4.3.4/1-5, 1983.
5. J. Etkin, D. Tabak, 'Distributed Communication Architecture: the Key Element of a Large Switching System', MELECON 83, Athens, Greece, pp. A5.07.1-2, 1983.
6. J. Etkin, D. Tabak, 'Method for Implementing Distributed Control of a Communication System', Microprocessing and Microprogramming, Vol. **14**, No. 3,4, pp.181-186, October 1984.
7. J. Etkin, D. Tabak, 'Communication Protocols of Distributed Systems: How Far Should We Follow the Standardization Trend?' Computer Networking Symposium, NBS, Gaithersburg, pp.113-121, 1984.
8. J. Etkin, D. Tabak, 'The Dynamic Aspects of a Distributed Communication System: Reconfiguration and User Friendly Approach', EUROMICRO, Brussels, Belgium, pp.1-10, Sept. 3-6, 1985.
9. J. Etkin - Ph.D thesis: 'Computerized Communication System: Distributed Architecture and Reconfiguration', Ben Gurion University, Beersheva, Israel, Aug. 1985.
10. W.R. Franta, E.D. Jensen, R.Y. Kain, G.D. Marshall, 'Real-Time Distributed Computer Systems', Advances on Computers, Vol. **20**, pp. 39-82, 1981.
11. W.R. Franta, I. Chlamtac, Local Networks: Motivation, Technology and Performance, Lexington Books, Lexington, MA, 1981.
12. W.N. Hsieh, I.G. Gitman, 'Routing Strategies in Computer Networks', Computer, Vol. **17**, No. 6, pp. 46-56, June 1984.
13. K. Hwang, F.A. Briggs, Computer Architecture and Parallel Processing, McGraw-Hill, N.Y., 1984.
14. S.P. Kartashev, S.I. Kartashev, LSI Modular Computer Systems, Vol. **1** : Designing and Programming Modern Computers and Systems, Prentice-Hall, Englewood Cliffs, N.J., 1982.
15. J.M. Kassan, 'The Rolm Computerized Branch Exchange, an Advanced Digital PBX', Computer, Vol. **12**, No. 6, pp. 24-31, June 1979.
16. L. Kleinrock, Queueing Systems, Wiley, N.Y., 1976.
17. B.W. Lampson et al, Distributed Systems-Architecture and Implementation, Springer-Verlag, Berlin, 1983.
18. D.A. Lawson, 'A New Software Architecture for Switching Systems', IEEE Trans. on Communication, Vol. **COM 30**, No. 6, pp. 1281-1289, June 1982.
19. B.H. Leibowitz , J.H. Carson, Multiple Processor Systems for Real Time Applications, Prentice-Hall, Englewood Cliffs, N.J., 1985.

20. J.M. McQuillen, 'The Evolution of Message Processing Techniques in the ARPA Network,' From Tutorial: A Practical View of Computer Communication Protocols, pp. 111-149, IEEE Computer Society Press, 1978.
21. G.L. Rainey, B.E. Voss, 'SL-1: A Business Communication System with Digital Switching and Stored Program Control', Proceeding International Switching Symposium, October 1976.
22. Report: Integrated Voice/Data PBXs, (pp.95-101: Mitel-SX2000, pp.143-47: Lessons not to be Forgotten: Data Point ISX), Architecture Technology Corporation, 1983.
23. H.J. Siegel, R.J. McMiller, R.T. Mueller, Jr., 'A Survey of Interconnection Methods for Reconfigurable Parallel Processing Systems', AFIPS Conference Proceedings, Vol. **48**, pp. 529-542, NCC 1979.
24. A.S. Tanenbaum, Computer Networks, Prentice-Hall, Englewood Cliffs, N.J., 1981.
25. M. Van Rumste, 'The iAPX432 - A Next Generation Microprocessor', Microprocessing and Microprogramming, Vol.**11**, pp. 69-106, 1983.
26. C.R. Vick, 'A Next Generation of Supercomputers: From Mainframes to Micros', EUROMICRO 80 Conference, pp.3-8, London, 1980.

CHAPTER 18

MICROPROCESSOR-BASED DATA REDUCTION AND COMPRESSION SYSTEMS

Spyros Tzafestas and George Papakonstantinou
Division of Computer Science
Department of Electrical Engineering
National Technical University
Zografou, Athens, Greece

Data compression and reduction is used whenever a large amount of data values need to be stored and processed many times before transmission [1-18]. The advantages that are offered by data compression/reduction include: (i) storage saving (since less coefficients/points of smaller word length need to be stored for representing the waveform, (ii) computing time saving (since data reduction can take place simultaneously with data acquisition and processing requirements are reduced), (iii)data processing in real time (using the reduced/compressed data).

One of the main approaches for data compression is using discrete orthogonal transforms such as Fourier, Walsh-Hadamard, Haar, etc. The data compression (or bit rate reduction) ability is due to the fact that the variances of the orthogonal transform coefficients are not the same for all coefficients. General theoretical aspects of data compression are given in [10-13], some applications of data compression to electrocardiographic signals (ECG) are described in [14,15,21-23], and some data reduction techniques are given in [16-20]. Among the data reduction techniques of ECG's the amplitude-zone-time epoch-coding (AZTEC) [16] is the most popular in commercial applications. It can provide a reduction of up to 20:1 and is very fast and thus appropriate for real-time applications. However, it introduces significant discontinuities and distortion of the reconstructed signal [18], and hence it is mainly used for rhythm analysis systems where the quality of the reproduced signal is not so important. For better signal reproducibility other methods need to be used requiring more complex algorithms and processing time. Piecewise linear approximation of digitized waveforms [2-9] is such a useful approach for data reduction.

The purpose of the present chapter is to give an introduction to the current methods of data reduction and data compression using piecewise linear approximation and discrete orthogonal transforms, respectively, and to study their implementation using microprocessor-based systems.

Here by data reduction we mean the process of reducing the number of the sample points in piecewise linear approximations of continuous waveforms, while the term data compression has, its standard meaning as the process of reducing the number of terms in finite orthogonal expansions of signals.

N. K. Sinha (ed.), Microprocessor-Based Control Systems, 369–399.

Specifically in Part I after an introduction to the piecewise linear approximation concept, the following algorithms are outlined:

- Rectified minimum perimeter polygon approximation
- Fast piecewise linear approximation
- Peak preserving piecewise linear approximation
- Optimal piecewise linear approximation

Then their microprocessor-based implementation is discussed and some systems actually built and tested by the authors and their co-workers are presented.

In Part II we study the data compression problems using discrete orthogonal transforms and give a review of their performance shown in many practical data compression applications. Then we concentrate on the fast Walsh transform (FWT) method and present a 2-μP FWT data compression system actually implemented and tested by our group.

I DATA REDUCTION VIA PIECEWISE LINEAR APPROXIMATION ALGORITHMS

1. Introduction

The problem of approximating functions or sampled data by piecewise linear functions, is called *"piecewise linear approximation"* (PLA) problem. When we approximate closed boundaries of objects in images by piecewise linear functions then we speak about the polygonal approximation problem. We also refer to the former case as the 1-space problem and to the later case as the 2-space problem.

Many algorithms have been proposed for the above problems [1],[2]. They can be classified according to specific criteria the most important of which are [2]:

1. *The object of minimization.* Some algorithms are trying to minimize an error measure, others the number of line segments or the total length keeping also the error below a certain threshold.
2. *The connectivity of the segments.* The ends of successive line segments may or may not coincide. In this last case there is discontinuity at the ends of the approximating line segments.
3. *The error measure.* Some algorithms are using the area between the approximated curve and the approximating line segments. Other algorithms are using the pointwise error i.e. the distance along the y-axis, between the original curve and the approximating line segments. This measure is suitable for waveforms (1-space problems). A third category of algorithms is using the Euclidean distance error i.e. the distance of an original point from the corresponding approximating line segment.
4. *The method used.* Some of these methods are reported as the least-square, the cone intersection, the parallel lines, the non-linear programming, the minimax etc.
5. *The number of passes through the sample points.* Some algorithms require many passes while others which are called "scan-along" can process the sample points sequentially and produce the line segments in a single pass. Hence space and time saving is obtained with possibly suboptimal results.
6. *The computational cost of the algorithms.* Obviously scan-along

algorithms will have smaller computational cost as compared to multi-pass algorithms.

When considering microprocessor-based systems which are limited in terms of speed and memory, scan-along algorithms are more suitable. Such algorithms are those of references [3]-[7]. We shall next describe some such scan-along algorithms for the PLA problem, suitable for microprocessor based systems as well as for real time systems.

2. The RMPP algorithm

The RMPP algorithm (Rectified Minimum Perimeter Polygon) was proposed by Sklansky and Gonzalez [3],[4]. It exhibits many interesting characteristics, such as:

- It is scan-along algorithm.
- It works both in 1-space or 2-space problems.
- It can accept digitized curves whose points are non-uniformally spaced.
- It gives as vertices of the resulting curve a subset of the sample data.
- It requires a fixed and very small amount of working space etc.

A brief informal description of the RMPP algorithm is as follows:

A waveform is defined as a sequence of the coordinates of its sample points $(x_1,y_1),(x_2,y_2),\ldots,(x_n,y_n)$, such that $x_i,y_i\geq 0$ for $1\leq i\leq n$.

The basic idea of the algorithm is shown in Fig. 1.

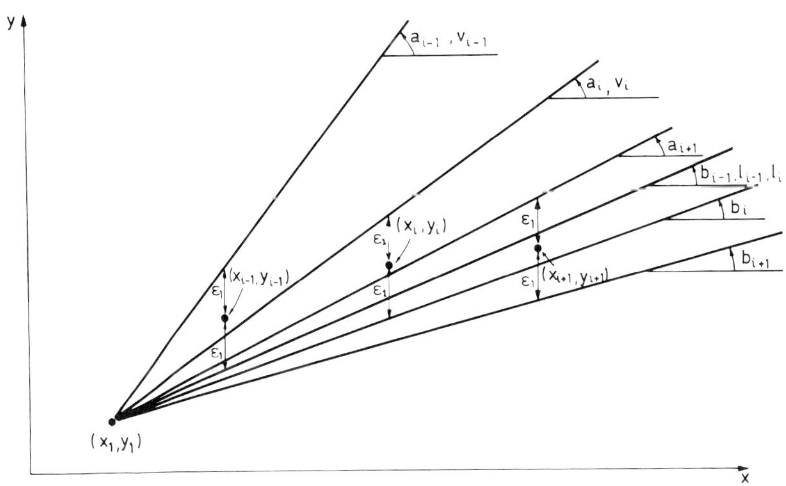

Fig. 1 An example of a piecewise linear approximation.

Starting with the point (x_1,y_1) of the waveform two angles v_1 and

l_1 with initial values $\frac{\pi}{2}$ and $-\frac{\pi}{2}$ respectively,are defined. The algorithm proceeds stepwise. For each sample point (x_i,y_i) two lines are drawn defined by the points $(x_1,y_1),(x_i,y_i+\varepsilon)$ and $(x_1,y_1),(x_i,y_i-\varepsilon)$ respectively, where ε is a small predetermined value. These lines form with the x-axis the angles a_i,b_i as shown in Fig. 1. For each point (x_i,y_i) we define the angles v_i and l_i so that

$$\begin{aligned} v_i &:= \text{minimum of } v_{i-1},a_i \\ l_i &:= \text{maximum of } l_{i-1},b_i \end{aligned} \qquad (1)$$

If $v_i > l_i$ and the point (x_i,y_i) is inside its current cone, i.e. the area defined by the lines $[(x_1,y_1), \tan v_i]$ and $[(x_1,y_1),\tan l_i]$, then the point is a valid one. The point (x_i,y_i) in Fig. 1 is a valid one, while the point x_{i+1},y_{i+1} is not since it is not inside its current cone. If a point (x_j,y_j) is not a valid one, then the line between the points (x_1,y_1) and (x_{j-1},y_{j-1}) is the piecewise linear approximation between these two points, (x_{j-1},y_{j-1}) become the starting point of the next line segment with $v_{j-1}:=\frac{\pi}{2}$ and $l_{j-1}:=\frac{\pi}{2}$. The algorithm guarantees that all points (x_i,y_i), $1\leq i\leq n$, of the waveform will have a Δy difference (pointwise error) from the line segment approximating the waveform such that $\Delta y\leq\varepsilon$.

In the above example the pointwise error was utilized. In 2-space problems the Euclidean distance error is more suitable. The above algorithm can also be used in this case, modifying only the angles a_i,b_i as it is indicated in Fig. 2.

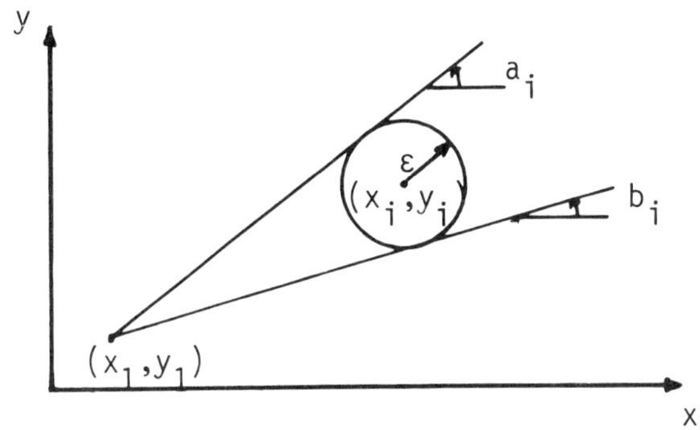

Fig. 2 Modification of a_i,b_i for 2-space problems.

The algorithm described above, requires at each step to compute the quantities a_i,b_i,v_i,ℓ_i, and determine if the current sample point

(x_i,y_i) is inside its current cone. Hence floating point operations are necessary for the computation of slopes, which is costly in micro-processor-based systems.

3. The FPLA algorithm

The FPLA (Fast Piecewise Linear Approximation) algorithm [5] is also a scan-along algorithm which avoids all floating point type operations as well as the integer type division operations, which are time consuming. It only requires integer type multiplications additions substractions and comparisons. The FPLA algorithm is basically a variant of the RMPP algorithm described previously. Instead of using the cone and comparing angles, we use line segments defined on a line perpendicular to the x-axis through the current sample point, and compare their lengths.

Let:

A_i be the point $(i,y_i+\varepsilon)$

B_i " " " $(i,y_i-\varepsilon)$

V_i,L_i be the points defined by the intersection of the current cone and the line C_i (Fig. 3)

P_i be the intersection of the lines OV_{i-1} and C_i

Q_i " " " " " " OL_{i-1} and C_i

y_D " " y-coordinate of a point D, assuming that the x-y axis have been moved to the starting point 0.

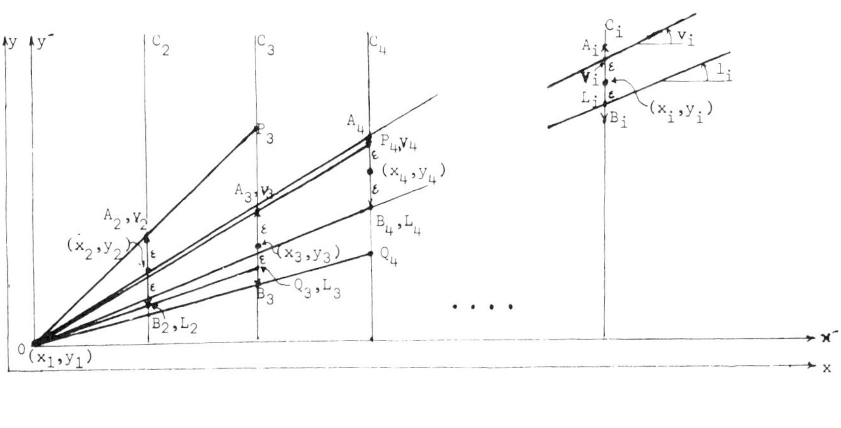

Fig. 3 Illustration of the proposed FPLA algorithm.

From Fig. 3 we can easily obtain:

$$y_{P_3} = \frac{2}{1}\, y_{V_2}\,, \qquad y_{Q_3} = \frac{2}{1}\, y_{L_2}$$

$$y_{V_3} = \mathrm{MIN}(y_{P_3}, y_{A_3}) = y_{A_3}\ ,\quad y_{L_3} = \mathrm{MAX}(y_{Q_3}, y_{B_3}) = y_{Q_3} = \frac{2}{1}\, y_{L_2}$$

$$y_{P_4} = \frac{3}{2}\, y_{V_3}\ ,\quad y_{Q_4} = \frac{3}{2}\, y_{L_3} = \frac{3}{2}\cdot\frac{2}{1}\, y_{L_2} = \frac{3}{1}\, y_{L_2}$$

$$y_{V_4} = \mathrm{MIN}(y_{P_4}, y_{A_4}) = y_{P_4} = \frac{3}{2}\, y_{V_3}$$

$$y_{L_4} = \mathrm{MAX}(y_{Q_4}, y_{B_4}) = y_{B_4} = y_4 - \varepsilon$$

$$\vdots$$

$$y_{P_i} = \frac{\lambda_i}{\mu_i}\cdot y_{V_j}\ ,\quad y_{Q_i} = \frac{\rho_i}{\sigma_i}\cdot y_{L_k}\ ,\quad y_{V_i} = \mathrm{MIN}(y_{P_i}, y_{A_i})$$

$$y_{L_i} = \mathrm{MAX}(y_{Q_i}, y_{B_i}) \text{ with } j,k < i \text{ and } \lambda_i, \mu_i, \rho_i, \sigma_i \text{ integers}$$

which can be calculated from $\lambda_{i-1}, \mu_{i-1}, \rho_{i-1}, \sigma_{i-1}$ as it will be shown in the description of the following FPLA algorithm.

We can now observe that instead of comparing y_{P_i} with y_{A_i} we could compare $\lambda_i\, y_{V_j}$ with $\mu_i y_{A_i}$. In the same way instead of comparing y_{Q_i} with y_{B_i} we could compare $\rho_i y_{L_k}$ with $\sigma_i y_{B_i}$. Hence divisions can be avoided.

The FPLA algorithm will be next described in Algol-like notation in its off-line version where:

i is the index of the current sample point
s is the index of the start point of a line segment
ys is the y coordinate of the start point of a line segment
$z_1, z_2, z_3, w_1, w_2, w_3, w_4$ are auxiliary variables.

Algorithm

```
Initialization: i: =2; s: =1, ys: =y1; yV: =yi-ys+ε;
                yL: =yi-ys-ε; λ: =2; μ: =1; ρ: =2; σ: =1;
                output (1,y1);
     step1:     i:=i+1; if i is the last point then
                        begin output (i,yi); stop; end;
```

```
z1:=yV*λ;z2:=yi-ys+ε;z3:=z2*μ;
if z1>z3 then
   begin yV:=z2;λ:=i-s;μ:=i-s; end;
z1:=yL*ρ;z2=yi-ys-ε;z3:=z2*σ;
if z1 <z3 then
   begin yL:=z2; ρ: i-s; σ:=i-s; end;
w1:=yV*λ;w2: =(yi-ys)*μ;
w3:=yL*ρ;w4:=(yi-ys)*σ;
if w1 <w2 or w3 >w4 then
begin s:=i-1; output (s,ys);ys:=yi-1;
      yV:=yi-ys+ε;yL:=yi-ys-ε;
      λ:=2;μ:=1;ρ:=2;σ:=1;
      goto step 1;
end
             else
begin λ:=λ+1;ρ:=ρ+1; goto step1; end;
```

We can see that if the coordinates y_i are integers then integer arithmetic is sufficient. Moreover at each step 8 multiplications are required plus additional additions, substructions and comparisons. The required amount of memory is a few places (including y_i, y_{i-1}), used as working space, and independent of the size of the data. Hence the above algorithm is suitable for real time applications and microprocessor-based applications.

4. The PPLA algorithm

The PPLA (Peak Preserving Piecewise Linear Approximation) algorithm [6] is also a scan-along algorithm suitable for applications where the peak points must be accurately preserved and incorporated as corners in the piecewise linear approximation of the curve. This is important in applications where the peak points are carrying important information, e.g. the QRS waves in ECGs.

The PPLA algorithm is a combination of two algorithms. The first may be either the RMPP or the FPLA algorithm and the second is a peak detection algorithm, described informally below.

An upgoing portion of the waveform is defined as a portion with positive slopes. A downgoing portion is defined as a portion with negative slopes. A noisy portion after an upgoing portion is defined as a portion with y_i coordinates such that $y_i - \varepsilon_2 \le y_i \le y_j$, where y_j is the y coordinate of the end point of the previous upgoing portion and ε_2 is a predetermined value approximately equal to the peak noise amplitude. A noisy portion after a downgoing portion is defined as a portion with y_i coordinates such that $y_i+\varepsilon_2 \geq y_i \geq y_j$, where y_j is the y coordinate of the end point of the previous downgoing portion.

The algorithm finds the different kinds of portions of the waveform and marks the following points as peaks:

(a) the end points of those upgoing portions which are followed by downgoing portions, ignoring possible noisy portions, and

(b) the end points of those downgoing portions which are followed by upgoing portions, ignoring possible noisy portions.

We shall now describe the proposed combination i.e. the PPLA algorithm, using the RMPP algorithm and the previous peak detection algorithm. The model of a finite state automaton will be utilized for the description. The constant ε_1 stands for the constant ε in the description of the RMPP algorithm.

The algorithm works as follows: for each sample point a PLA, started at a peak point is continued until another peak point is found or until a possible peak point is found. In the latter case another PLA starts which is ignored if the possible peak point was not really a peak point but which becomes the continuation of the previous PLA if the possible peak point was really a peak point.

The increment at the sample point (x_i,y_i) is defined as $\Delta y_i = y_i-y_{i-1}$ for $1<i\leq n$. We classify the increments into five categories, namely:

small positive (sp) for $0<\Delta y_i \leq \varepsilon_2$,

large positive (lp) for $\varepsilon_2<\Delta y_i$,

small negative (sn) for $-\varepsilon_2 \leq \Delta y_i<0$,

large negative (ln) for $\Delta y_i<-\varepsilon_2$,

zero (z) for $\Delta y_i = 0$.

The states of the automaton are:

1. for an upgoing portion,
2. for a noisy portion after an upgoing one,
3. for a downgoing portion,
4. for a noisy portion after a downgoing one.

The transition diagram of this automaton is given in Fig. 4. The

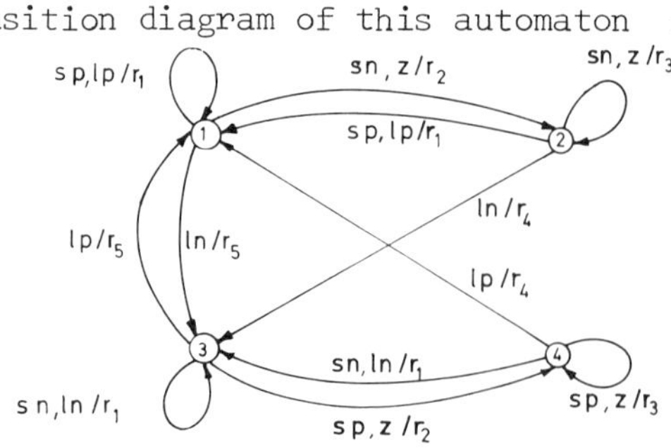

Fig. 4. Transition diagram of the finite state automaton.

input to the automaton consists of the increments sp, lp, sn, ln, z, and the output consists of the semantic routines r_1, r_2, r_3, r_4, r_5, which are described later.

Let k be the index for the quantities of the kth line segment, with $k \geq 0$. The zeroth line segment is an auxiliary one, not belonging to the resulting piecewise curve. Let (X_k, Y_k) be the starting point of the kth line segment, and let $v_i^k, l_i^k, a_i^k, b_i^k$ be the angles as defined in Fig. 1 for the sample point (x_i, y_i) and with (X_k, Y_k) the starting point of the line segment.

The following procedures described in an Algol-like notation will be used in the semantic routines r_1, r_2, r_3, r_4 and r_5:

(a) Initialize (x_i, y_i, k): for the initialization of the kth line segment.

$X_k := x_i;\quad Y_k := y_i;\quad v_i^k := \frac{1}{2}\pi,\quad l_i^k := -\frac{1}{2}\pi;$

(b) Continue (x_i, y_i, k): for the continuation of the PLA.

$v_i^k :=$ minimum of $v_{i-1}^k, a_i^k;$

$l_i^k :=$ maximum of $l_{i-1}^k, b_i^k;$

<u>if</u> (x_i, y_i) is not a valid point of line segment k

<u>then</u> <u>begin</u> $k:=k+1;\quad X_k:=x_{i-1};\quad Y_k:=y_{i-1};$

$v_i^k:=a_i^k;\quad l_i^k:=b_i^k;$ <u>end</u>

(c) Peakreg(k): for the inclusion of a peak point as a corner of the resulting piecewise line curve.

<u>if</u> $X_k < X_0$ <u>then</u> $k:=k+1;$

$X_k:=X_0;\quad Y_k:=Y_0;\quad v_{i-1}^k:=v_{i-1}^0;$

$l_{i-1}^k:=l_{i-1}^0.$

The initialization of the automaton is done by setting the initial state to 1, the index k to 1, the index i to 2, the X_1, Y_1 to x_1, y_1, respectively, and by calling initialize (x_1, y_1, k).

The semantic routines r_1, r_2, r_3, r_4 and r_5 can now easily be described as follows:

r_1: continue (x_i, y_i, k); $i:=i+1;$

(for the continuation the kth line segment).

r_2: continue (x_i, y_i, k); initialize $(x_{i-1}, y_{i-1}, 0)$;

continue $(x_i, y_i, 0)$; $i:=i+1;$

(for the continuation of the kth line segment and the start of an auxiliary line segment with 0 index).

r_3: continue (x_i,y_i,k); continue $(x_i,y_i,0)$; $y_i:=y_{i-1}$; $i:=i+1$;
(for the continuation of both the kth and the 0th line segments if (X_0,Y_0) is a possible peak point).

r_4: peakreg(k); continue (x_i,y_i,k); $i:=i+1$,
(for the inclusion of the peak point (X_0,Y_0) as a corner of the resulting piecewise line curve and the continuation of the next line segment).

r_5:k: =k+1; initialize (x_{i-1},y_{i-1},k);
continue (x_i,y_i,k); $i=i+1$;
(for the inclusion of the peak point (X_{i-1},Y_{i-1}) as a corner of the resulting piecewise line curve and the continuation of the next line segment).

We have assumed that $\varepsilon_1 \geq \varepsilon_2$, hence the auxiliary PLA which starts at a possible peak point has only one segment, and the line index k may be increased only by 1, while the automaton is in the state 2 or 4.

The quantity ε_2 must be small enough so that the peaks will not be distorted and at the same time it must be large enough so that small peaks due to noise will be eliminated.

The proposed algorithm guarantees that each point of the waveform will have a Δy difference from the resulting piecewise linear curve (pointwise error) smaller or equal to ε_1. Moreover the peak points of the waveform will be preserved in the resulting piecewise line curve.

5. The OPLA algorithm

Scan-along algorithm may lead to suboptimal results. The OPLA (Optimal Piecewise Linear Approximation) algorithm [9] is not a scan-along algorithm but it gives an approximation with the minimum possible number of line segments, keeping the error measure below a certain threshold ε. It can be used in 2-space or in 1-space problems, using as error measure the Euclidean distance or the pointwise error.

A dynamic programming approach is utilized in order to find the optimal solution. The time cost of the algorithm is inversely proportional to the reduction of the digitized curve. More specifically, the time required for finding the optimal solution is approximately k times the time required for finding a non-optimal solution (e.g. by the previous algorithms), where k is the reduction ratio i.e.:

k - number of sample points/number of the resulting segments

This makes the OPLA algorithm practically 10 times slower while the improvement in the number of line segments is of the order of 15-40%. Hence, the use of the OPLA algorithm is justified in cases where the data reduction will be done once but the processing of the reduced data repeatedly.

We shall first describe the OPLA algorithm for 2-space problems. The method of cone as defined in section I.2 is not sufficient in order to obtain optimal solutions, as it will be shown later with a counter example. Hence we must generalize it as described in the following definitions.

Let us introduce the following symbols:

ε: an error bound expressing the maximum allowable Euclidean distance of the sample points from their approximating line segments.

C_i: the "reference circle" i.e. the circle with center (x_i,y_i) and a fixed radius which can be chosen arbitrarily.

$r_i(k)$: the arcs of the reference circle defined as in Fig. 5 for a "source point" (x_i,y_i) and a sample point (x_k,y_k) $i<k$.

$R_i(j)$: the common part of all $r_i(k)$, $i+1\leq k\leq j$. If there is not common part then $R_i(j)=\underline{nil}$

valid (i,p,j): a predicate which takes the value true if:

a) $R_i(j)\neq\underline{nil}$ and b) the line $[(x_i,y_i),(x_p,y_p)]$, $p\geq i+1$ intersects $R_i(j)$.

T_i: the set $\{j, j\in(i+1,\sigma],$ such that valid $(i,j,j)=\underline{true}\}$ (2)

where σ is an index such that $R_i(\sigma)\neq\underline{nil}$ and $R_i(\sigma+1)=\underline{nil}$

$d(i,j,\varepsilon)$: the "distance" of the sample point i from the sample point j, $i\leq j$ i.e the minimum number of line segments approximating the sample point $i,i+1,\ldots,j$ within ε. The distance $d(i,i,\varepsilon)$ is defined as zero.

We shall use throughout this paper the word "longer" with the following meaning. A line segment $[(x_i,y_i),(x_p,y_p)]$ $i<p$, is longer than the line segment $[(x_i,y_i),(x_q,y_q)]$ $i<q$, if $p>q$.

Lemma 1: If the Euclidean distance of a sample point (x_k,y_k) from the line $[(x_i,y_i),(x_\tau,y_\tau)]$, with $i<k<\tau$, is within ε then the above line intersects $r_i(k)$ and vice versa.

Proof: The proof is obvious according to Fig. 5.

Theorem 1. The index $\tau=MAX(T_i)$ defines the longest line segment $[(x_i,y_i),(x_\tau,y_\tau)]$ approximating the sample points j, $i\leq j\leq\tau$, such that the Euclidean distance error for these points is within ε.

Proof: Let us assume that an index $\tau'\in T_i$ exists such that the Euclidean distance of each sample point j, $i\leq j\leq\tau'$ is within ε. Then according to Lemma 1 the line $[(x_i,y_i),(x_{\tau'},x_{\tau'})]$ intersects $r_i(j)$ for $i\leq j\leq\tau'$ hence $R_i(q)\neq\underline{nil}$ for $i<q\leq\tau'$ and valid $(i,\tau',\tau')=\underline{true}$. Since valid

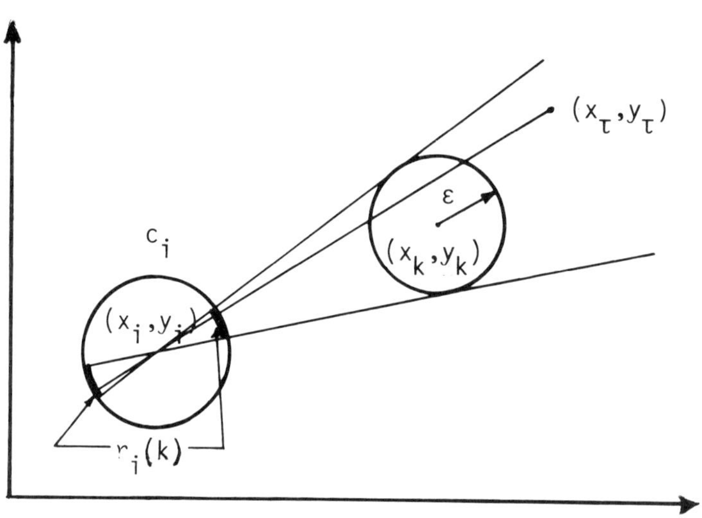

Fig. 5 Definition of $r_i(k)$.

$(i,\tau',\tau')=$true, $\tau' \in T_i$ which contradicts with the assumption that $\tau' \notin T_i$ q.e.d.

The following algorithm 0 in Algol-like notation will give index τ when the source point is i.

Algorithm 0:

```
begin
    τ:=i+1; j: i+1;  calculate  R_i(j);
    while  j<n-1  do
        begin
            j:=j+1;
              calculate R_i(j) and valid (i,j,j);
              if R_i(j)=nil then return;
              if  valid (i,j,j)=true then τ:=j
        end
end
```

In Fig. 6 a counter example is given to show that if the cone method is used we shall not obtain the longest line segment. Using the cone method the longest line segment is 1-2, while using the arc method as defined previously the longest line segment is 1-4, for the same value of ε.

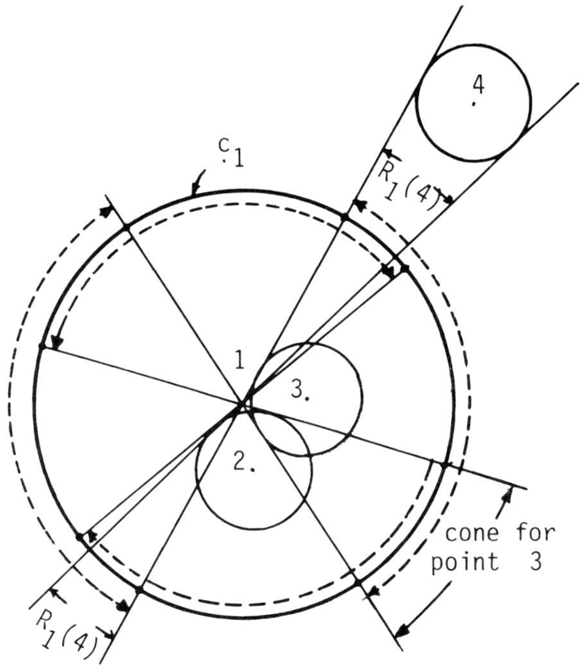

Fig. 6 A counter example.

Lemma 2: All possible line segments starting from the sample point i that approximate sample points j, $i \le j \le n$ with Euclidean distance error within ε, are:

$$[(x_i, y_i), (x_p, y_p)] \quad \text{with} \quad p \in T_i \quad \text{and} \quad T_i \quad \text{as defined in (2).}$$

Proof: It follows the proof of theorem 1.

Lemma 3: The distance $d(i,n,\varepsilon)$ of a sample point i from the last sample point n with a given ε is:

$$d(i,n,\varepsilon) = \text{MIN}(D_i)+1 \text{ , with } D_i' = \{d(j,n,\varepsilon), \text{ such that } j \in T_i\} \quad (3)$$

Proof: The first segment of the piecewise linear approximation of the sample points p, $i \le p \le n$ will end at a point j, $j \in T_i$ according to lemma 2. Obviously, $d(i,n,\varepsilon)=d(j,n,\varepsilon)+1$ for $j \in T_i$, hence $d(i,n,\varepsilon)=\text{MIN}(D_i)+1$ q.e.d.

We shall next describe informally using Algol-like notation, an algorithm for determining the distance $d(i,n,\varepsilon)$, $1 \le i \le n$ and an optimal solution, based on lemma 2 and lemma 3.

OPLA Algorithm:

```
begin
    d(n,n,ε): = 0;
    for  i:=n-1 step -1 until 1 do
       begin
          calculate T_i {T_i as defined in (2)};
          d(i,n,ε):=MIN(D_i)+1 {D_i as defined in (3)};
       end;
    k:=1; τ_k:=1; e:=d(1,n,ε);i:=1;
    while  i<n  do
    begin
       find T_i;  k:=k+1; e:=e-1; τ_k:=p; i:=p;
          { where p ∈ T_i  and  d(p,n,ε)=e }
    end
end
```

The algorithm is based on the simple idea that if we know the distances $d(j,n,\varepsilon)$ for $i \leq j \leq n$ then we can, according to lemma 2 and lemma 3 find the distance $d(\overline{i-1},n,\varepsilon)$. The initial condition is that $d(n,n,\varepsilon)=0$.

The OPLA algorithm can be modified to start from the sample point 1 and move to the sample point n, i.e. scan the sample points in the other direction. It can also be modified to give all optimal solutions.

The OPLA algorithm can also be used, utilizing the pointwise error norm. In this case the following changes in the definitions must be done:

1) The reference circle C_i and the arcs $r_i(k)$ do not hold.
2) $R_i(j)$ is the "cone" angle $v_j - l_j$ as defined in section I.2 when x_i, y_i is the starting point $\{(x_1,y_1)$ in Fig. 1)}. If $v_j - l_j < 0$ then $R_i(j)=$nil by definition.
3) valid (i,p,j) is a predicate which takes the value true if:
   a. $R_i(j) \neq$ nil and
   b. the point (x_p,y_p) is inside the cone $R_i(j)$.

6. Microcomputer-Based Implementation

The algorithms previously described were first implemented in FORTRAN in their off-line version using the CDC-3300 computer. Some of them were then transfered to a microcomputer-based data acquisition system as shown in Fig. 7 [20].

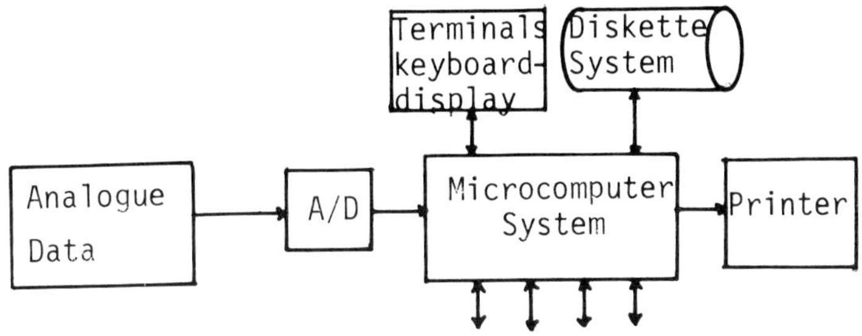

Fig. 7 Data Acquisition System.

It is based on the Data Applications International S.A. (DAI) microcomputer system [51]. This system provides the digital computing power of the 8-bit microprocessor 8080A, enhanced by fully implemented serial and parallel Input-Output capability, program and data memory, independent interval timers and a fully vectored interrupt facility. The system consists of eurocard modules which peform a variety of functions and they have plug-to-plug as well as software compatibility. Parallel connection of additional microcomputer eurocards to the system bus is possible through the I/O ports, for distributed operation. Finally a utility software package, as well as an Assembler/Editor and a FORTRAN package allow for an easier program development.

The system of Fig. 7 was used in the development of a real-time data reduction microcomputer system, using the FPLA algorithm. The analogue data were real electrocardiograms (ECGs). The ECG Hittman CDC-70 system was used to obtain the ECG input signal in analogue form. This was converted to digital form using the SDM853 Burr-Brown data acquisition module which was directly coupled to the system bus of the microcomputer. The signal must be practically sampled at a rate greater than or equal to 200 samples/sec. Using this sampling rate the time left for data reduction for each sample point was less than or equal to 5msec.

The output of the implemented system was the coordinate pairs $[(i-s),y_i]$ $s<i$, where $i-s$ is the distance in sample points of two consecutive corners of the approximating polygon and y_i the y coordinate of the corner point i. This pair was sent to the system bus for further manipulation.

The FPLA algorithm in its off-line version was first tested on the CDC-3300 mainframe computer. The above program was then transfered to our μP system, using a subset of ANSI FORTRAN, acceptable by the μP system. Finally, with some changes to this program another program was developed for the on-line acquisition and reduction of ECG-signals. For the acquisition of the signals, the last program cooperates with a sampling program named SAMPLE in which only the 8-most significant bits of the A/D converter are used.

During the development of the system the output, consisting of the coordinate pairs $[(i-s),y_i]$, was stored in the memory of the computer for a 5 sec time period. The final memory required was approximately

2k bytes, including FORTRAN program, assembly language program, system routines and scratch pad memory.

The reduced data were sent to the system bus for further manipulation, e.g. for the QRS detection.

In Fig. 8 an illustrative example is given of an ECG signal and its piecewise linear approximation obtained by the proposed algorithm for $0 \le y_i \le 255$ and $\varepsilon=4$. The reduction was about 10-to-1 points.

Fig. 8a depicts the input ECG signal and Fig. 8b shows the output reduced signal.

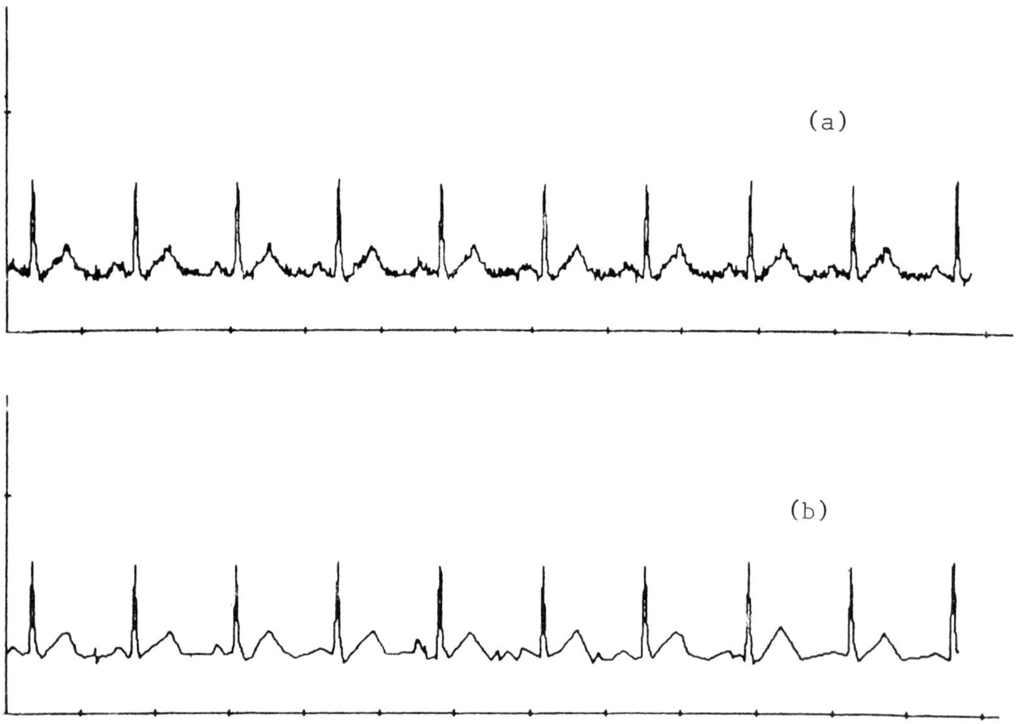

Fig. 8(a). The sampled ECG input signal.

Fig. 8(b). The processed with the FPLA algorithm, output ECG singal.

The OPLA algorithm was implemented in FORTRAN in the CDC-3300 computer using the Euclidean distance error norm. It was tested using artificial data (animal contours), taken from a CALCOMP 942 digitizer and real ECG waveforms, taken from the CSE ECG library [52]. It was compared with the algorithm RMPP and the algorithm PPLA which besides the piecewise linear approximation it represents accurately the peak points of the curve. The last algorithm utilizes the pointwise error norm. Hence OPLA was modified in these cases to utilize the pointwise error norm, in order to be able to do the comparison.

In Table 1 some experimental data are given for a typical example with the same number of input data and the same error bound.

The time required by the algorithm OPLA compared to that of the RMPP can theoretically be estimated. If the time required for the completion of one step in RMPP algorithm (i.e. the computation of v and l) is T_{RMPP}, then approximately $n \cdot T_{RMPP}$ steps are required, where n is the number of the sample points. If the time required for the completion of one step in the OPLA algorithm (i.e. computation of $r_i(p)$) is T_{OPLA} then approximately $n \cdot (T_{OPLA}(n/s))$ steps are required, where s is the number of segments of the resulting piecewise linear approximation. Hence the RMPP algorithm is approximately $n \cdot T_{OPLA}/sT_{RMPP}$ times faster. Since T_{OPLA} and T_{RMPP} are comparable then RMPP is approximately $k = n/s$ times faster.

Table 1. Experimental data

Algorithm X	$\frac{N_x - N_{OPLA}}{N_x}$	$\frac{T_{OPLA}}{T_x}$
RMPP	0.35	18
PPLA*	0.40	14

*The results depend upon the noise threshold ε_2 utilized in PPLA. N stands for the number of line segments and T for the CPU time. K_{OPLA} was approximately 16.

In order to (visually) have a good approximation, about 10:1 data reduction was sufficient which makes OPLA about 10 times slower.

The improvement in the number of line segments is shown in Table 1. The improvement seems to increase with k for the case of the RMPP algorithm. It also seems to improve with more noisy data.

A practical data acquisition system should be: a) capable of co-operating with any computer systems, b) flexible so that it can be adjusted to the needs of specific problems, c) capable to accurately reproduce the analogue system e.t.c. In order to achieve all these characteristics a general purpose computer system is not suitable. Instead a special interface card [53] was developed based on an 8085A microprocessor which is shown in Fig. 9.

The card communicates with any system which has an RS-232C port, via a suitable software protocol. The card can be in two modes namely the data Acquisition mode or the Monitor mode. The external system sends an appropriate ASCII character ("A" or "M") upon initialization, to bring the card in one of these modes. In the monitor mode the user can use the card as a development system and develop his own signal processing algorithms. In the data Acquisition mode the external system (using the above mentioned protocol) sends information concerning the *Baud rate*, the number of required sampling periods etc. and receives the

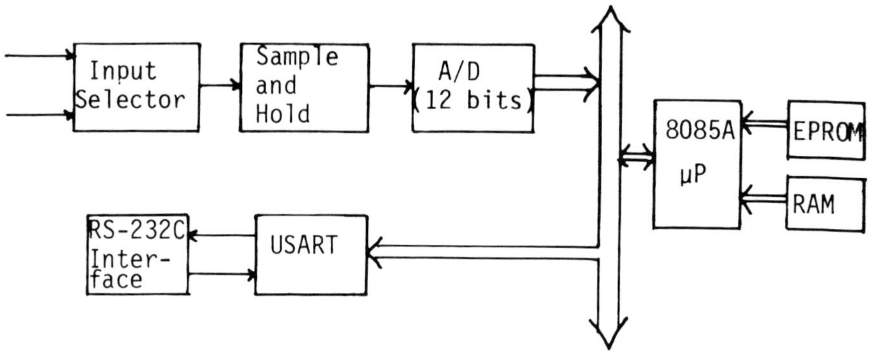

Fig. 9 Block diagram of special interface system.

sampled data as ASCII characters. When the external system sends the character S then the communication stops.

II. DATA COMPRESSION VIA ORTHOGONAL TRANSFORMS: THE FAST WALSH TRANSFORM CASE

1. Orthogonal Transforms and Data Compression

Consider a discrete data vector $X^T = [x_0, x_1, \ldots, x_{N-1}]$ and an orthogonal transform $P = [\psi_0, \psi_1, \ldots, \psi_{N-1}]^T$, where $\psi_0, \psi_1, \ldots, \psi_{N-1}$ are orthonormal real-valued vectors, i.e. vectors having the property $\psi_i^T \psi_j = \delta_{ij}$ (Kronecker delta) or $P^T P = I$ (unit N×N matrix). Applying P upon X gives the transformed vector components $y_i = \psi_i^T X$ or in matrix form

$$Y = PX, \quad Y^T = [y_0, y_1, \ldots, y_{N-1}] \tag{4}$$

Premultiplying (4) by P^T and using the orthogonality of P gives the inverse transform

$$X = P^T Y = \sum_{i\ 0}^{N-1} y_i \psi_i \qquad (P^T = P^{-1}) \tag{5}$$

Clearly, P and P^{-1} are linear inverse transforms and are used in both the communication and control fields. In both fields the basic problem is to select a family of vector functions $\psi_0, \psi_1, \ldots, \psi_{N-1}$ such that the signal X can be reconstructed or estimated using a smaller number of terms $y_i \psi_i$, $i = 0, 1, \ldots, M-1$. This implies that only $M < N$ data values y_i have to be used or transmitted for reconstructing X as

$$\hat{X} = \sum_{i=0}^{M-1} y_i\psi_i + \sum_{i=M}^{N-1} c_i\psi_i \tag{6}$$

where c_i are preselected constants.

Using fewer data points or transform coefficients y_i we have what is known as "data compression" which allows to transmit the information signal X approximately with reduced information rate, a result which is mostly important in practice.

The optimum transform (or expansion) with respect to the mean square criterion is that in which ψ_i, $i = 0,1,\ldots, N-1$ are the eigenvectors of the data covariance matrix $\Sigma_x = E\{(X-\bar{X})(X-\bar{X})^T\}$, where $\bar{X} = E[X]$ is the mean value of X. This is the well known Karhunen-Loève expansion. Indeed, the mean square error V is given by

$$V = E\{(X-\hat{X})^T(X-\hat{X})\} = \sum_{i=M}^{N-1} E\{(y_i-c_i)^2\} \tag{7}$$

and the optimality conditions, $\partial V/\partial c_i = 0$, $\partial V/\partial\psi_i = 0$ subject to $\psi_i^T\psi_i = 1$ give, respectively:

$$c_i = E[y_i] = \psi_i^T\bar{X}, \qquad \Sigma_x\psi_i = \lambda_i\psi_i \tag{8}$$

where λ_i are the Lagrange coefficients for the constraints $\psi_i^T\psi_i = 1$. This shows that λ_i is an eigenvector of Σ_x and λ_i is the corresponding eigenvalue. Now from $Y = PX$ we obtain

$$\Sigma_y = P\Sigma_x P^{-1} = P\Sigma_x P^T = \text{diag.}[\lambda_0, \lambda_1, \ldots, \lambda_{N-1}], \quad \text{i.e. } \Sigma_y$$

is diagonal which implies that the components of y_i are uncorrelated.

The minimum value V^0 is obtained by introducing (8) into (7), i.e.

$$V^0 = \sum_{i=M}^{N-1} \psi_i^T\Sigma_x\psi_i = \sum_{i=M}^{N-1} \psi_i^T\lambda_i\psi_i = \sum_{i=M}^{N-1} \lambda_i$$

This shows that V^0 is equal to the sum of the eigenvalues corresponding to the data components y_i which are discarded. Thus in practice one must keep the M data components having the largest eigenvalues. The discarded data components are replaced by the constants $c_i = \psi^T\bar{X}$ which can be reduced to zero by preprocessing the given data such that $\bar{X} = 0$.

This data compression criterion for selecting the set of y_i components with the largest variances is known as "variance data compression criterion". Note that for all other orthogonal transforms the covariance matrix Σ_y is not diagonal.

In the case of discrete-time signals we have $x_k = x(kT)$, where k is the sample index and T is the time between samples, Eq. (5) gives

$$x(kT) = \sum_{i=0}^{N-1} y_i \psi_i(kT) \tag{9}$$

where $\psi_i(kT)$ is the kth component of the vector ψ_i, and actually is the value of $\psi_i(t)$ at the kth sampling instant t=kT. Here NT is the window size. Some common types of $\psi_i(kT)$, besides the Karhunen-Loève vectors, are the *discrete Fourier, Slant,* and *Walsh-Hadamard* vectors.

2. Performance of Orthogonal Transforms in Data Compression Applications

In [29] a comparison is made of the Fourier, Walsh-Hadamard and Karhunen-Loève transforms regarding their ability to reduce the bit rate in speech signal transmission. To this end, a general expression is derived for the information rate needed for the transmission of the coefficients y_i such that the reconstructed signal x(kT) has the same signal-to-quantizing noise ratio as that of the standard PCM transmission. If N samples are required to determine the expansion coefficients (or, equivalently, if N coefficients are used to represent a set of time series samples of size N), then the quantized noise (error) due to the quantized transmission (using n_i bits) of the coefficients $y_i, i = 0,1,\ldots,N-1$ is given by

$$Q = \frac{1}{N^2} \sum_{i=0}^{N-1} C_i \sigma_i^2 / 2^{2n_i} \tag{10}$$

where σ_i^2 is the variance associated with y_i, and C_i is a constant expressing the consequence of the scaling and companding associated with the coefficient quantization. The quantization noise of the PCM transmission of the same information is $Q_{PCM} = C_{PCM}\sigma^2/2^{2M}$, where M is the total number of bits used to quantize each sample, σ^2 is the signal power, or variance in the zero bias case, and C_{PCM} is the scaling-companding constant. The quantization noise Q in (10) is minimized when all terms of the summation are equal. Hence, to ensure that $Q = Q_{PCM}$ one must select the number of bits n_i such that $C_i\sigma_i/N^2 2^{2n_i} = C_{PCM}\sigma^2/N2^{2M}$, i.e. $n_i = M + \log_2(\sigma_i^2/N\sigma) + \frac{1}{2}\log_2(NC_i/C_{PCM})$

The total bit rate for the transform (coefficient) transmission is $R_0 = (1/NT) \sum_{i=0}^{N-1} n_i$, and for the PCM transmission is $R_{PCM} = M/T$. Therefore, to reduce the bit rate by the orthogonal transform (i.e. to ensure that $R_0 < R_{PCM}$) one must have $(1/N) \sum_{i=0}^{N-1} n_i < M$.

For a window size NT the frequency resolution in the Fourier transform case is 1/NT. Thus, for N = 16 and T = 125μsec, the frequency resolution is 500Hz. This resolution is sufficient to resolve long-terms spectral distribution of speech in terms of the variance of the Fourier coefficients (components). Higher values of N give higher resolution by which short-term spectral distribution of speech can be resolved. Due to the similarity of the spectral distributions of the Fourier, Hadamard and Karhunen-Loeve transforms the above procedure of determining the sample window holds approximately for the Hadamard and Karhunen-Loeve transforms as well.

Experimental work on both man and woman speeches of duration 65sec [29], sampled every T=125 μsec (8000 samples per second) has showed that to obtain a S/Q_n ratio (Q_n denotes quantized noise) equal to that of a standard 56 kbits/sec PCM, the bit rates required in kbits/sec are as follows:

Processing Transform	Fourier	Hadamard	Karhunen-Loève
Bit rate	46	48.5	42.5
Bit rate reduction	10	7.5	13.5

As expected, the best performance is obtained using the Karhunen-Loève processing transform. The behaviour of the above transforms was also studied by comparing their S/Q_n ratios with that obtained using logarithmically companded PCM at bit rates 56,28 and 14 kbits/sec. The S/Q_n improvements are 5db for Fourier, 3db for Hadamard, and 9db for Karhunen-Loève processing, averaged over the three bit rates used. A somewhat independency of the S/Q_n improvement of each transform on bit rate was also observed.

In [30] the Hadamard-Haar processing scheme was applied on speech signals sampled against a rate of 8000 samples/sec, and broken down into blocks of N=64 samples. Each block of samples was Walsh-Hadamard transformed yielding the corresponding N=64 Hadamard coefficients. The Hadamard coefficients were then Haar transformed yielding N=64 Haar coefficients which were to be transmitted in digitally coded form.

To realize the above processing scheme they passed the input speech signals through a lowpass filter (cutoff frequency 3.4 kHz), and used

10 bit A/D and D/A converters for signal conversion, and a PDP 15 computer for processing. By examining the p.d.f. of the Haar coefficients for the 64-point transformation, they found that there is a small group of M=6 coefficients that tend to predominate for both male and female speakers. This shows that a strong data compression can be used, and that it may be possible to transmit intelligible speech using fewer than N=64 coefficients. Actually, a good quality speech was reconstructed (tested by natural listening) using M=6 coefficients which corresponds to a transmission bit rate of 10kbits/sec. Using 3 coefficients the quality of reconstructed speech was again intelligible but much more noisy. Finally, using 2 coefficients the resulting speech was again intelligible with "metallic" sound.

Similar results were obtained in [31] where it was shown that an intelligible speech transmission can be made using Hadamard processing at bit rates of about 13bits/sec, and in [32] where by an appropriate coefficient labelling technique the bit rate of transmission was reduced to 8bits/sec (using 6-8 dominant Hadamard coefficients) without any loss of speech intelligibility.

Some experimental results on ECG signals using the discrete cosine transform (DCT) and the Haar transform (HT) compared with those corresponding to the Karhunen-Loève transform (KLT) and the identity transform (IT) were derived in [14]. Three hundred ECG signals, recorded using a limb lead system with three leads, were digitized using a sampling rate of 400 samples per second (assuming that the ECG signal bandwidth is much less than 200 Hz), and represented by 128 samples each, such that the QRS complex and T wave would always appear. Half of the 300 ECG's were normal and half of them abnormal.

The overall data covariance was computed by $\Sigma_x = \frac{1}{2}\{\Sigma_{x_1} + \Sigma_{x_2}\}$ where Σ_{x_1} (Σ_{x_2}) is the covariance matrix of the normal (abnormal) ECG waveforms. The transform covariance matrix was computed as $\Sigma_y = P\Sigma_x P^T$. The variance distribution Σ_y was plotted against the number of components for KLT, DCT and IT. area under each curve for a given number of components is a measure of variance energy contained in them. It has been verified that the order of effectiveness for data compression is KLT>DCT>IT. Note that IT leaves the data unaltered.

It was found that the whole area under the KLT and DCT curves was contained in 45 components. Thus a data compression 150:45=3:1 was used. Actually 43 DCT components with the largest variances were used to represent each ECG in the transform domain (the remaining components were set equal to zero). The inverse DCT, was taken and the resulting ECG's were plotted using a CALCOMP plotter. The same was repeated using the HT.

Typical results for normal and abnormal ECG's are shown in Fig.10 (a,b). One observes that the information loss resulting from the 3:1 data compression adopted is not diagnostically significant. However a premature beat could be lost in such reconstructed data.

We close this section with a few words on the important work of

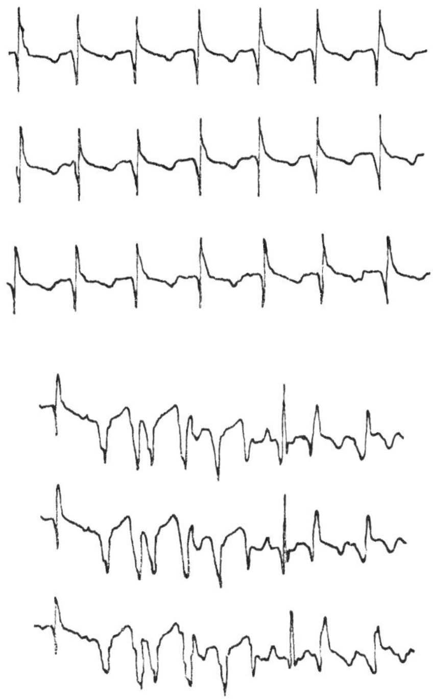

(a)

(b)

Fig. 10 ECG signals reconstructed using 3:1 data compression

(a) Normal ECG's,
(b) Abnormal ECG's

Kuklinski [15] where the feasibility of using a fast Walsh transform (FWT) algorithm to implement a microprocessor-based real-time E.C.G. data-compression system was demonstrated. However, the results were using a minicomputer FORTRAN implementation. The recursive sequenc - ordered FWT algorithm of Brown [38] was used.

Two methods of data compression were tried. In the first, the Walsh coefficients were quantized using 3 to 11 bits per coefficient (whereas the original samples were represented using 12 bits) and in the second, only a fraction (1/16) of the Walsh spectrum was used for reconstructing the ECG's. To improve the reconstructed ECG's a 9th-order FIR low-pass filter was used, which reduces the existing high-frequency noise components which are due to the data compression.

3. The Fast Walsh Transform

The Walsh functions constitute a family of binary orthogonal functions $W(k,t) = 0,1,\ldots,n-1$, $n = 2^m$ in the interval $t \varepsilon [0,1)$. There are several types of ordering for the Walsh functions, but in digital signal processing applications, the sequency-ordered Walsh functions (Fig. 11) are the most useful. The Walsh functions can be generated by hardware [46] or software.

A continuous-time signal $x(t)$, which is absolutely integrable in the time interval $[0,1)$, has the Walsh series expansion:

$$x(t) = \sum_{n=0}^{\infty} a_n W(n,t) \tag{11}$$

where a_n is the coefficient of Walsh function $W(n,t)$, and can be determined by

$$a_n = \int_0^1 x(t)W(n,t)dt \tag{12}$$

Equations (11) and (12) constitute the "Walsh Transform pair". For digital applications, in discrete time, Eqn. (12) can be approximated as:

$$a_n = \frac{1}{N} \sum_{i=0}^{N-1} x(i)W(n,i), n = 0,1,2,\ldots,N-1 \tag{13}$$

and the corresponding inverse Walsh transform is:

$$x(i) = \sum_{n=0}^{N-1} a_n W(n,i),\ i = 0,1,2,\ldots,N-1 \tag{14}$$

A direct computation of the Walsh transform, given by Eqn (13), requires N^2 operations. A number of algorithms have been developed that can compute a N-point Walsh transform in $N\log_2 N$ rather than N^2 operations. Such algorithms which are known as *fast Walsh transforms* (FWT) include [33-38]. Actually all FWT algorithms are modifications of the well known Cooley-Tukey fast Fourier transform (FFT) algorithm [33].

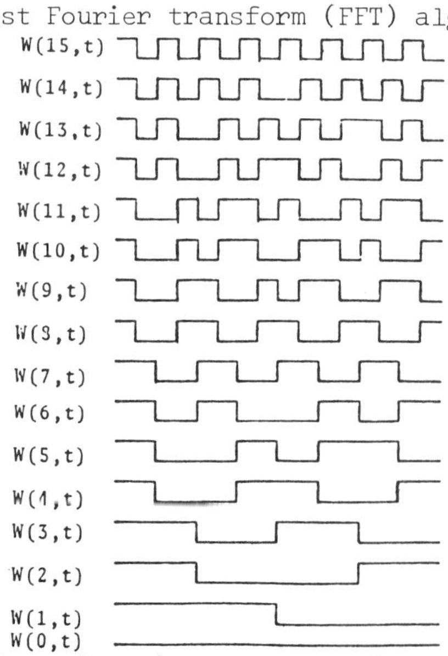

Fig. 11 Walsh Functions in sequency order.

The flow diagram of such a Fast Walsh Transform (FWT) algorithm is shown in Fig. 12 with normal input and sequency ordered output [50]. This algorithm offers considerable advantages and being an in-place computation, requires about half the memory of other algorithms.

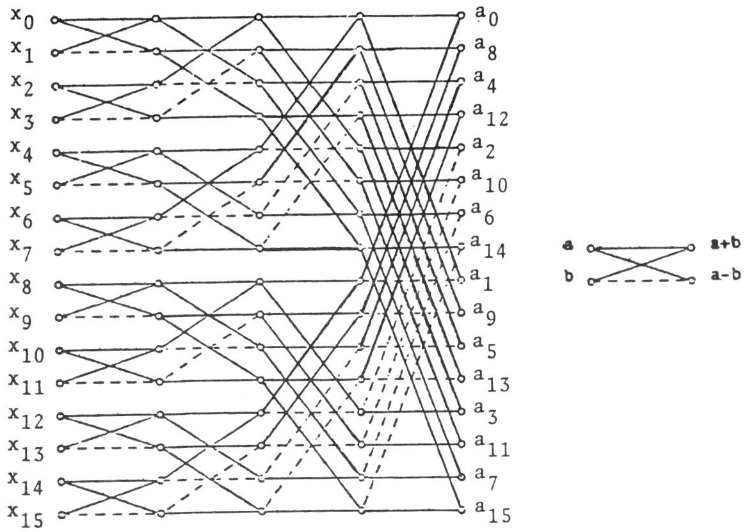

Fig. 12. Flow graph of FWT. The input data is in normal order. The Walsh Coefficients are in bit reversed sequency order.

Now, using the data compression principle, outlined in section II.1 and on the basis of the results of [14,15,47,48] we can use only a set of M most dominant terms (in absolute value order) among the complete set of N terms (M<N). Of course the value of M depends on the type of signals and the desired accuracy of the reconstruction.

4. Using a 2-μP System

The block diagram of a data compression-transmission and retrieval system for Electrocardiogram input signal (ECG), which uses the algorithm of Fig. 12 for the FWT computation, and selects and transmits the M-dominant terms rather than N terms, is shown in Fig. 13.

For the implementation of the transmitter section of Fig. 13, a distributed μP system using two μC's working in parallel was used (Fig. 14). It functions as follows. At the beginning the μC-1 samples and converts into digital form the first group (in our case 64 points) of the input points.

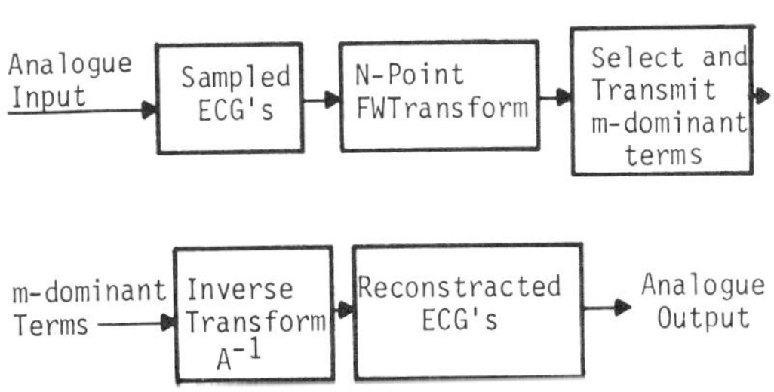

Fig. 13. A data compression-transmission and retrieval system.

At the end of this sampling period it informs the μC-2 to start sampling the next input points of the second group while μC-1 continuous with the application of FWT for the determination of the Walsh coefficients, the calculation of the most dominant terms and the transmission of them. When the sampling by μC-2 is completed, the μC-1 is informed by the μC-2 to continue with the sampling of the 3rd group while μC-2 determines the Walsh coefficients of the points already obtained, calculates the most dominant terms and transmits them. This changing of roles between μC-1 and μC-2 is continued, until all the input points have been processed.

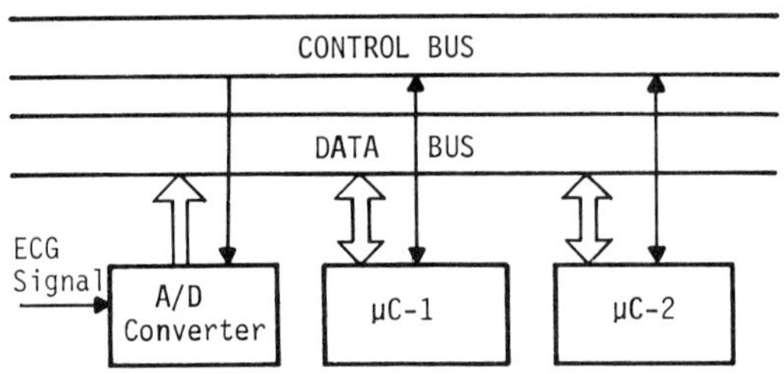

Fig. 14. Block diagram of the system.

In Fig. 15, the timing function of the operation of μC-1 and μC-2 is given.

The reconstruction of the original signal at the receiver section applying the reverse fast Walsh transform, can be accomplished by hardware [49] or software methods.

The analogue ECG input signal was obtained by the Hittman CDC-70 system, while the SDM-853 Burr-Brown data acquisition module was used for the convertion of this analogue signal into digital form.

Each μC is based on the 8080A μP, and the required memory was about 3K-ROM and 0.5K-RAM memory for each μC.

The sampling rate which can be achieved by the system was found to be at least 500 samples/sec, without optimization.

In Fig. 16(a,b) the reconstructed ECG signal is shown on the original signal, using 8 and 16 most dominant terms.

We can see that the reconstruction is not so accurate in the QRS complex but this is not so important in the case of arrythmia analysis in contrast to the complete analysis of the ECG, where accurate measurements are required.

The quickshort algorithm, for sorting the Walsh coefficients can be used which is very fast. The system can be easily extended by using more μC's. Additional data compression can also be achieved by using a limited number of bits in order to represent the Walsh coefficients.

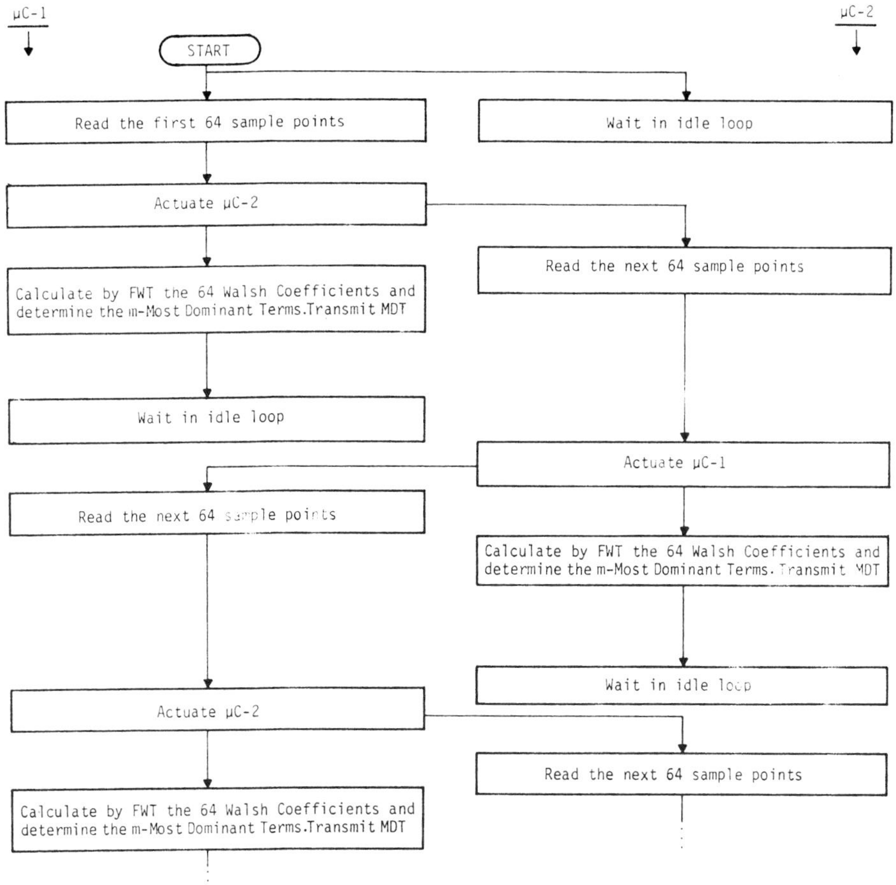

Fig. 15 Timing diagram of the system.

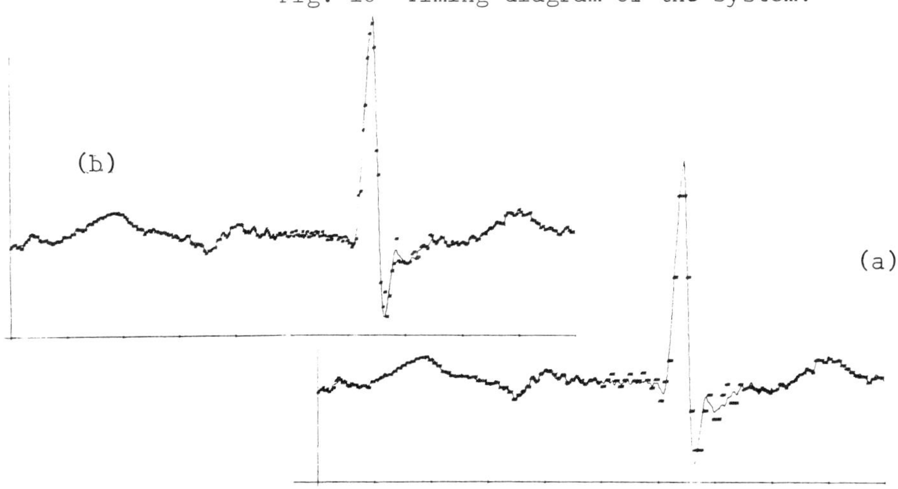

Fig. 16 (a) Reconstruction of ECG signal, using 8 most dominant terms.
(b) Reconstruction of ECG signal, using 16 most dominant terms.

CONCLUSIONS

In communications, remote control biomedical signal processing systems, etc., a large amount of input data must be processed and transmitted. Many methods and algorithms have been proposed to perform data compression and reduction operations. One of the methods for data compression consists of encoding the input digital signal by means of a linear orthogonal transformation. A well known transformation of this type is the Walsh-Hadamard transform. On the other hand one of the main tools for data reduction is the piece wise linear approximation.

Microprocessor (μP) based systems can be used for the implementation of such approximation and transformation algorithms, since they have great flexibility and can replace hardwired circuits, increasing the capability and lowering the cost. But in general, the speed of operation is not high and the accuracy is relatively low. To overcome these disadvantages, efficient algorithms for parallel or distributed processing must be applied, especially in real-time applications.

In this chapter a survey of piecewise linear approximation and discrete transformation algorithms for data reduction and compression with emphasis on the FWT algorithm, was presented. Some microprocessor-based systems implemented by the authors and their co-workers were described.

REFERENCES

1. T. Pavlidis, "Structural Pattern Recognition", *Springer*, New York(1977).
2. Y. Kurozumi and W. Davis, "Polygonal Approximation by the Minimax Method, *Computer Graphics and Image Processing*, Vol. 12, 248-264 (1982).
3. J. Sklansky and V. Gonzalez, "Fast Polygonal Approximation of digitized curves, *Pattern Recognition*, Vol. 12, 327-331, (1980).
4. V. Gonzalez and J. Sklansky, "Fast Polygonal Approximations of digitized signals", *TP-79,14*, School of Engineering, University of California, Irvine (1979).
5. G. Frangakis and G. Papakonstantinou, "A Real Time Reduction Microcomputer System" in Methods and Applications of Measurement and Control (MECO'83) S. Tzafestas and M. Hamza, Eds), Vol. 2, 606-608, ACTA Press, (1984).
6. F. Gritzali and G. Papakonstantinou, "A Fast Piecewise Linear Approximation Algorithm" *Signal Processing*, Vol. 5, 221-227(1983).
7. I. Tomek, "Two Algorithms for Piecewise Linear Continuous Approximations of Functions of one variable", *IEEE Trans Comput.*, C-23,445-448 (1974).
8. C. Williams, "An Efficient Algorithm for the Piecewise Linear Approximation of Planar Curves, *Computer Graphics and Image Processing*, Vol. 8, 286-293,(1978).
9. G. Papakonstantinou, "Optimal Polygonal Approximation of Digital Curves, *Signal Processing*, Vol. 8, 131-135 (1985).
10. G. Benelli, V. Gappellini and F. Lotti, "Data compression techniques and applications, *The Radio and Electronic Eng.*, Vol. 50, No.1/2, Jan./Feb. 1980, pp. 29-53.

11. T. Berger, "Rate distortion theory - A mathematical basis for data compression (Prentice Hall, Englewood Cliffs, N.J., 1971).
12. L.D. Davisson, The theoretical analysis of data compression systems; *Proc. IEEE,* Vol. 56, No. 2, Feb. 1968, pp. 176-186.
13. L.D. Davisson, Theory of adaptive data compression, in *"Advances in Communication",* Vol. 2 (Academic Press, New York, 1966).
14. N. Ahmed, P.J. Milne and S.G. Harris, Electrocardiographic data compression via orthogonal transforms, *IEEE Trans.Biom.Eng.*, Vol.BME-22, No. 6, Nov. 1975, pp. 484-487.
15. W.S. Kuklinski, Fast Walsh transform data-compression algorithm: ecg applications, *Med. and Biol.Eng. and Comput.,* Vol. 21, 1983, pp. 465-472.
16. J.H. VanBemmel and J.L. Willems (Eds), *Trends in Computer-processed electrocardiographs,* (North-Holland, Amsterdam,1977).
17. T. Shibahara et al., CCA-A knowledge based system with causal knowledge to diagnose rhythm disorders in the heart, *Proc.Int.Conf. CSCSI/SCEIO,* 1982, pp. 71-78.
18. A. Lestenberg and Kunt M., A robust-digital QRS-detection algorithm for arrythmia monitoring, *Comp. and Biomed.Research,* Vol. 16, 1983, pp. 273-286.
19. M.L. Ahlstrom and W.J. Tomkins, Automated high-speed analysis of Holter tapes with microcomputers, *IEEE Trans.Biom.Eng.,* Vol.BME-30, No. 10, 1983, pp. 651-657.
20. G. Frangakis, "Digital and μP-Based Techniques in Signal Processing and System Simulation, Ph.D. Thesis, Southampton Univ.(1985).
21. B.H. Jansen, J.R. Bourne and J.W. Ward, Spectral decomposition of e.c.g. intervals using Walsh and Fourier transforms, *IEEE Trans. Biom.Eng.,* Vol. BME-28, 1981, pp. 836-838.
22. B.W. Weide, L.T. Andrews and A.M. Tannone, Real-time analysis of e.c.g. using Walsh transforms, *Comput.Biol.Med.,* Vol. 8, 1978, pp. 225-263.
23. M.E. Womble et al, Data compression for storage and transmitting e.c.g.s./v.c.g.s., *Proc. IEEE,* Vol. 65, 1977, pp. 702-706.
24. H. Harmuth, Transmission of information by orthogonal functions (Springer, 1969).
25. R. Lackey, The wonderful world of Walsh functions, *Proc.1972 Symp. on Appl.of Walsh functions,* The Catholic Univ. of America, Washington D.C., pp. 2-7 (1972).
26. C. Chen, and C. Hsiao, Walsh series analysis in optimal control, *Int.J.Control,* Vol. 21, pp. 881-897 (1975).
27. P. Stavroulakis and S. Tzafestas, Walsh series approach to observer and filter design in optimal control systems, *Int.J.Control,* Vol. 26, pp. 721-736,(1977).
28. S. Tzafestas, Walsh series approach to lumped and distributed system identification, *J.Franklin Inst.,* Vol. 305, pp. 199-220,(1978).
29. S. Campanella and G. Robinson, A comparison of orthogonal transformations for digital speech processing, *IEEE Trans.on Commun.,* Vol. COM-19, pp. 1045-1050 (1971).
30. E. Frangoulis and L. Turner, Speech coding by the Hadamard-Haar transform, *Computers and Digital Techniques,* Vol.1,pp.106-112,(1978).

31. Y. Shum, A. Elliot and W. Brown, Speech processing with Walsh-Hadamard transforms, *IEEE Trans.on Audio and Electroacoustics*, Vol. AU-21, pp. 174-179, (1973).
32. E. Frangoulis and L. Turner, Hadamard-Transformation Technique of Speech Coding: Some further Results, *Proc.IEE*, pp. 845-852, (Oct. 1977).
33. J.E. Whechel and D.F. Guinn, The fast Fourier-Hadamard transform and its use in signal representation and classification, *EASCON'68 Rec.*, 1968, pp. 561-573.
34. J.L. Shanks, Computation of the fast Walsh-Fourier transform, *IEEE Trans.Comput.*, Vol.C.18, May 1969, pp. 457-459.
35. K.W. Lindenberg, A Sequency-ordered in-place fast Walsh transform algorithm, *Proc.San Diego Biomedical Symp.*, V13, 1974, pp.495-498.
36. J.W. Manz, A sequency ordered fast Walsh transform, *IEEE Trans.Audio Electroacoust.*, Vol. AU-20, 1972, pp. 204-205.
37. H. Larsen, An algorithm to compute the sequency ordered Walsh transform, *IEEE, Trans. Acoust., Speech, and Signal Processing*, Aug. 1976, pp. 335-336.
38. R.D. Brown, A recursive algorithm for sequency-ordered fast Walsh transforms, *IEEE Trans.on Computers*, Vol. C-26, 1977, pp.819-823.
39. S.H. Fuller et al., Multi-Microprocessors: An overview and working example, *Proc.IEEE*, Vol. 6, No.2, Feb. 1978, pp. 216-218.
40. Ramamoorthy et al., Hardware-software issues in multimicroprocessor computer architectures, *Proc.Microcomputers*, Fort Collins, Colo., Aug.-Sept. 1977, pp. 73-79.
41. P.H. Enslow, Multiprocessor organization, A survey, *Computing surveys*, Vol. 9, No. 1, March 1977, pp. 103-129.
42. L.H. Anderson, The microcomputer as distributed intelligence, Proc. *Int.Symp. Circuits and Systems*, Boston, Mass., April 1975, pp. 337-340.
43. M. Krieger, Task driven multimicroprocessor system, *Proc. 1st Canadian Workshop Design and Develop. of Computer Systems*, May 1979, pp.81-88.
44. B.A. Bowen and R.J.A. Buhr, The logical design of multiple microprocessor systems (Prentice Hall, Englewood Cliffs, N.J.,1980).
45. E.T. Fathi and M. Krieger, Multiple Microprocessor systems: What, Why, and When, *Computer*, March 1983, pp. 23-32.
46. S. Tzafestas, G. Frangakis and Pimenidis T., Global Walsh function generators, *Electronic Engineering*, 1976, pp. 45-49.
47. R. Kitai, Synthesis of Periodic sinusoids from Walsh waves, *IEEE Trans. on Instr. and Measur.*, Vol. IM24, pp. 313-317.
48. W.O. Brown and A.R. Elliot, *IEEE Proc. Symp. Applications of Walsh functions*, Wahsington, D.C., AD7444650, 1972.
49. S. Tzafestas, G. Frangakis, A fast Walsh Hadamard signal processing system, *Mathematics and Computers in Simulation*, Vol. 23, 1981, pp. 163-169.
50. H. Larsen, An algorithm to compute the sequency ordered Walsh transform, *IEEE Trans.Acoust., Speech and Signal Processing*, Vol.24, 1976, pp. 335-336.
51. DCE Microcomputer Systems Designer's Handbook, *Data Applications International S.A.*, Bruxelles 1977.

52. J.L. Willems et al, Common standards for quantitative electrocardiography, The CSE pilot study, *Proc. Medical Informatics Europe 81* (F. Cremy et al, eds) Springer-Verlag, Berlin, pp. 319-326, 1981.
53. J. Sarris, A microcomputer card for AID convertion, *Tech. Report*, National Technical University of Athens, Computer Science Division, Athens, 1985.

INDEX